The Benefits of Climate Change Policies

ANALYTICAL AND FRAMEWORK ISSUES

ORGANISATION FOR ECONOMIC CO-OPERATION AND DEVELOPMENT

ORGANISATION FOR ECONOMIC CO-OPERATION AND DEVELOPMENT

Pursuant to Article 1 of the Convention signed in Paris on 14th December 1960, and which came into force on 30th September 1961, the Organisation for Economic Co-operation and Development (OECD) shall promote policies designed:

– to achieve the highest sustainable economic growth and employment and a rising standard of living in member countries, while maintaining financial stability, and thus to contribute to the development of the world economy;

– to contribute to sound economic expansion in member as well as non-member countries in the process of economic development; and

– to contribute to the expansion of world trade on a multilateral, non-discriminatory basis in accordance with international obligations.

The original member countries of the OECD are Austria, Belgium, Canada, Denmark, France, Germany, Greece, Iceland, Ireland, Italy, Luxembourg, the Netherlands, Norway, Portugal, Spain, Sweden, Switzerland, Turkey, the United Kingdom and the United States. The following countries became members subsequently through accession at the dates indicated hereafter: Japan (28th April 1964), Finland (28th January 1969), Australia (7th June 1971), New Zealand (29th May 1973), Mexico (18th May 1994), the Czech Republic (21st December 1995), Hungary (7th May 1996), Poland (22nd November 1996), Korea (12th December 1996) and the Slovak Republic (14th December 2000). The Commission of the European Communities takes part in the work of the OECD (Article 13 of the OECD Convention).

FOREWORD

The OECD initiative on the Benefits of Climate Policies was launched in 2002. The overall aim of this initiative is to improve information on the benefits of climate policies for policymakers. A modest first step has been to advance a conceptual framework for the assessment of climate policy benefits. The approach of the initiative has been consultative, drawing on papers and ideas from different experts and government representatives, through a first Workshop in December 2002 which was followed by an informal expert meeting in September 2003. The Workshop agenda and discussion spanned a wide range of topics related to mitigation policy benefits, including ancillary benefits and potential linkages with the adaptation measures. A full set of the original working papers from the Workshop can also be found on the OECD climate change website: www.oecd.org/env/cc.[1] The September 2003 expert meeting focussed on the emerging framework and helped to develop recommendations for future work in this area (see Chapters 1 and 9).

This book presents a selection of papers, which were originally prepared for the December 2002 Workshop.[2] They focus on different aspects of the benefits of mitigation policy; however they also include perspectives on how such benefits are intertwined with adaptation and adaptation benefits. Overall the book suggests a rich research agenda, which if further developed and followed would improve information on the avoided (or delayed) impact benefits of GHG mitigation.

---

1 A workshop report is also available on-line – see OECD, 2003.

2 In addition to this volume, another complementary selection of papers from the Workshop was also released in the journal of Global Environmental Change (Corfee Morlot and Agrawala, eds., Volume 3, October 2004).

ACKNOWLEDGEMENTS

The OECD would like to thank the partners that have made this book possible. In particular thanks are due to individual authors whose contributions make up the core of this volume. In addition to the authors featured here, a number of participants in the December 2002 Workshop and/or the September 2003 expert meeting have actively contributed to the rich discussions that have taken place on these issues, including: Philippe Ambrosie, Sebastian Catovsky, Lynda Danquah, Hadi Dowlatabadi, Enno Harders, Geoff Jenkins, Eberhard Jochem, Rik Leemans, Jane Leggett, Reinhard Madlener, Petra Mahrenholz, Robert Nicholls, Merylyn McKenzie Hedger, Andre van den Moor, Anand Patwardhan, Martin Parry, Ademar Ribeiro Romeiro, Rich Richels, Dale Rothman, Jan Rotmans, Ken Ruffing, Tana Stratton, Dennis Tirpak, Richard J.S. Tol, Farhana Yamin, Gary Yohe.

The Secretariat would also like to thank the Governments of Canada, Finland, Germany and the United States for their financial support for this initiative.

The views presented in this volume represent those of the authors. Some of the papers presented here have benefited from the oversight, input and suggestions gathered from OECD government representatives through the OECD Working Party on Global and Structural Policy. However, the papers do not represent the views of OECD Member countries nor of the OECD. The book is published under the responsibility of the Secretary-General of the OECD.

Jan Corfee Morlot and Shardul Agrawala of the OECD Environment Directorate edited this book and oversaw the project leading up to it. Tom Jones, Head of the Global and Structural Policies Division of the OECD Environment Directorate provided useful suggestions at all stages of the project and Martin Berg, Travis Franck, Carolyn Sturgeon and Jane Kynaston provided invaluable staff support for the book and the project.

TABLE OF CONTENTS

This book is available in printed form with black and white graphics; however, it is also available as an e-book or electronic file (pdf) with colour, higher resolution graphics.

EXECUTIVE SUMMARY

In recent climate policy assessments and debate, too little attention has been given to estimation of the *direct benefits* of greenhouse gas mitigation – that is, the benefits of avoiding climatic change and reducing the likelihood of any ensuing net adverse impacts. The problem is partly relative lack of research and partly lack of synthesis of research into some coherent measure or set of measures for policymakers and the public to understand and weigh benefits.

What can be meaningfully conveyed to policymakers about the direct benefits of climate policy? This volume considers this question through a series of review papers. The goal was not to come up with new, monetised or even physical estimates of direct benefits, but rather to survey available information to work towards an eventual framework and set of priorities for future work, which over time could improve accounting for benefits to facilitate decision-making on international policies.

A number of specific challenges are underscored in this collection of papers, which points to large uncertainties in estimates of impacts or of monetised benefits. There are several reasons for this, including that many categories of impacts have not been researched at a global scale. In addition, socio-economic baselines for impact studies sometimes are not consistent with those emissions driving the climate change projections and adaptation is sometimes not included, or may be assumed to be unrealistically effective and the costs of adaptation are sometimes not tabulated. Further, impact assessments generally only examine responses to changes in mean climate, not those associated with changes in variability or extreme events, or with the risk of non-linearities, abrupt changes and "surprises." Finally, different types of impacts are fundamentally incomparable, such as changes in human health risks versus species extinctions, and monetizing and aggregating them may be misleading.

Another challenge is that impacts vary across economies, and across market and non-market systems and a range of subjective and technical judgments are embedded in any choice of assumptions to monetise and aggregate benefits across time and space. Any choice of assumptions may be controversial, if not carefully constructed to reflect the views of those affected.

In addition, the benefits of mitigation policies are likely to be experienced by different populations than those that pay for the mitigation, with the differences of distribution spread over both time and space. These differences will affect how various people view what policies are appropriate.

A broad conclusion is that sound summary estimates of benefits in a single (monetary) measure, as might be sought to compare with aggregate costs, may not be adequate on their own to inform policy decisions, especially given the incommensurable nature of benefits. Thus, benefit-cost methods alone may be inadequate to resolve many of these problems and would be usefully complemented with risk-based methods, such as probabilistic approaches to consider climate change and related impacts across a range of possible futures. Such a dual approach also calls for the presentation of benefits information in at least two different forms, using different monetary and non-monetary metrics of change: monetised estimates and physical impact estimates.

To improve information for policymakers it is also desirable to develop a coherent set of indicators that present a balance of the physical and economic metrics of change. Preferably this would include information at the local, regional and global scales, and would be structured to provide transparency about embedded assumptions when viewing any particular set of estimates. More systematic research and discussion of benefits would allow more explicit, transparent consideration of them in policy dialogue and decisions. However, much work will be needed to make available reliable global, aggregated estimates of the benefits of climate policies. A more modest and preliminary goal should be to have some consistent and comparable regional information against which to assess impacts associated with various levels of global mitigation.

Despite the uncertainties and incommensurable nature of benefits and impacts, some general patterns emerge when looking across the literature on global impacts. Some sectors, such as agriculture, may experience net positive impacts globally of a small amount of climate change. However, no research for any sector suggests positive impacts from climate change as temperatures increase beyond certain levels. A consistent pattern of marginal adverse impacts emerges across all sectors for which data were available beyond a 3-4 °C increase in global mean temperature – translating into possible large and positive net benefits to mitigation policies that can limit climate change to this level or possibly below it.

In addition, results from a number of studies suggest that accounting for the risks of irreversible, abrupt change – risks that grow with forcing of the climate system and with the pace of climate change – is likely to increase the economically "optimal" level of mitigation, calling for more investment in abatement in the near-term.

Looking forward, a conceptual framework for future work emerges here with the aim to help improve information on global and regional avoided impact benefits and to support mitigation policy decisions. The main elements of the framework include a portfolio of indicators of change, first in physical units and at the sub-global scale, before moving onto monetised and aggregated benefits assessment. The framework suggested here is necessarily partial, with emphasis on mitigation and direct climate impacts elements of any more comprehensive framework. By setting out an initial framework to structure further work, it is hoped that impacts research can be used to inform not just adaptation policy but also mitigation policy decisions by helping to assess the trade-offs associated with different global mitigation pathways.

Chapter 1

OVERVIEW

by Jan Corfee Morlot and Shardul Agrawala,

Environment Directorate, OECD, France

This chapter underscores a number of specific challenges in the estimation of the direct, avoided climate change impact benefits of mitigation policies. A broad conclusion is that sound summary estimates of climate benefits in a single (monetary) measure, as might be sought to compare with aggregate mitigation costs, may not be adequate on their own to inform policy decisions. Risk-based tools can usefully complement cost-benefit assessment, relying upon probabilistic approaches to help manage uncertainty as well as on a rich presentation of physical and economic benefits information. A portfolio of indicators of change is called for here, first in physical units and at the sub-global scale, before moving onto monetised and aggregated, global estimates. The suggested framework is necessarily partial, focusing on mitigation and direct climate impacts elements. Through this initial framework it is hoped that impacts research can be used to better inform policy decisions over time by helping to assess the trade-offs associated with different global mitigation pathways.

Despite uncertainties and the incommensurable nature of impacts, some general patterns emerge looking across the global literature. Though positive impacts appear at lower levels of change in some sectors and regions, no research suggests positive impacts as global mean temperatures (GMT) increase beyond certain levels. Marginal adverse impacts emerge, across all sectors with available data, beyond a 3-4 °C increase in GMT. This indicates possible large and positive net benefits to mitigation policies that can limit climate change to this level or possibly below it. Additionally, the risks of irreversible, abrupt climate change increases the risks and the likely costs of rapid climate change, calling for more investment in abatement in the near term.

ISBN 92-64-10831-9
THE BENEFITS OF CLIMATE CHANGE POLICIES

CHAPTER 1. OVERVIEW

by Jan Corfee Morlot and Shardul Agrawala

1.1 Introduction

Human activities have already had a discernible impact on the earth's climate (IPCC, 1995, 2001a). There is growing evidence of observable impacts of climate change on physical and biological systems. While developing countries might be particularly vulnerable to climate change impacts due to their limited coping capacities and natural resource dependence, the mid- to high-latitudes, where many OECD countries are located, have experienced significantly higher rates of recent warming, and, in the northern hemisphere, such regions have also experienced an increase in heavy precipitation events (IPCC, 2001a).

With considerably more significant and widespread biophysical, social and economic impacts being projected as climate change unfolds over time, views vary widely amongst governments, business and other parts of civil society about how we should deal with climate change. Mitigation – or reduction in greenhouse gas (GHG), emissions and concentrations – is the key pillar of international climate policy negotiations and of national climate policies. Recently interest has grown in adaptation to projected climatic changes as a complementary policy response as it has been recognised that we will not be able to avoid all adverse effects of climate change even under aggressive mitigation (IPCC, 2001d). Nevertheless, much of the policy and analytical discourse to date has been characterized by asymmetric attention to the costs of mitigation commitments on the one hand, and, more recently, the potential benefits of adaptation on the other, with only limited analysis of the costs of adaptation and of the benefits of GHG mitigation. Analysis of the benefits mitigation to date has been dominated by attention to near-term secondary or ancillary benefits in related domains such as air pollution and public health.

Comparatively little attention has been given to estimation of the *direct benefits* of greenhouse gas mitigation – that is, the benefits of avoiding climatic change and reducing the likelihood of any ensuing net adverse impacts. The problem is not so much the lack of research, but the difficulty of, and lack of, efforts synthesizing available work into some coherent measure or set of measures that aid policymakers and the public in thinking about potential mitigation benefits. Critical policy decisions – with regard to how much and how fast to cut greenhouse gas emissions – would be better informed by careful assessment of what is at stake for natural and human systems. Increased attention to the direct avoided impact benefits would provide policymakers with a more complete picture about the full range of mitigation policy benefits.

Key questions would seem to include: what are the cost and benefit trade-offs between alternative mitigation strategies, or in moving from today's more limited climate policies towards significant and rapid emission reductions in the coming decades? While noting that complete, quantitative estimates of benefits do not exist, the IPCC's Third Assessment Report (IPCC, 2001a-d), began to assess links between climate impacts – or "reasons for concern" – and global mean temperature change in 2100 (Figure 1), which in turn can be linked to different emission pathways or mitigation scenarios.

Figure 1. Relating global mean temperature change to reasons for concern

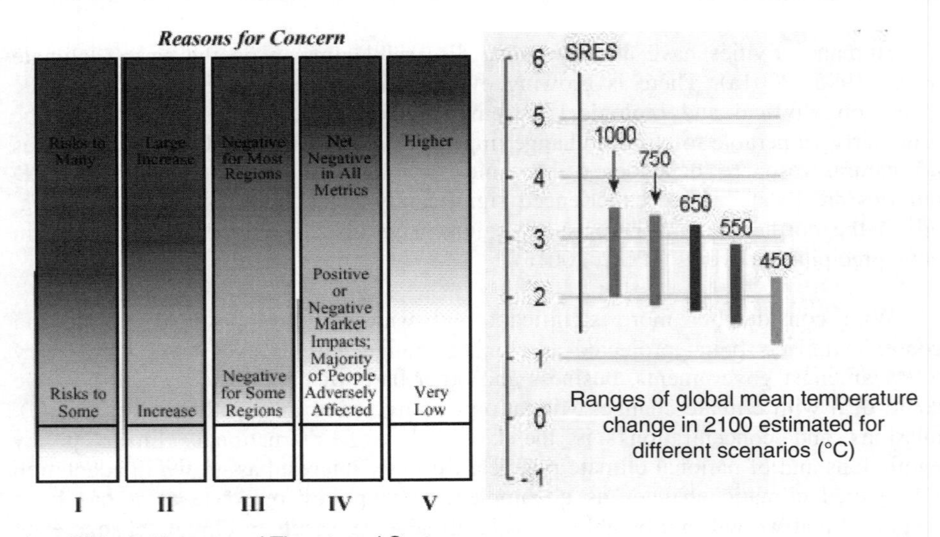

I Risks to Unique and Threatened Systems
II Risks from Extreme Climate Events
III Distribution of Impacts
IV Aggregate Impacts
V Risks from Future Large-Scale Discontinuities

Source: IPCC, 2001d, Synthesis Report.

Aim of this volume

This book is part of an ongoing OECD initiative aiming to improve information on the benefits of climate policies for policymakers. The volume examines a number of questions, including:

- What are the likely climate change benefits of increasingly aggressive or alternative types of mitigation policies that move us from one level of climate change to another?
- How can these benefits be assessed and communicated, taking into account associated uncertainties?
- How do benefits of mitigation interact with autonomous and planned adaptation?
- What is the nature of the risks of climate change and how well are these represented in the models and analytical tools, including data, of which we have access to today?

1.2 Analytical issues and challenges in assessing mitigation policy benefits

Several factors complicate the assessment of global mitigation policy benefits to render it difficult at best and controversial at worst. Mitigation costs and ancillary benefits of mitigation, or adaptation benefits and costs, typically (but not always), accrue at the same location (or region), as those in which mitigation or adaptation actions are undertaken. However, there is no such link between where actions to mitigate climate change and the related set of avoided climate change impacts occur. This is because mitigation measures undertaken anywhere in the world will influence global greenhouse gas concentrations, which will then translate into changes in climate and associated impacts. Local or regional actions to reduce emissions result in dispersed benefits. Although the views of any population are likely to be formed by the set of mitigation costs and benefits that they would experience directly, assessment of economic efficiency can only be considered by comparison of costs and direct benefits of mitigation at a global level.

Even looking at benefits of mitigation at the global level, a number of other analytical issues bring into question the straightforward application of standard methods to aggregate and monetise the benefits of avoided impacts. Benefits of avoided climate change impacts accrue much later than the costs of mitigation, raising the issue of the choice of appropriate discount rate. In addition, impacts vary across different world regions, and across market and non-market systems, requiring methods to monetise and aggregate these widely different types of impacts across different locations and populations. Subjective judgements are embedded in any choice of assumptions to monetise and aggregate across time and space and any choice of assumptions may be controversial, especially if not carefully constructed to reflect the views of those affected.[3]

---

3 It may not be practical to consult stakeholders prior to conducting relevant analyses however it is one way to ensure that assumptions made reflect the views of those affected. Sensitivity assessment across a wide range of possible assumptions or techniques (e.g. for aggregation), is another way to handle this analytically. The point

Figure 2. Mitigation and adaptation policy benefits over space and time

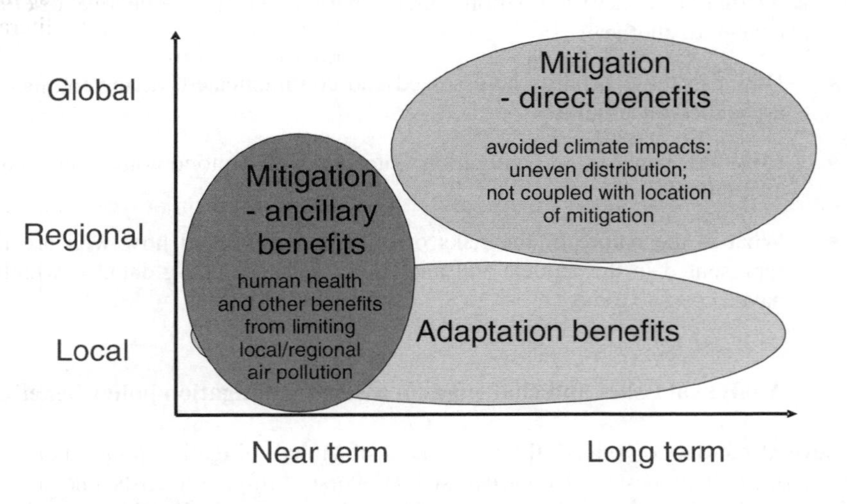

Another complicating factor is that impact assessments may not be measurably sensitive to relatively small changes in the climate system from mitigation actions that occur over limited time frames, and there are very large path dependencies that make outcomes dependent upon assumptions about future actions. Furthermore, impact assessments generally only examine responses to changes in mean climate, and not those associated with changes in variability or extreme events, or with the risk of non-linearities, abrupt changes and "surprises". Finally, the inherent uncertainty embedded in any prediction of the chain of events that starts with GHG emissions leading to changes in the atmosphere and eventually to climate change and climate impacts raises questions about how much can be said about benefits with confidence (Figure 3). Working with the ubiquitous uncertainty in a policy context suggests that standard benefit-cost methods may be inadequate to address climate change and could be usefully complemented with risk-based assessments of benefits.

The concept of risk provides a means to combine ideas on uncertainty with those on adverse consequences, where risk assessment generally refers to an evaluation of adverse consequences and their likelihood (or probability) of occurring. Risk management, on the other hand, refers to decisions that can limit or control the risk of adverse consequences (Jones, this volume; Shlyakhter *et al.,* 1995). Climate change is not the only policy problem characterised by widespread uncertainty. In other policy areas, such as traffic safety, chemical exposure or hazardous waste management, policymakers routinely make decisions in the face of sometimes large uncertainty (Morgan and Henrion, 1990). Where consequences are potentially large but low probability, decisions on whether or not to forestall action until further information is available are valuable decisions in and of themselves (Morgan and Henrion, 1990; Shlyakhter *et al.,* 1995; Webster, 2002).

---

is that a wide range of views and assumptions may be valid (see also Pittini and Rahman – this volume).

In the case of climate change, the evidence that human-induced climate change is already occurring with potentially irreversible impacts on natural and human systems, is already well-established by the IPCC in its recent assessments of the scientific literature (IPCC, 1995; IPCC, 2001a). Also through the Framework Convention on Climate Change, the international community has already begun to work together to manage the risks of climate change by agreeing to work together to bring about a wide range of mitigation and adaptation actions at the national and local levels.

Figure 3. From emissions to impacts: the cause-effect chain and cascading uncertainty

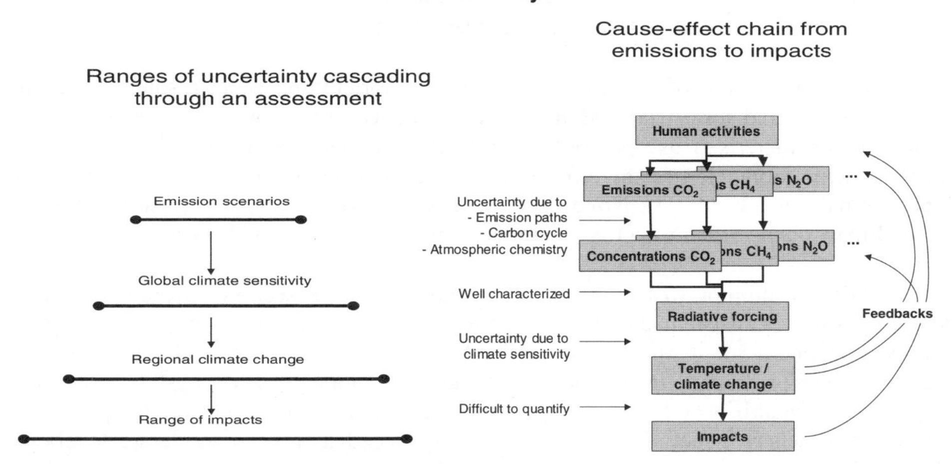

Sources: (left panel), adapted from Jones, 2000; and (right panel), Corfee and Höhne, 2003.

1.3 Structure and contents of this book

The remainder of this volume is a selection of papers that begins to review the state of current knowledge and methods to assess global risks of climate change and damage costs and how these might be affected through climate change policies (Hitz and Smith and Schellnhuber, and Schneider and Lane, this volume). The papers also explore a number of key issues in some depth, including adaptation and abrupt climate change (Callaway, Schneider and Lane this volume), as well as the critical issue of valuation of impacts more generally such that they become damage costs, which can be used to estimate the economic benefits of avoided climate impacts from alternative mitigation strategies. Finally the last three papers deal with framing issues, providing specific suggestions for how to frame future assessments of climate policy benefits so that they are more useful to key decision-makers (Wigley, Jones, Jacoby, this volume).

The remainder of this chapter outlines key findings from these papers and the others that were released through this project. It also sets out an agenda for ongoing work to continue to improve information on the benefits of climate change policies.

1.4 Key findings

1.4.1 What can we learn from global impacts literature?

A review of the global impacts literature was conducted to assess the general shape of the damage curve, expressing globally aggregated impacts by sector as a function of changes in climate, expressed as increase in global mean temperature (Hitz and Smith, this volume). The analysis considered the question of what is the magnitude of avoided damage (benefits), in going from one level of climate change to lesser one. Are incremental reductions of impacts constant, decreasing, increasing, or do they change in sign (from positive to negative), at some point?

Based on a review of studies of sea level rise, agriculture, water resources, human health, energy, and terrestrial and marine ecosystems among others, impacts in each sector are characterised as "parabolic" or U-shaped (decreasing initially, shifting to increase with more significant climate change), increasing with climate change, or indeterminate (see Figure 4). None of the available studies suggested positive impacts from climate change in any sector as temperatures increased beyond certain levels.[4]

While no attempt was made to aggregate across the sectors covered, Hitz and Smith do find a consistent pattern of progressively adverse impacts across all sectors analysed beyond a 3-4 °C increase in GMT. At lower levels of climate change however, the relationships range from increasing adverse impacts (in coastal resources, biodiversity, health[5], and possibly marine ecosystem productivity), to parabolic relationships where beneficial impacts are experienced at low to moderate levels of climate change (agriculture, terrestrial ecosystem productivity), to no consistent pattern (water, energy, aggregate costs). Converting these findings into the "benefits" framework suggests that at high levels of climate change or beyond 3-4 °C, the literature suggests that marginal benefits of mitigation would be positive and perhaps large. As mitigation reduces climate change, the marginal benefits would appear to stay constant or decline until some relatively low level of climate change. However the relationships that emerge from the global impacts literature at low levels of climate change are less clear: there are some marginal "negative" benefits in some sectors and regions, so the global results will depend upon aggregation methods and assumptions.[6]

---

4 Also while this review considered marginal impacts, it is important to note the need to consider the possibility of non-marginal change and irreversibilities, for example, through abrupt climate change and or through loss of historical and culturally significant sites. No global studies exist in a number of relevant areas, including tourism and energy use, thus these results are still partial.

5 The literature is not clear about health, but based on their knowledge of impacts, Hitz and Smith believe health will show increasing damages.

6 Another way to see this is that the estimated net benefits in some regions from low levels of climate change in agriculture and forestry sectors would be incrementally lost through mitigation.

Figure 4. Sector damage relationships with increasing global mean temperature[7]

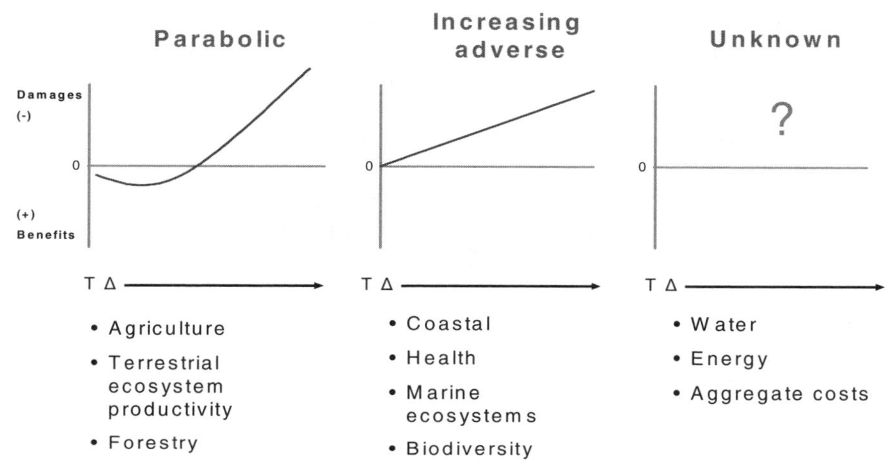

Note:

For some systems/sectors few studies were available; the authors point to uncertainties in the characterisation of health as adverse and increasing, where the results of studies examined are inconsistent. For marine ecosystems and forestry, this relationship is estimated of the basis of only one study. No global studies exist for the following sectors/systems: recreation and tourism; transport; buildings; insurance; and human amenities.

On biodiversity and other ecosystem effects, global impact studies are particularly weak. However, Leemans and Eickhout (2004), use a series of different indicators of change in ecosystem categories across the world's terrestrial surface area, to assess global impacts on ecosystems at both low and moderate to high levels (as well as implied rates), of climate change. By presenting a multi-dimensional set of indicators of change, the Leemans and Eickhout study varies significantly from the single indicator studies covered in the Hitz and Smith survey (Chapter 2). Looking across the range of indicators at different level and rates of climate change, Leemans and Eickhout's conclusions are consistent with the findings of recent meta-analyses (Root *et al.,* 2003; Parmesan and Yohe, 2003): even small levels of climate change will have significant impacts on temperature-limited ecosystems, such as the tundra, and on diversity of species within ecosystems.

Leemans and Eickhout also conclude that risks to many regional and global ecosystems increase rapidly above a 1 to 2°C change in global mean temperature over the course of this century, mainly due to the inability of forest ecosystems to adapt to such rapid rates of temperature increase. The authors note that mitigation may prove to be the most effective of policy options to limit ecosystem stresses from climate change,

---

7 Based on Hitz and Smith in this volume. These graphs are illustrative only, and do not attempt to summarise the variety of relationships that were found in this study. The results are based on global impact assessments only and do not take into account a much larger and richer literature that exists at the regional scale.

where every degree of global mean temperature increase avoided will yield clear benefits in terms of avoiding ecosystem disruption. Relating their assessment back to the five "reasons of concern" identified by the IPCC, the authors propose adding another reason of concern related to adaptive capacity of regional and global ecosystem (Figure 5).[8]

Figure 5. Risk to regional and global ecosystems by global mean temperature increase

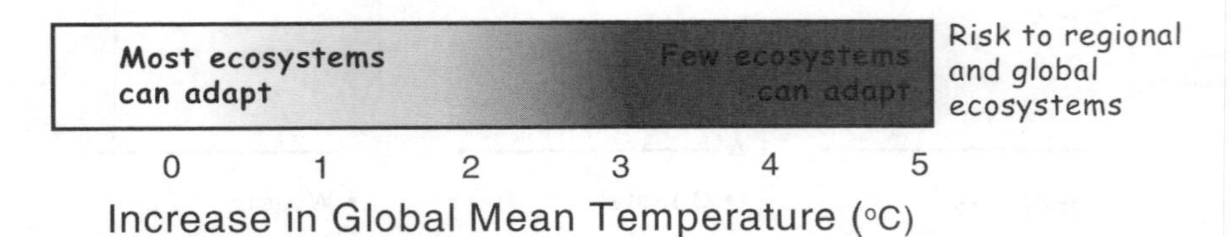

Source: Leemans and Eickhout, 2004.

Note:

This figure does not capture the timing of changes in GMT, yet timing or rates of temperature change are an important determinant of potential impacts on ecosystems. Some ecosystems may be able to adapt to gradual change but not to rapid change. These points are highlighted in the Leemans and Eickhout, 2004.

More work is still needed to provide a comprehensive understanding of impacts in this area, including decisions on standard metrics for monitoring change, assessing non-linear change in eco-systems and path dependency of such change (see also Schneider and Lane, this volume), as well as the economic implications of these changes (Gitay *et al.,* 2001; Smith *et al.,* 2003).

Turning to global impacts in coastal zones, Nicholls and Lowe (2004 and 2003), examine the possible benefits of GHG mitigation in this impacts sector. A key conclusion is that there is a long-term commitment to sea level rise due to thermal lags in the ocean system. This is not to say that near-term mitigation will not generate avoided impact benefits for coastal zones, just that these benefits will manifest themselves late in this century and beyond (Figure 6).

Also, the coastal zone benefits of mitigation will be largely in the form of "delayed" rather than "avoided" impacts, providing more time for the planning and implementation of effective adaptation measures.

---

8 The main difference between this risk area and the first bar in Figure 1 is the focus in this bar ecosystems, whereas "unique and threatened systems" of Figure 1 refers both to human systems, e.g. island communities, and to specific and unique ecosystems such as coral reefs or mangroves.

1.4.2 Abrupt climate change, distributional issues, and adaptation

This volume also examines a number of key issues relevant to the discussion on benefits, but which are not explicitly accounted for in the global impact assessments outlined above.

Figure 6. Additional people flooded in coastal storm surges

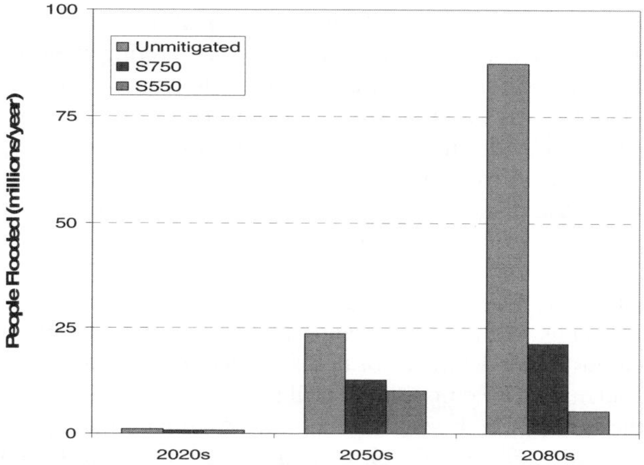

Source: Arnell *et al.,* 2002 as cited in Nicholls, 2003.

Note:

Unmitigated (IS92a), versus mitigated emissions; based on climate scenarios from the HadCM2 model. Assumptions as described in Arnell *et al.*, 2002.

Two papers bring attention to the issue of low probability, high consequence events which are almost always left out of standard impact assessments (Schneider and Lane, this volume; Schellnhuber *et al.,* this volume). Schneider and Lane uses two sets of examples from ocean circulation and atmosphere-biosphere interactions to demonstrate that the response of coupled systems to external forcing, such as increases in GHG emissions and climate change, can push the system from one equilibrium to another – thereby leading to non-linear, abrupt changes. Schellnhuber *et al.* note a variety of "switch and choke" elements in the earth's bio-geophysical systems that might be activated or deactivated by human interference with the global climate. Analysis of the drivers and thresholds for these "switch and choke" elements is still in its infancy, however results from a number of studies suggest that accounting for irreversible, abrupt change is likely to shift the economically "optimal" level of mitigation, calling for more investment in abatement today.[9] Schellnhuber *et al.* go

[9] This issue is addressed in Schneider and Lane (this volume), as well as in Narain and Fisher, 2000; Baranzini *et al.,* 2003; and Mastrandrea and Schneider, 2001; Yohe, 2003 –based on Roughgarden and Schneider, 1999; Azar and Lingren, 2003. In contrast, Nordhaus and Boyer (2000), ignore irreversibility of ecosystem effects and

further, suggesting that any human alteration of these ecosphere modes would have severe impacts and "should be avoided by all reasonable means available."

Schneider and Lane (this volume), argues that the harder and faster a system is disturbed, the higher is the likelihood of such abrupt events, which could be catastrophic. Thus, among the benefits of early and stringent GHG mitigation could be a reduction in the likelihood of such high consequence events.

Another important issue missing from most global assessments of impacts is the distribution of those impacts, their relationship to underlying patterns of wealth and poverty, and the implications of climate change for social justice and economic development patterns across different world regions. Tol *et al.* (2004), conclude that the distribution of adverse climate impacts is heavily skewed towards the poor in the near future, and will deteriorate for more than a century before becoming more egalitarian. Therefore another metric of assessing the benefits of mitigation policies could be based on their equity or social justice implications.

Regarding the interface between adaptation and mitigation benefits and costs, Callaway (2004), points to the lack of consistent approaches incorporating adaptation and its associated costs into estimates of damages due to climate change (and thus, into the benefits of reducing GHG emissions). Callaway (this volume) presents a framework for estimating the benefits and costs of adapting to climate change under uncertainty and illustrates the role of reliable climate information to shift to partial and full adjustment to climate change. His framework points to the need to assess the costs and benefits of adaptation in different contexts. More systematic attention to adaptation would consider how climatic changes intersect with investment, production, and consumption decisions in climate-sensitive sectors to mediate the net observed impacts and provide a better basis for estimates of the benefits of mitigation.

1.4.3 Valuation of impacts and estimating economic benefits of mitigation

The valuation of impacts is a critically important step for economic analyses of policy options to address climate change. Pittini and Rahman (this volume), review recent literature on the social costs of carbon, and identify a number of reasons for the wide variation among published estimates of damage costs. Firstly, the key economic assumptions being used vary widely, including discount rates as well as use of the notion of equity weighting to account for distributional differences in wealth and well-being in different parts of the world. Yet the authors note that a wide range of choices is valid, as these choices depend on one's ethical point of view. Secondly, some key issues are not dealt with consistently (or not addressed at all), in many economic assessments; these include treatment of adaptation, non-market impacts, abrupt climate change and "socially-contingent" impacts.[10] Along with Schneider and Lane (this

---

suggest that catastrophic change would make no difference to the timing of abatement (though it might raise the costs in the distant future).

[10] This could include forced migration due to extreme weather events.

volume), and others, they suggest the possibility of combining economic assessments with complementary approaches that can accommodate different (non-monetary), metrics, such as multi-criteria assessment, so as to take better account of distributional and ethical concerns.

1.5 Towards an improved assessment framework

Beyond summarising what is known today about the avoided impact benefits of climate policies, a number of papers have helped us to broadly reflect on how to set out an improved assessment framework for these benefits.

1.5.1 Bottom-up and top-down perspectives on risk

Wigley (this volume), demonstrates the value to decision makers of probabilistic assessment when considering a wide range of possible climate outcomes in the distant future. By assessing the probability of global mean temperature change over different time frames and under "no policy" and "policy" emission scenarios, he suggests that this type of analysis can be used to enhance our understanding of global impacts in exploring the question: "what is dangerous climate change?"

Any stabilisation target might require large reductions in emissions relative to today (and hence significant costs). However, the choice of target also depends on what types and levels of risk are acceptable. Wigley shows it is possible to identify a range of global emission pathways that limit "unacceptable" risk. Where wide uncertainties exist, probabilistic analysis provides a means to look at how policy can shift the distribution of outcomes and narrow the risk of more extreme outcomes, e.g. what might be clearly dangerous change (see also Schneider and Lane, this volume; Jones, this volume).

Jones (this volume), also underscores that climate change is best managed through a risk-based framework, and highlights the need to include both "bottom-up" and "top-down" perspectives on risk. Considering the likelihood of exceeding critical thresholds, he works with examples of possible local or regional climate impacts (i.e. coral reefs and water resources), to consider, first, local thresholds for coping ranges and vulnerability, then global thresholds for mitigation. Using generally accepted regional thresholds for risk management, such as those in use for basin-wide water resource management in Australia, he connects regional scale impacts to a discussion about global target setting. He also considers the question of how to hedge risks at the global level, asking "What is the level of greenhouse gas stabilisation needed to stay below critical outcomes?"[11]

Both Jacoby (this volume), and Yohe (2004), raise the concern that sound estimates of benefits in a single (monetary), measure are not feasible, given the

---

[11] Jones develops this probabilistic assessment based on information initially provided in the paper by Wigley (this volume).

incommensurable nature of benefits. Yohe suggests the need for a risk-based precautionary approach, while Jacoby observes that in any policy choice, explicit or implicit benefit-cost considerations are inescapable. Decisions about a climate response – stringent or relaxed policies now, tight or loose constraints in the future – do imply some weight of the likely climate benefits from these various strategies. Jacoby therefore recommends the construction of a "portfolio of benefits measures" to include physical and economic metrics of change and structured to provide transparency about embedded assumptions when viewing any particular set of estimates. Jacoby and Yohe also endorse the use of a probabilistic or risk-based framework as a means to manage uncertainty and risk to shape decisions about climate policies.

1.5.2 Conceptual framework

A conceptual framework is proposed here which includes a portfolio of benefits measures.[12] It is composed of three specific elements: global physical variables of climate change; regional-scale impacts of climate change at different levels of global warming; and regional-scale valuation.

The main elements of such a framework include a portfolio of indicators of change, first in physical units and at the sub-global scale, before moving onto monetised and aggregated benefits assessment. The framework suggested here is necessarily partial, with emphasis on the mitigation and direct climate impacts elements of any more comprehensive framework. It is intended to make incremental progress to improve information for policymakers on the benefits of mitigation.

Global physical variables of climate change

The first component of the framework focuses on global physical variables of climate change, expressed in variables that can be analysed in probabilistic terms. At a minimum, this includes reporting on global mean temperature (GMT) change; however, the use of other possible global indicators to report on climate change will also be explored, including regional mean temperature changes, sea level rise, or changes in precipitation. These indicators should be expressed in physical and probabilistic terms – e.g., there is a 10% chance that a given emission pathway would lead to GMT change of 4°C or more by 2080 or 2100. This level of assessment is free of the difficulties of valuation and/or aggregation that arise with economic impacts measures. Though difficult to quantify, it would be useful to qualitatively capture any differences that might arise in the risk of abrupt climate events (e.g. breakdown in thermohaline circulation), under different emission pathways or climate scenarios.

---

12 This is based on Jacoby (this volume), also drawing on inputs and conversations from the OECD Workshop (2002), and expert meeting (2003), (see Appendices, this volume).

Global mean temperature change serves as a proxy for levels of climate change and as a means to integrate both regional and global benefits. Thus on the one hand, GMT change is a main indicator of how a particular mitigation strategy affects climate change. It reveals the change in global warming avoided under different mitigation strategies, for example, that a particular mitigation strategy is expected to limit global mean temperature change in the 21st century to 2°C rather than 3°C or 4 °C compared to a reference scenario without mitigation policy. On the other hand, GMT change can also be used to scale and compare climate change impacts, thus as a common denominator to indicate how climate benefits vary with different mitigation strategies. In this way, the likelihood of a move from one level to another of global warming is used as both an indicator of the benefits of mitigation and as a scale against which other benefits can be assessed.

1.5.3 Regional-scale impacts of climate change at different levels of global warming

A second component of the framework is the characterisation of regional-scale impacts of climate change. These should also be stated in natural/physical units for most impacts, with monetary estimates to be included where the market exists to provide monetary units of measurement. To the extent possible, these impact indicators should also be presented with quantified uncertainty ranges. Uncertainty or risk measures might be presented in relationship to impact thresholds where these have been identified by stakeholders or analysts in the region (Jones, 2001; Pittock *et al.,* 2001; Jones this volume). Should resources permit, it would be desirable to extend the regional scale non-market impacts estimates to monetary estimates, based on the regional valuation literature (see 3.3).

Regional-scale valuation

A third component of the framework is bottom-up, regional scale valuation and regional or global aggregation, as methods and data allow. A principal challenge in implementing this last tier is non-market or natural system impact valuation, as this area is clearly the weakest in the current literature on climate damages (see Tol, 2002a, 2002b; Hitz and Smith, this volume). Nevertheless without valuation and aggregation, the work to improve indicators of benefits from mitigation policies will remain disconnected from wider modelling efforts. Such modelling is needed to show how mitigation today is connected to a longer term pathway leading from emissions to possible climate change effects. Emission pathways will need to shift over time to limit climate change and what is feasible in the future will depend upon what is done today.[13] The path-dependence feature of climate change – with extremely long time lags

13 Path-dependency is found in all "stock" pollutants where the overall quantity of pollution over time is what triggers impacts. It connects near-term pollution-levels to long-term effects such as climate change.

between the time of emission and the time during which the impact is observed – presents a main challenge to today's policymakers.

Improving inputs for economic and integrated assessment models in the form of damage cost curves is an inevitable and essential use for the types of improved regional impact information. A side benefit then of work in this area should be to help to improve damage cost curves, or at a minimum, to test the shapes of the curves already in use or identified in the global impacts literature as highlighted by Hitz and Smith (this volume). Thus exercises in valuation are a critical step and should focus initially on the sectors and/or systems where data and methods exist to test and extend existing knowledge.

1.5.4 Structuring assessment

The framework above suggests at least three layers of assessment to monitor change associated with different levels of climate change in any particular system:

- Mean physical values of the climate change impact at different levels of climate change, with baselines held constant.
- Mean physical values of the climate change impact with dynamic baselines for adaptation, socio-economic, infrastructure and bio-geophysical variables over time.
- A risk measure of climate change impact, taking account of natural variability and uncertainty in other key drivers, and other change over time.
- Where data permit, extension of physical values to economic values for each of the three layers of results outlined above, with consistent economic assumptions for aggregation and discounting.

In addition, a number of criteria have been suggested for use in selection of impact areas of study:[14]

- The significance of climate change as a stress on the system relative to other stress factors.
- Sensitivity of the impacted sector/system to mitigation.
- Existence of well-accepted (no climate policy), baselines for comparison.

---

14 These are based on an exchange among experts in an OECD-hosted meeting, September 2003.

The advocated approach is necessarily partial rather than comprehensive, aiming to improve understanding about a few key impact areas. The framework is designed to generate improved information on the effects of climate change, taken from different perspectives and using the best available information. Once the framework has been tested in a few areas, the work might be expanded to be more comprehensive in its coverage. This is a first essential step required to support the development of improved monetised regional benefits estimates over time as well as aggregate or global level benefits estimates. Providing suggestions for a limited set of (global and regional), impact indicators would not provide the means to tell the complete, complex story about climate change, but it would be an improvement over what is available today to support policy deliberations.

1.6 Next steps

Ideally, the framework will facilitate better communication between impact researchers, integrated assessment and other modellers and climate policymakers on questions surrounding the benefits of strategic global mitigation options. It is not the final solution but merely a start to identify and advance work on an important set of questions.

The framework should therefore be considered a "living document" which is tested and refined over time. An initial starting point would be to identify sensitive sectors or impact areas and, for these, explore selected indicators of impacts and risk as well as what is known about uncertainties in input assumptions. Using risk assessment methods, further work would assess whether it is feasible to analyse and communicate climate change policy benefits through the selected indicators. The work would also explore appropriate temporal and spatial dimensions for assessment, dimensions which might vary by sector or impact area. Testing the framework should be done in collaboration with experts in relevant impact fields but should also inform policy makers of interim results and gather feedback about their usefulness. Ultimately, it will also be important to use insights gained to refine the framework.

This initial round of work has produced a number of new insights about what we know and what we do not know about the benefits of climate change policies. Beyond suggesting a need to test and refine a conceptual framework of a "portfolio of benefits measures", a number of additional priorities for future work have also emerged in our expert-policymaker dialogue (see Box 1). New research is required to answer many of the outstanding questions, including in-depth work on selected impact sectors or systems most vulnerable to climate change. OECD governments are well-placed to sponsor such research. The long-term nature of the climate change problem suggests that such an investment could, over time, provide invaluable information about benefits of global greenhouse gas emission reductions.

Box 1. Additional priorities for future work

Beyond the "portfolio of benefits measures" agenda outlined above, a number of other policy-relevant priorities for the broader research community include:

- a concise review of which impacts have been quantified/modelled in which regions and over which timescales;
- identification and elaboration of selected impacts numeraires for physical impacts, organised by type of impact, level of detail and region;
- probabilistic characterisation of uncertainty and further work on how risk-based approaches can be used in climate policy analysis and research;
- better characterization of eco-system and other natural system impacts, in non-monetary and non-monetary terms;
- assessment of "hot spots" with top-down and bottom-up tools;
- the characterisation of extreme events as drivers for impacts;
- identification of socio-economic constraints and evaluation of eco-system thresholds that might correspond to long term climate predictions, including better characterisation of abrupt, non-linear change;
- improved damage cost functions for use in integrated assessment and other modelling exercises, based on improved physical impact estimates;
- new work to address "science of integration" questions and science-policy networks to improve integrated assessment modeling;
- integrated approaches to mitigation and adaptation.

Source: OECD, 2003.

References

Arnell, N.W., M.G.R. Cannell, M. Hulme, R.S. Kovats, J.F.B. Mitchell, R.J. Nicholls, M.L.Parry, M.T.J.Livermore and A.White (2002), "The Consequences of CO_2 Stabilisation for the Impacts of Climate Change", *Climatic Change,* Vol. **53,** 413-446.

Azar, C. and K. Lindgren (2003), "Catastrophic Events and Stochastic Cost-Benefit Analysis of Climate Change – Editorial", *Climatic Change,* **56**, 245-255.

Baranzini, A., M. Chesney, J. Morisset (2003), "The impact of possible climate catastrophes on global warming policy", *Energy Policy*, **31**, 691-701.

Callaway, J.M. (2004), "Assessing and linking the benefits and costs of adapting to climate variability and climate change", in *The Benefits of Climate Change Policies*, OECD, Paris.

Callaway (2004), "Adaptation benefits and costs: Are they important in the global policy picture and how can we estimate them?" *Global Environmental Change: Special Edition on the Benefits of Climate Policy*, **14**, 273-282.

Corfee Morlot, J. and S. Agrawala (eds.) (2004), *Global Environmental Change: Special Edition on the Benefits of Climate Policy,* volume **14**.

Corfee-Morlot, J. and S. Agrawala (2004), "Overview", in *The Benefits of Climate Change Policies*, OECD, Paris.

Corfee Morlot, J. and N. Hoehne (2003), "Climate change: long-term targets and short-term commitments", in *Global Environmental Change,* **13**, 277-293.

Gitay, H., S. Brown, W. Easterling, B. Jallow (2001), "Ecosytems and their Goods and Services", in *Climate Change 2001: Impacts, Adaptation, and Vulnerability,* A Report of the Working Group II of the Intergovernmental Panel on Climate Change, edited by J.J.McCarthy, O.F.Canziani, N.A.Leary, D.J.Dokken, K.S.White, Cambridge University Press, Cambridge.

Hitz, S. and J. Smith (2004), "Estimating global impacts from climate change", in *The Benefits of Climate Change Policies*, OECD, Paris.

IPCC (2001a), *Climate Change 2001: The Scientific Basis*, A Report of the Working Group I of the Intergovernmental Panel on Climate Change, edited by J.T.Houghton, Y.Ding, D.J.Griggs, M.Noguer P.J.van der Linden, X.Dai, K.Maskell, C.A.Johnson, Cambridge University Press, Cambridge.

IPCC (2001b), *Climate Change 2001: Impacts, Adaptation, and Vulnerability,* A Report of the Working Group II of the Intergovernmental Panel on Climate Change, edited by J.J.McCarthy, O.F.Canziani, N.A.Leary, D.J.Dokken, K.S.White, Cambridge University Press, Cambridge.

IPCC (2001c), *Climate Change 2001: Mitigation,* edited by B. Metz, O.Davidson, R. Swart and J.Pan, Cambridge University Press, Cambridge.

IPCC (2001d), *Climate Change 2001, Synthesis Report*, Cambridge University Press, Cambridge.

IPCC (1995), *Climate Change 1995: The Science of Climate Change*, Contribution of Working Group I to the Second Assessment of the Intergovernmental Panel on Climate Change, JT Houghton, LG Meira Filho, BA Callender, N Harris, A. Kattenberg and K. Maskell (eds.), Cambridge University Press, UK, p. 572.

Jacoby, H. (2004), "Toward a framework for climate benefits estimation", in *The Benefits of Climate Change Policies,* OECD, Paris.

Jochem, E. (2003), paper prepared for the OECD Workshop on Benefits of Climate Change Policies, Paris [OECD/EPOC/GSP(2003)16/FINAL].

Jones, R.N. (2000), "Managing Uncertainty in Climate Change Projections – Issues for Impact Assessment", *Climatic Change,* **45**, 403-419; Part 3/4.

Jones, R.N. (2001), "An Environmental Risk Assessment/Management Framework for Climate Change Impact", *Natural Hazards,* **23**, 197-230.

Jones, R. (2004), "Managing climate change risks", in *The Benefits of Climate Change Policies*, OECD, Paris.

Krupnick, A., D. Burtraw, and A. Markandya (2000), "The Ancillary Benefits and Costs of Climate Change Mitigation: A Conceptual Framework", in OECD *et al.*, *Ancillary Benefits and Costs of Greenhouse Gas Mitigation,* Proceedings of an Expert Workshop, Paris.

Leemans, R. and B. Eickhout (2004), "Another reason for concern: regional and global impacts on ecosystems for different levels of climate change", *Global Environmental Change: Special Edition on the Benefits of Climate Policy* **14**: 219-228.

Mastrandrea, M. and S. Schneider (2001), "Integrated assessment of abrupt climatic changes", *Climate Policy.* **1**, 433-449.

Mendelsohn, R. and M. Schlesinger (1999), "Climate Response Functions", *Ambio,* Vol. 28(4), 362-366.

Morgan, M.G. and M. Henrion (1990), *Uncertainty: a guide to dealing with uncertainty in quantitative risk and policy analysis*, Cambridge University Press, Cambridge.

Narain, Urvashi and Anthony Fisher (2000), "Irreversibility, Uncertainly, and Catastrophic Global Warming", Gianinni Foundation Working Paper 843, Berkeley: University of California, Department of Agriculture and Resource Economics, Berkeley.

Nicholls, R. and J. Lowe (2004), "Benefits of mitigation of climate change for coastal areas", *Global Environmental Change: Special Edition on the Benefits of Climate Policy* **14**: 229-244.

Nicholls, R. and J.A. Lowe (2003), "Case Study on Sea Level Impacts" paper prepared for the OECD Workshop on Benefits of Climate Policy: Improving Information for Policymakers, OECD ENV/EPOC/GSP(2003)9/FINAL.

Nordhaus, W.D. and J.G. Boyer (2000), *Warming the World: Economic Models of Global Warming*, MIT Press, Cambridge MA.

OECD, IPCC, RFF, WRI, The Climate Institute (2000), *Ancillary Benefits and Costs of Greenhouse Gas Mitigation,* Proceedings of an Expert Workshop, Paris.

OECD (2003), "Summary Report" on the OECD Workshop on Benefits of Climate Policy: Improving Information for Policymakers, held 12-13 December 2002, ENV/EPOC/GSP(2003)3/REV1, OECD, Paris.

Parmesan C. and G. Yohe (2003), "A globally coherent fingerprint of climate change impacts across natural systems" in *Nature*, **421**, 37-41.

Pittini, M. and M. Rahman (2004), "The social cost of carbon: key issues arising from a UK review", in *The Benefits of Climate Change Policies*, OECD, Paris.

Pittock, A.B., R. N. Jones, C.D. Mitchell (2001), "Probabilities will help us plan for climate change", *Nature*, 413:249.

Root, T., J. Price, K. Hall, S. Schneider, C. Rosenzweig and J.A. Pounds (2003), "Fingerprints of global warming on wild animals and plants", *Nature*, 421, 57-60.

Rothman, D., B. Amelung and P. Polmé (2003), "Estimating Non-Market Impacts of Climate Change Policy", paper prepared for the OECD Workshop on Benefits of Climate Change Policies, Paris (ENV/EPOC/GSP(2003)12/FINAL).

Roughgarden T. and S.H. Schneider (1999), "Climate change policy: quantifying uncertainties for damages and optimal carbon taxes", in *Energy Policy,* **27**(7), pp.415-429.

Schneider, S.H. and J. Lane (2004), "Abrupt non-linear climate change and climate policy", in *The Benefits of Climate Change Policies*, OECD, Paris.

Schellnhuber, J., R. Warren, A. Haxeltine, and L. Naylor (2004), "Integrated assessment of benefits of climate policy", in *The Benefits of Climate Change Policies*, OECD, Paris.

Shlyakhter, A., L.J. Valverde A. Jr., R. Wilson (1995), "Integrated Risk Analysis of Global Climate Change", *Chemosphere*, **30**, No. 8, 1585-1618.

Smith, Joel B., Hans-Joachim Schellnhuber and Monirul Q. Mirza (2001), "Lines of Evidence for Vulnerability to Climate Change: A Synthesis (Chapter 19)", in *Climate Change 2001: Impacts, Adaptation, and Vulnerability,* A Report of the Working Group II of the Intergovernmental Panel on Climate Change, edited by J.J.McCarthy, O.F.Canziani, N.A.Leary, D.J.Dokken, K.S.White, Cambridge University Press, Cambridge.

Smit, B., I. Burton, R.J.T. Klein, R. Street (1999), "The Science of Adaptation: A Framework for Assessment", in *Mitigation and Adaptation Strategies for Global Change*, **4** (3-4), 199-213.

Tol, R. (2004), "Distributional Aspects of Climate Change Impacts", Special Issue, *Global Environmental Change,* (based on OECD ENV/EPOC/GSP(2003)5/FINAL).

Tol, R. (2002a and b), "Estimates of the Damage Costs of Climate Change, Part 1: Benchmark Estimates" and "Part II: Dynamic Estimates", in *Environmental and Resource Economics*, **21**, 47-73 and 135-160 (respectively).

Webster, M. (2002), "The Curious Role of 'Learning' in Climate Policy: Should We Wait For More Data?" *The Energy Journal,* **23**(2), 97-119.

Wigley, T.M.L. (2004), "Modelling climate change under no-policy and policy emission pathways", in *The Benefits of Climate Change Policies*, OECD, Paris.

Yohe, G. (2003), "Estimating Benefits: Other Issues Concerning Market Impacts", OECD, Paris (ENV/EPOC/GSP(2003)8/FINAL).

Yohe, G. (2004), "Some thoughts on perspective", *Global Environmental Change: Special Edition on the Benefits of Climate Policy* **14**: 283-286.

Chapter 2

ESTIMATING GLOBAL IMPACTS FROM CLIMATE CHANGE

by Sam Hitz and Joel Smith,

Stratus Consulting Inc., United States

We surveyed the literature to assess the state of knowledge with regard to the (presumed) benefits or avoided damages of reducing atmospheric concentrations of greenhouse gases to progressively lower levels. The survey included only published studies addressing global impacts of climate change; studies that only addressed regional impacts were not included. The metric we used for change in climate is increase in global mean temperature (GMT). The focus of the analysis centred on determining the general shape of the damage curve, expressed as a function of GMT. Studies in sea level rise, agriculture, water resources, human health, energy, terrestrial ecosystems productivity, forestry, biodiversity, and marine ecosystems productivity were examined. In addition, we analysed several studies that aggregate results across sectors. Results are presented using metrics as reported in the surveyed studies and thus are not aggregated.

We found that the relationships between GMT and impacts are not consistent across sectors. Some of the sectors exhibit increasing adverse impacts with increasing GMT, in particular coastal resources, biodiversity, and possibly marine ecosystem productivity. Some sectors are characterised by a parabolic relationship between temperature and impacts (benefits at lower GMT increases, damages at higher GMT increases), in particular, agriculture, terrestrial ecosystem productivity, and possibly forestry. The relationship between global impacts and increase in GMT for water, health, energy, and aggregate impacts appears to be uncertain. One consistent pattern is that beyond an approximate 3 to 4°C increase in GMT, all of the studies we examined, with the possible exception of forestry, show increasing adverse impacts. Thus, in total, it appears likely that there are increasing adverse impacts at higher increases in GMT. We were unable to determine the relationship between total impacts and climate change up to a 3 to 4°C increase in GMT. There are important uncertainties in the studies we surveyed that prevent us from a precise identification of 3-4°C as the critical temperature transition range, beyond which damages are adverse and increasing. We are confident in general however, that beyond several degrees of GMT, damages tend to be adverse and increasing. We conclude by suggesting some priorities for future research that, if undertaken, would further our understanding of how impacts are apt to vary with increases in GMT.

ISBN 92-64-10831-9
THE BENEFITS OF CLIMATE CHANGE POLICIES

CHAPTER 2. ESTIMATING GLOBAL IMPACTS FROM CLIMATE CHANGE

by *Sam Hitz and Joel Smith*[15]

1. Introduction

The Intergovernmental Panel on Climate Change (IPCC) has concluded that the costs of reducing emissions to stabilize atmospheric greenhouse concentrations rise with successively lower levels of stabilization. Costs rise as concentrations are decreased from 650 ppm to 550 ppm CO_2 and then rise more sharply as concentrations are decreased from 550 ppm to 450 ppm (Metz *et al.*, 2001). An important question is how the *marginal* benefits, or avoided damages, associated with controlling climate vary with particular levels of mitigation. In other words, what are the (presumed) benefits or avoided damages of reducing atmospheric concentrations of greenhouse gases to progressively lower levels?[16] Do the marginal benefits increase or decrease at successively lower levels of greenhouse gas concentrations? A number of previous studies have attempted to address these questions. Some have focused on quantifying the benefits of stabilizing climate at particular levels, typically expressing those benefits in terms of a single metric, most often dollars, which allows for a direct comparison of the benefits of controlling climate change to the greenhouse gas emission control costs necessary for doing so (e.g., Fankhauser, 1995; Nordhaus and Boyer, 2000; Tol, 2002a). Some use non-monetary units (Alcamo *et al.*, 1998). Others have sought to identify important climate thresholds (e.g., Smith *et al.*, 2001). However, the approaches employed in these studies have some important limitations.

While studies that aggregate impacts from climate change in terms of a single metric provide useful insight about how marginal impacts change, especially at higher levels of climate change, there are a number of concerns with them. One is that the common metric, particularly if it is dollars, may be difficult to apply to sectors that involve services that are not traded in markets and can also undervalue impacts in developing countries. A second is that it may actually be more useful for policy

---

15 Stratus Consulting Inc., PO Box 4059, Boulder, CO 80306-4059, United States, Tel. 1 (303) 381-8000.

16 See Questions 3 and 6 in the IPCC Synthesis Report (Watson and the Core Writing Team, 2001).

purposes to express results sector by sector rather than as a single aggregate, to show how the response to climate change can vary across sectors.

In this study, we identified the global marginal benefits associated with different levels of climate change in a sector-by-sector fashion. We did so based on a survey of primarily sectoral studies that have attempted to quantify global impacts of climate change. Instead of converting impacts to a common metric such as dollars, we retained the different metrics reported by the authors. Our goal was not to develop a single estimate of global benefits across sectors, but to examine the relationships between climate change and impacts in particular sectors to discern any general patterns.

2. Method

We examined the following sectors:

- coastal resources;
- agriculture;
- water resources;
- human health;
- energy;
- terrestrial ecosystems productivity;
- forestry;
- terrestrial biodiversity;
- marine ecosystems productivity

To the extent of our knowledge, no published studies investigated global recreation, tourism, human amenity value, or migration; also, local and regional impacts in these sectors could be substantial (e.g., Lise and Tol, 2002). We also examined recent studies that estimated aggregate impacts (cross sectoral) on a global scale.

We present results using the metrics as they are reported in these studies, which is a broad range of units (e.g., change in GDP, number of people affected, agricultural production, and primary productivity). Each of these metrics has advantages and disadvantages, many of which are discussed in the studies. We note above some limitations associated with using a monetary metric, but also affirm that this sort of metric is appropriate to measure impacts on markets. Number of people at risk is similarly a sensible numeraire in sectors in which ultimately impacts on people are of greatest concern (agriculture, coastal resources, health), and it also has the advantage of allowing for cross-sectoral and regional comparisons to some extent. This metric counts all individuals the same and in some sense is more equitable than the monetary numeraire. But it does not measure intensity of impact (see Schneider *et al.*, 2000 on multiple numeraires).

We used global mean temperature (GMT) as the index for measuring change in climate. For any concentration of greenhouse gases, there is a range of potential changes in climate (Houghton *et al.*, 2001). Furthermore, for any change in GMT there is a range of concomitant changes in global precipitation and other meteorological variables. A wide range of potential regional patterns of climate change is also associated with a particular change in GMT. Variation in these regional patterns can have a profound effect on regional impacts and even net global impacts. Thus, one would expect an examination of the type we undertook to yield a wide range of potential impacts for any given GMT. We used GMT because it is the most feasible index of climate change, but note its limitations (see Smith *et al.*, 2001). Regional impacts are discussed in only limited fashion to highlight the point that they often differ substantially from global impacts.

Our analysis focussed on determining the general shape of the damage curve, expressed as a function of GMT. We attempted to determine whether impacts appear with a small amount of warming and increase with higher levels of warming. If they did, we sought to determine if they would increase linearly or exponentially with increasing GMT, or whether they would stabilise at a particular level. We also looked for thresholds below which there are no impacts and cases where the relationship between impacts and climate change might be parabolic (e.g., net benefits and then damages). These questions are important because their answers determine whether there are benefits associated with lower GMT and whether those benefits remain constant, decrease, or increase as GMT rises.

Most of the studies we examined used output from general circulation models (GCMs) for simulating future climate (typically equilibrium model runs of doubled CO_2 in older studies and transient model runs in more recent studies). We took a cross-model approach, comparing impacts simulated by climate input from different GCMs. One difficulty with such an approach is that not only can factors such as precipitation be drastically different from model to model, but also regional patterns of temperature may differ (making it more challenging to compare regional impacts). A further limitation is that most studies use only a few GCMs, limiting the output we could analyse. Elucidating the relationship between impacts and GMT is not always straightforward given the few data points that most studies provide and we note this as a central limitation of our analysis. Where possible, we used literature on the underlying biophysical relationships with climate to bolster our conclusions regarding the shapes of damage curves.

We also examined the studies to determine how they differed from one another in several important elements. These differences point to some of the limitations of our approach of comparing results across studies. First, the scenarios of climate change that the various studies examined are often quite different. Houghton *et al.* (2001) concluded that GMT could increase by 1.4 to 5.8°C above 1990 levels by 2100. Few of the studies we examined encompass this full range. Furthermore, few studies also considered the impacts from changes in climate other than gradual increase average conditions, such as changes in extreme events or climate variance. Rate of climate change is also an important dynamic that has generally not been examined. Similarly, the time frames examined by most studies typically differ. As noted above, some studies examined results from different climate models in what is essentially a single point in time. Others examined time slices from a dynamic climate model run. Comparing the results from the two different approaches can be problematic, not least of so because of differences in socio-economic variables at different points in time. We also noted differences in the studies with respect to treatment of key factors such as adaptation, socio-economic baseline changes, sectoral interactions (water availability on agriculture for instance), and assumptions concerning biophysical processes such as carbon dioxide fertilization. Studies differ significantly in the role or influence they posit these factors have or the realism with which they are modelled. Finally, while we were interested primarily in global results, the spatial and distributional scales at which studies estimate impacts are often different.

An appendix to this document provides a summary of some key features of the studies we examined.

3. Results

3.1 Coastal resources

We examined two studies that investigated the effects of rising sea level: Fankhauser (1995) and Nicholls *et al.* (1999). A key difference in how adverse impacts from sea level rise were estimated in each of these studies has to do with what was assumed in terms of adaptation. With sea level rise, adaptation typically refers to the decision of whether or not to protect coastal development. Fankhauser assumed an economic paradigm of optimal protection, based on benefit-cost analysis, while Nicholls *et al.* used a more arbitrary approach based on observed practices. The Fankhauser study minimised the discounted sum of three streams of costs — protection costs, dryland loss, and wetland loss — for each region it considered. Central to this effort, Fankhauser estimated the optimal degree of coastal protection, where protection efforts would be undertaken if the benefits from avoided damage were estimated to exceed the incremental costs of additional action. Fankhauser presented the direct costs of sea level rise as a function of the assumed magnitude of that rise.

Nicholls *et al.* (1999) used a flood model algorithm similar to that employed by Hoozemans *et al.* (1993). This algorithm uses transient output from two GCMs along with results from an ice melt model to derive global sea level rise scenarios. Storm surge flood curves are then raised by relative sea level rise scenarios. Nicholls *et al.* estimated land areas threatened by different probability floods arising from several scenarios. These land areas were then converted to *people in the hazard zone* (the number of people living below the 1000-year storm surge elevation). Lastly, the standard of protection was used to calculate *average annual people flooded* (the average annual number of people who experience flooding by storm surge) and *people to respond* (the average annual number of people who experience flooding by storm surge more than once per year).

The results from both Fankhauser (Figure 1a) and Nicholls *et al.* (Figure 1b) suggest that adverse impacts increase linearly with sea level rise. As Fankhauser pointed out, one might expect protection costs to rise nonlinearly with sea level rise, because construction costs of sea walls increase with required height. This might well be the case, but costs of land and wetland loss dominate Fankhauser's bottom line. Ultimately, were wetland loss the only damage associated with sea level rise, this might suggest a levelling off of adverse impacts, since there is a finite area of wetlands to be lost. Fankhauser's results are sensitive to choice of discount rate, and he assumed a discount rate of zero. Nicholls *et al.* projected that the number of additional people in the hazard zone also increases linearly as a function of sea level rise. The results displayed in Figure 1b assume protection standards increase as incomes rise, though not in response to sea level rise. The second curve, which displays the results for people to respond as a function of sea level (those who are apt to migrate out of the coastal zone because of repeated flooding), exhibits a somewhat steeper increase after a 2°C increase in GMT (roughly 0.25 m sea level rise), which is assumed to occur by the 2050s. Nicholls *et al.* indicated that this is due mainly to the increased frequency of flooding within the existing flood plain as sea level rises. The expansion of the size of the flood plain is a smaller effect.

Figure 1a. Coastal resources

Costs of sea level rise in OECD countries

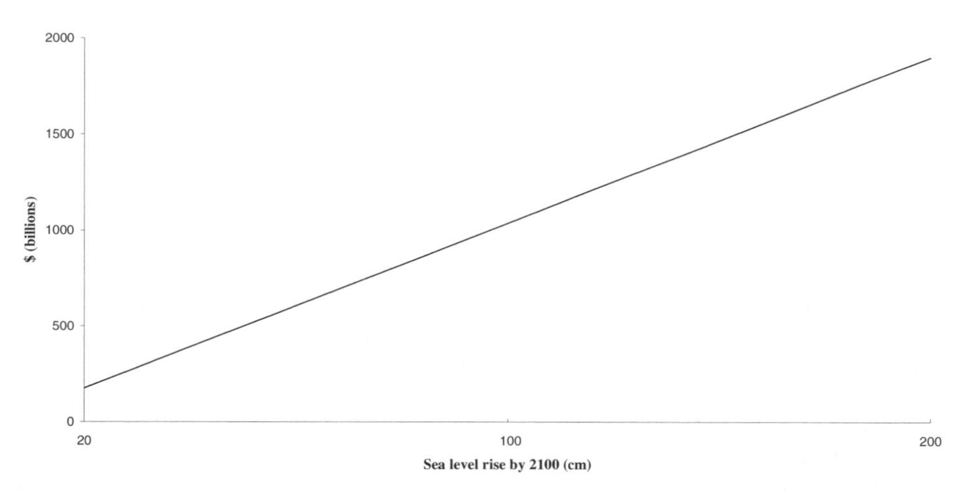

Source: Fankhauser, (1995).

In general, based on these results and the underlying relationship between sea level and impacts, we are highly confident that adverse impacts will increase with GMT increase and sea level rise. While it is impossible to determine whether the relationship between impacts and sea level is a straight line or exponential, the studies we examined are consistent with this more general conclusion; more land will be inundated as sea level rises, damages from higher storm surges will mount, and costs will increase as coastal defences are raised or lengthened to provide necessary additional protection. In addition, there will be other adverse impacts such as increased saltwater intrusion.

Figure 1b. Coastal resources

Additional people in the hazard zone as well as people to respond as a function of temperature

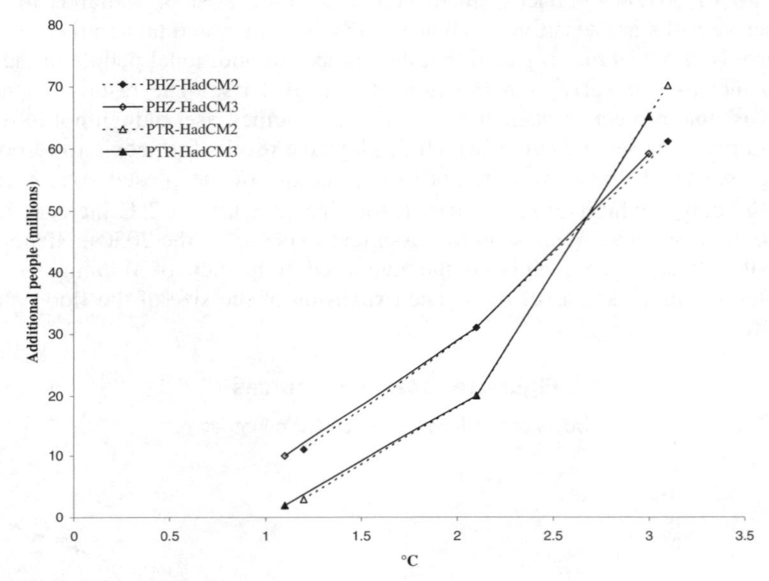

Source: Nicholls *et al.,* 1999.

3.2 Agriculture

We examined five studies that investigated the possible effects of climate change on global agricultural production: Darwin *et al.* (1995), Rosenzweig *et al.* (1995), Parry *et al.* (1999), Fischer *et al.* (2002), and Parry *et al.* (2004). Rosenzweig *et al.* (1995), Parry *et al.* (1999), Fischer *et al.* (2002), and Parry *et al.* (2004) generated estimates of the number of people at risk of hunger (defined as those with an income insufficient to either produce or procure their food requirements). Darwin *et al.* (1995) and Fischer *et al.* (2002) also examined changes in the global production of agricultural commodities.

Rosenzweig *et al.* (1995) used a crop yield model linked to a world food trade model. The Parry *et al.* (1999) model system, like Rosenzweig *et al.*, relied on two main steps, estimating potential changes in crop yields and estimating world food trade responses. Darwin *et al.* (1995) used a framework composed of a geographic information system (GIS) and a computable general equilibrium (CGE) economic model. The basic premise is that climate change would affect not only agriculture but also all manner of production possibilities associated with land and water resources throughout the world, including livestock, forestry, mining, and manufacturing, among others. The resultant shifts in regional production possibilities would alter patterns of world agricultural output and trade. Fischer *et al.* (2002) took a somewhat different approach, developing a global spatial data base of land resources and associated crop production potentials. Current land resources were characterised according to a number of potential constraints, including climate, soils, landform, and land cover. Potential output was determined for each land class for different varieties of crop. Future output was projected by matching the characteristics and extent of future agricultural land to this inventory. The economic implications of these changes in agro-ecology and the consequences for regional and global food systems were explored using a world food trade model, the Basic Linked System. Parry *et al.* (2004) is an extension of the work in the 1999 study and examined impacts on agriculture in light of linked socioeconomic and climate scenarios.

The results of the studies paint a fairly consistent picture of how agriculture might be affected by changes in temperature, with the possible exception of the Parry *et al.* (2004) study. Rosenzweig *et al.* (1995) (Figure 2a) suggests a steeply increasing trend in adverse impacts, measured as a percentage change in the number of people at risk of hunger above about 4°C. In contrast, the results of the low temperature (GISS-A) scenario in the Rosenzweig *et al.* study suggest that benefits might actually exist at lower temperatures. This GISS-A scenario, unlike the other Rosenzweig *et al.* (1995) scenarios, does not incorporate farm level adaptation. Accordingly, benefits at low temperatures might be larger than the Rosenzweig *et al.* (1995) results indicate. It is also clear from the plot that at each level of temperature change, the more optimistic (level 2) scenario of adaptation reduces adverse impacts. While only one low temperature point indicates initial benefits, the results do seem to suggest a parabolic damage curve.

Parry *et al.*'s (1999) results, also shown in Figure 2a, indicate adverse impacts at Parry *et al.* (1999) results, also shown in Figure 2a, indicate adverse impacts at approximately 1°C, and the impacts increase sharply above approximately 2°C. HadCM2, with higher levels of CO_2, seems to lead to predictions of lower risk of hunger in the 2050s and 2080s relative to HadCM3. The fact that these curves become steeper over time may well result as much from a larger, more vulnerable exposed population in 2080 as from increases in temperature.

Darwin *et al.*'s (1995) results (Figure 2b) are more ambiguous, but do indicate a decrease in production in non-grain crops above 4°C. Production in total crops may also begin to decrease above this 4°C threshold. This reduction in total crops is offset by a sharp increase in the production of wheat above 4°C, driven by increases in wheat production in Canada and the United States. Nevertheless, the overall effect is pronounced. The basic trend, with the specific exception of wheat production, remains the same: increasing adverse impacts and increasingly steep impact curves.[17]

---

17 Arnell *et al.* (2002), in a study that in part provides the basis for Parry *et al.* (2001), presented results that are quite similar to these. Though the method is nearly identical

Figure 2a. Agriculture

Percent change in number of people at risk of hunger as a function of temperature

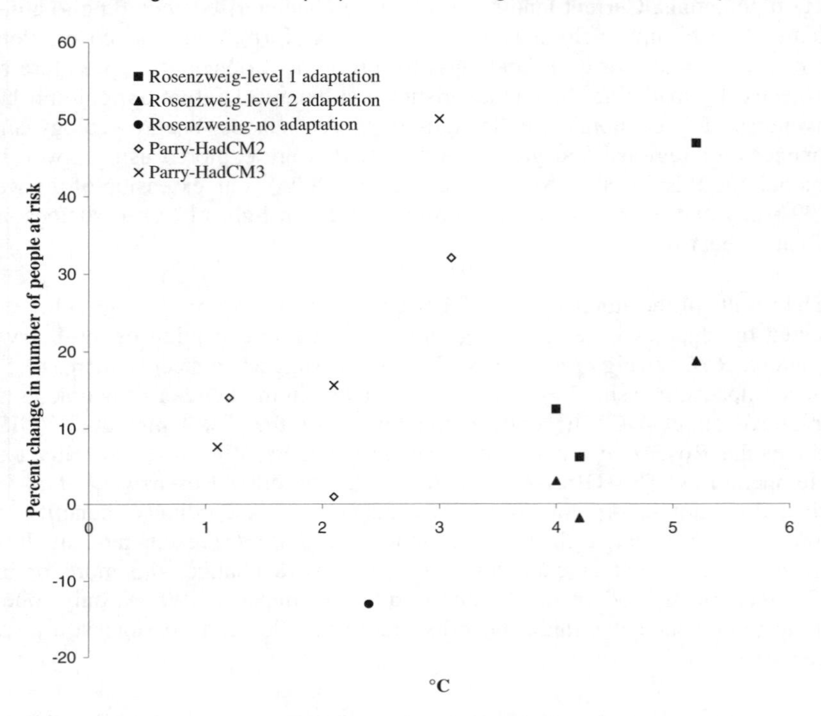

Sources: Rosenzweig *et al.*, 1995 and Parry *et al.*, 1999.

Note: Figure 2a shows results derived from both Rosenzweig *et al.* (1995) and Parry *et al.* (1999). Impacts from Rosenzweig *et al.* represent cross GCM comparisons for an equilibrium doubling of CO_2 and are shown for three different levels of adaptation. Impacts from Parry *et al.* are taken from transient runs of the HadCM2 and HadCM3 GCMs and shown as averages for the decades of the 2020s, 2050s and 2080s.

---

to that employed by Parry *et al.* (1999), the results rely on a single GCM, as do those for the other sectors that Arnell *et al.* (2002) and Parry *et al.* (2001) modelled. Because of the similarity of results, method, and the reliance on a single GCM, we do not discuss either study in detail here but do touch on some aspects of the general method in the Conclusions and Discussion section.

Figure 2b. Agriculture

Percent change in agricultural production as a function of temperature

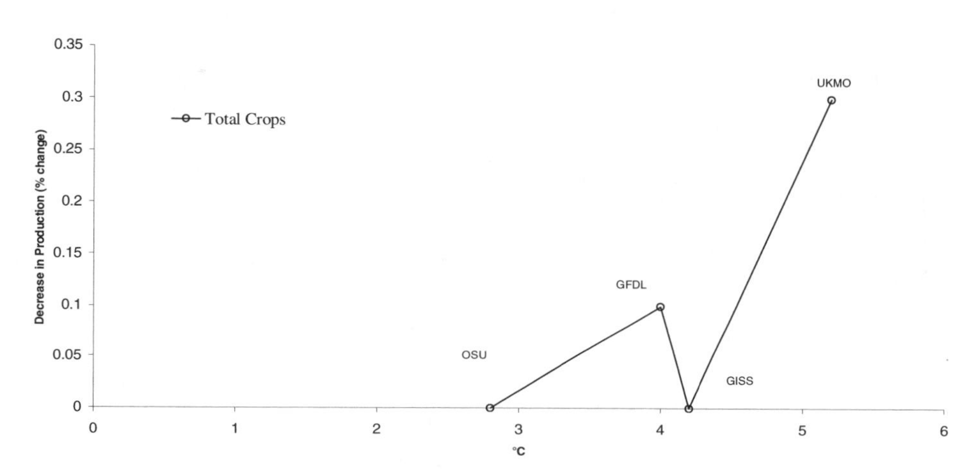

Source: Darwin *et al.*, 1995.

Note: Figure 2b plots results from Darwin *et al.* (1995) that show the change in production for various categories of crops.

Figure 2c. Agriculture

Increase in number of people at risk of hunger due to climate change in the 2080s

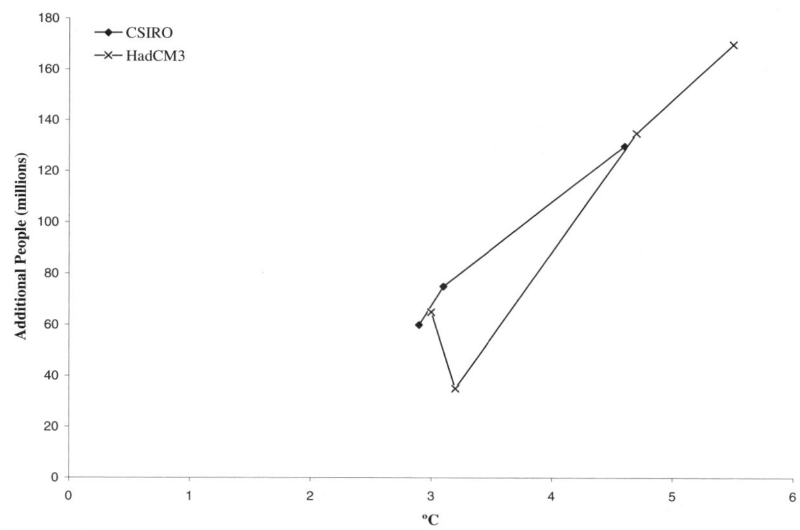

Source: Fischer *et al.*, 2002.

Figure 2d. Agriculture

Agricultural production as a function of GMT for SRES scenarios

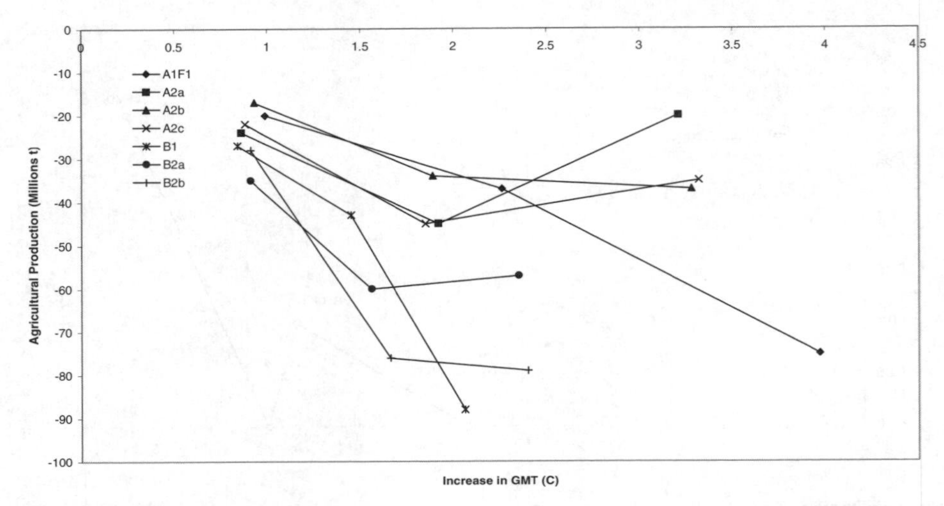

Source: Parry *et al.*, 2004.

Figure 2e. Agriculture

Additional number of people at risk of hunger as a function of increase in GMT

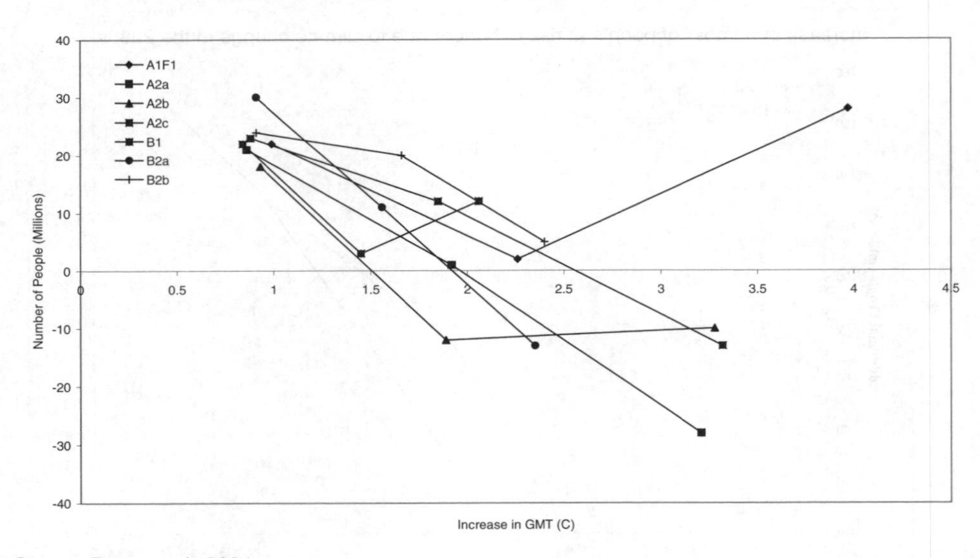

Source: Parry *et al.*, 2004.

Fischer *et al.* (2002) did not present results as a function of global mean temperature. However, by examining temporal results across various scenarios and knowing how temperature changes for the various GCMs and forcing scenarios, we were able to deduce such results. Figure 2c shows the increase in the number of people at risk of hunger as a function of global mean temperature. Results are shown for two GCMs. It should be noted that because presenting results in this fashion relies on looking across scenarios, neither CO_2 nor precipitation is constant. This may help to explain the downturn in number of people at risk in the HadCM3 results. In general, however, both models show that as GMT increases beyond 3°C, the number of people at risk of hunger increases steadily.

The results of the Parry *et al.* (2004) differ from those of the previous work or the other studies examined here. The picture is complicated when a full battery of potential economic changes is considered, as this study did. Changes in production and number of people at risk of hunger because of climate change are shown in Figure 2d and e. The nature of impacts varies with different socioeconomic scenarios; some result in negative impacts while others suggest benefits as temperature rises. Production generally experiences slight to moderate decreases, yet the number of people at risk of hunger actually decreases as temperature rises. The exceptions are A1 and B1, where production decreases more severely relative to other scenarios. In A1, this decrease in production is most likely due to its larger temperature increase, and in B1, the decrease is due to its low levels of CO_2 and consequently low fertilization effect. In both scenarios, the number of people at risk of hunger increases as temperatures rise more, due not only to these decreases in production but also to an increase in the number of poor people in each scenario as the century wears on.

It is important to note that although baseline (socioeconomic impacts decoupled from climate) results are not shown here, in absolute terms, climate change has a much smaller impact compared to socioeconomic change. For instance, the baseline reference indicates increases in production of 4,000 million metric tonnes by 2080 for all scenarios, while climate change produces decreases in production in the range of 20 to 90 million metric tonnes by the same time. Because of the complexity of interpreting the results of a study aimed at investigating the sensitivity of impacts to coupled socioeconomic and climate scenarios and the fact that climate seems to have a small impact on agriculture when set against these larger trends, the inconsistency of the results across scenarios is not surprising. This inconsistency does not change our basic conclusions.

All the studies indicated tremendous variation in regional results for agriculture, which we do not show here. One generalization is that, in most cases, the existing disparities in crop production between developed and developing countries were estimated to increase. These results are a reflection of longer and warmer growing seasons at high latitudes, where many developed countries are located, and shorter and drier growing seasons in the tropics, where most developing countries lie. Results in mid-latitude regions are mixed.

On the whole it appears uncertain whether global agriculture experiences benefits, adverse impacts, or virtually no effect for increases in GMT up to approximately 3° to 4°C. The first four studies, however, estimated increasing adverse global impacts beyond this level. These observations are consistent with the broader literature on agriculture, which shows crop yields declining beyond a global mean increase of approximately 3°C (see Gitay *et al.*, 2001). This phenomenon reflects the knowledge that grain crops, which represent the vast majority of crop revenues, have temperature thresholds beyond which yields decline. Farmers can grow crops at higher latitudes and altitudes to maintain production within optimal temperature ranges, but eventually this geographical shifting cannot compensate for higher temperatures. In addition, carbon dioxide spurs plant growth, but the effect saturates at higher concentrations of CO_2 (Rosenzweig and Hillel, 1998). The combination of higher temperatures exceeding thresholds and saturation of CO_2 can result in declines in crop yields (see Section 3.6). It is also possible that future research and development will result in crops with even higher temperature thresholds or improved ability to take advantage of high CO_2 concentrations. However, if climate change results in increased climate variance, greater threat of pests, substantial reductions in irrigation supply, or less efficient or effective adaptation, the threshold could be lower.

3.3 Water resources

We examined four studies that assessed the potential impacts of climate change on water resources: Alcamo *et al.* (1997), Arnell (1999), Vörösmarty *et al.* (2000), and Döll (2002).

Arnell (1999) used a macro-scale hydrological model to simulate river flows across the globe, and then calculated changes in national water resource availability. These changes were then used with projections of future national water resource use to estimate the global effects of climate change on water stress, and to estimate the number of people living in countries that experience water stress or in counties that experience a change in water stress. Vörösmarty *et al.* (2000) used a water balance model that is forced offline with GCM output to estimate the number of people experiencing water stress. Alcamo *et al.* (1997) used a global water model that computes water use and availability in each of 1,162 watersheds, taking into account socio-economic factors that lead to domestic, industrial, and agricultural water use as well as physical factors that determine supply (runoff and ground water recharge). Some aspects of the model's design and data came from the IMAGE integrated model of global environmental change (Alcamo *et al.*, 1994). The study relied on two GCMs for physical and climatic input. Alcamo *et al.* (1997) estimated the scarcity of water by means of a criticality index, which combines the criticality ratio (ratio of water use to water availability) and water availability per capita in a single indicator of water vulnerability. Döll (2002) used a global model of irrigation requirements, reporting changes in net irrigation requirements. Net irrigation was computed as a function of climate and crop type, with climatic input generated by two transient climate models.

The results from the water studies are far less consistent and conclusive than those of other sectors. Figure 3, based on Arnell's (1999) results, indicates the changes in the number of people living in countries experiencing water stress with increasing temperature. Arguably, it is impacts to this category of people that are most important. However, establishing what constitutes water stress is ultimately a rather subjective step. Nevertheless, there is not much change in water stress by this measure between the 2020s and the 2080s (increases in GMT of roughly 1°C and 3°C, respectively). As might be expected, the relatively wetter HadCM2 model predicts fewer people living in water stressed conditions. Figure 2e also shows the difference between the total population of countries where stress increases and the total population of countries where stress decreases. This measure gives a better sense of the total number of winners versus losers (though one could argue that the gains of winners do not really offset the losses of losers) with regard to changes in water stress, regardless of arbitrary thresholds. The trend is still ambiguous, since one model predicts net loss (HadCM2) and another predicts net gain (HadCM3). Counter to what one might expect, it is the drier model (HadCM3) that predicts a larger population of people in countries where water stress decreases. This is driven mainly by the fact that in the HadCM2 scenario, stress increases in the populous countries of India and Pakistan, while in the HadCM3 scenario, stress decreases in these countries. In both figures, the results are sensitive to large countries flipping from one situation to another. Regionally, the countries where climate change has the greatest adverse impact on water resource stress are located around the Mediterranean, in the Middle East, and in southern Africa. Significantly, these countries are generally least able to cope with changing resource pressures. Overall, these results indicate the importance of the regional distribution of precipitation changes to estimates of water resource impacts.

Vörösmarty *et al.*'s (2000) results indicate that climate change has little effect globally on water resource pressure. The effects of increased water demand due to population and economic growth eclipse changes due to climate. Here again it is important to note regional changes, which are masked by global aggregates. Vörösmarty *et al.* predicted significant water stress for parts of Africa and South America. This is offset by estimated decreases in water stress resulting from climate change in Europe and North America. In general, climate change produces a mixture of responses, both positive and negative, that are highly specific to individual regions. Of course, there is only a limited amount of climate change by 2025, the date at which the Vörösmarty *et al.* analysis ended.

Alcamo *et al.* (1997) presented results that highlight the impact of climate change on future water scarcity for only one point in time, 2075, and for one of the two GCMs that the study employed. The study suggested that, globally, overall annual runoff increases and water scarcity is somewhat less severe under climate change. In a world without climate change, 74% of the world's population is projected to live in water scarce watersheds by 2075. However, with climate change, this figure is reduced to 69%. These results are consistent with those of Vörösmarty *et al.* (2000), suggesting that climate change is not the most important driver of future water scarcity. Growth in water use due to population and economic growth is the decisive factor. Though Alcamo *et al.* (1997) suggested that climate change may ameliorate water scarcity globally, regionally the picture is quite different. Some 25% of the earth's land area experiences a decrease in runoff in the best guess scenario (which combines moderate estimates of future intensity and efficiency of water use) according to Alcamo *et al.* (1997), and some of this decrease is estimated to occur in countries that are currently facing severe water scarcity. The Alcamo *et al.* (1997) results also point to the possibility that industry will supersede agriculture as the world's largest user of water.

Figure 3. Water resources

Impacts on water resources as a function of temperature

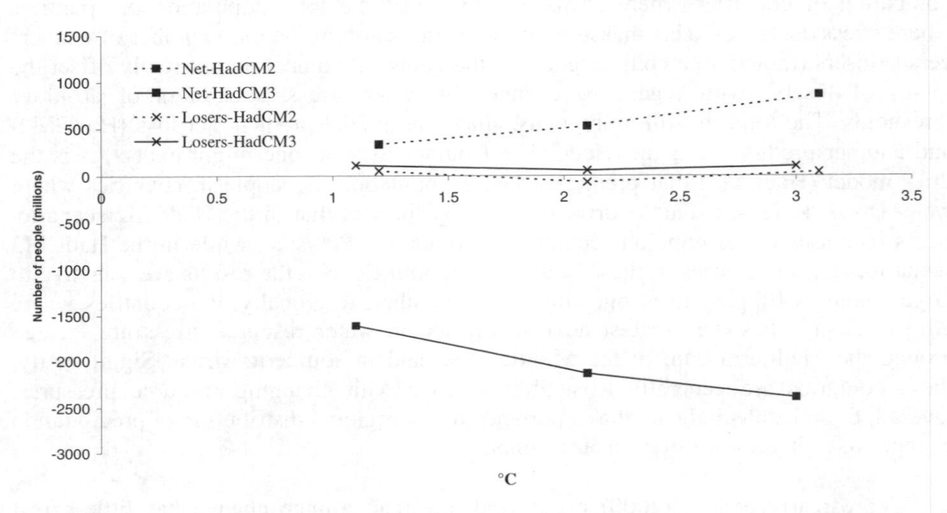

Source: Arnell, 1999. In both cases results are shown as averages for the decades of the 2020s, 2050s, and 2080s.

Note: Figure 3 shows two measures of the impact of climate change on users of water resources, both derived from Arnell (1999). Data represented by an "x" are changes in the number of people in countries using more than 20% of their water resources. This measure focuses on impacts on those people who live in or near a state of water stress. Data represented by a solid square are the difference between the total population in countries where water stress increases and countries where water stress decreases. This measure looks at the number of winners versus losers regardless of the whether they live in a state of water stress or not.

Döll's (2002) results mirror those of Vörösmarty *et al.* When cell-specific net irrigation requirements are summed over world regions, increases and decreases of cell values caused by climate change average out. Irrigation requirements, however, increase in 11 out of 17 of the world's regions by the 2020s, but not by more than 10%. By the 2070s, increases occur in 12 of these regions, 10 of which also show an increase in the 2020s.

The relationship between water resources and climate change appears to be inconclusive. A clear trend did not appear in the studies, perhaps because of the methods used and because of inconsistent changes in regional precipitation patterns across the climate models. Averaging world regions or even countries presents many problems. The water basin is the critical unit for analysis of water resources. Changes in one part of a basin, such as increased or decreased runoff, will affect other parts of the basin. Such changes have little effect outside the basin unless one basin feeds into another or is connected to another via water transport infrastructure. Since basins and transport infrastructure do not necessarily conform to national borders, an analysis based on estimating a uniform change for individual countries may not capture realistic impacts on water resources.

A second critical reason why we do not see a clear relationship between increases in GMT and effects on water resources appears to be inconsistent estimates of changes in regional precipitation. An increase in would increase global mean precipitation. However, the nature of regional changes in precipitation is quite uncertain and varies considerably across climate models. Differences in precipitation patterns from one climate model to another are probably more important than differences in mean temperature in terms of effect on estimates of impacts on water resources. Beyond this, the impacts on water resources are extremely complicated and can depend on such factors as how water is consumed, the ability to adjust uses, legal and institutional constraints, and the capacity to build or modify infrastructure.

Nevertheless, an argument can be made that adverse impacts to the water resources sector will probably increase with higher magnitudes of climate change.[18] This argument is based on two considerations. One is that water resource infrastructure and management are optimised for current climate. The more future climate diverges from current conditions, the more likely it is that thresholds related to flood protection or drought tolerance will be exceeded with more frequency and with greater magnitude than they currently are. The second consideration is that more severe floods and droughts are expected to accompany higher magnitudes of climate change. Some regions might benefit from a more hydrologically favourable climate, but it seems unlikely that the majority of the world's population would see improved conditions, especially since systems are optimised for current climate.

3.4 Human health

The effects of climate change on human health could find expression in numerous ways. Some health impacts would doubtless result from changes in extremes of heat and cold or in floods and droughts. Others might result indirectly from the impacts of

---

18 Parry *et al.* (2001) and Arnell *et al.* (2002) both presented results that suggest steadily increasing numbers of people at risk of water shortage as global mean temperature increases, for both the 2050s and the 2080s. However, they considered only the numbers of people already living with water stress who would experience an increase in stress due to climate change. This approach neglects those people for whom water stress decreases and in general neglects the impacts, negative or positive, on those people who do not currently live in water stressed countries. Essentially, this study considered losers only and provided no sense of net impacts.

climate change on ecological or social systems. Assessing the impacts of climate change on human health in any comprehensive way is extraordinarily difficult. Health impacts are complex and owe their causes to multiple factors. They may lead to increases in morbidity and mortality for some causes and decreases for other causes. Vulnerability will differ from one population to another and within every population over time (McMichael *et al.*, 2001). In general, there is insufficient literature to begin to form other than the most rudimentary conclusions concerning overall health impacts.

Malaria transmission is the only impact category with several studies with good global and temporal coverage. The impacts of climate change on vector-borne disease are unlikely to be limited to malaria (dengue and schistosomiasis are likely possibilities), but malaria might be representative of how climate change may affect the risks of vector-borne diseases in general. Consequently, we focused on four studies that assessed the possible impacts of climate change on the global transmission of malaria: Martin and Lefebvre (1995), Martens *et al.* (1999), van Lieshout *et al.* (2004), and Tol and Dowlatabadi (2002).

Climate change is likely to lead to increased water stress and deteriorate water quality in some areas, which in turn might well increase the incidence of water-borne diseases. Several studies suggest a correlation between average annual temperature and the incidence of diarrhoeal diseases. However, these studies are limited in the range of temperatures they examine or are not yet published. We present the results of one such study, Hijioka *et al.* (2002).

We also examined Tol's (2002a and b) results of how mortality is influenced directly by changes in temperature, both high and low.

3.4.1 Malaria

Martin and Lefebvre (1995) used a relatively simple model of malaria that predicts potential transmission, which occurs when environmental conditions are favourable at the same time and place to both malaria parasites and malaria vectors. The model also makes prediction based on endemicity, distinguishing between seasonal and perennial transmission. They presented results in terms of area of potential transmission.

Martens *et al.* (1999) is based on a model of malaria that is part of the MIASMA model (e.g., Martens *et al.*, 1997; Martens, 1999). This model is more sophisticated than that of Martin and Lefebvre in that it includes estimates of the distribution of 18 different malaria vectors, species-specific relationships between temperature and transmission dynamics, and a more realistic approach on malaria endemicity (epidemics versus year-round transmission). Results were presented in terms of changes in the number of people at risk of malaria infection.

Tol and Dowlatabadi (2002) transformed the results from several studies predicting risk of malaria transmission to actual mortality by assuming that the current regional death tolls from malaria increase as the risk of potential transmission increases with temperature. They also explored the importance of access to public health services on malaria mortality by assuming a linear relationship between regional per capita income and access to public health services and relating the latter to reductions in mortality.

Van Lieshout *et al.* (2004) is a follow-on study to Martens *et al.* (1999) and relies on the same MIASMA model and a similar methodology. However, the study sought to determine the additional population at risk of malaria because of climate change while incorporating a range of projections of socioeconomic development and greenhouse gas emissions. Also, the more recent study disaggregated results by subjectively classifying countries according to their vulnerability to malaria impacts.

In general, the studies portrayed an increase in health risks with increasing temperature. Martin and Lefebvre (1995; Figure 4a) suggested that a global increase of seasonal potential malaria transmission zones is caused by the encroachment of seasonal zones on perennial ones and by the expansion of seasonal malaria into areas formerly free of malaria. The increase in area of potential transmission in all malarious zones seems to be linear and increasing with temperature.

Figure 4a. Human health

Percent change in the extent of malaria transmission as a function of temperature, by type of transmission

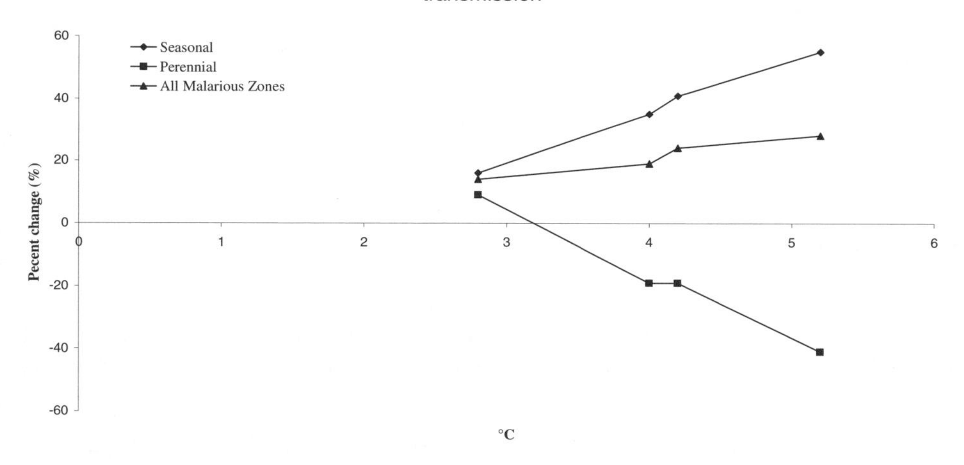

Source: Martin and Lefebvre, 1995.

The results from Martens *et al.* (1999) are shown in Figure 4b. The trends depict additional people at risk for *vivax* and *falciparum* malaria, for both the HadCM2 and HadCM3 models and for different types of transmission. Year-round transmission appears to increase linearly with temperature for both types of malaria parasite. However, the risk of epidemics is reduced and in both cases decreases gradually with temperature. It is more difficult to draw conclusions about seasonal transmission, though for *falciparum*, at least, risk also seems to decrease with rising temperature. In both cases, these measures risk missing potential increases in the actual disease burden. The portion of the year during which transmission can occur might increase, but if the increase is not enough to trigger a change in risk category, as defined in the study, this increase will not register. The results could, however, indicate an expansion of year-round transmission at the expense of seasonal and epidemic transmission, coupled with an expansion at the fringes of malarious zones, mostly likely in the form of epidemic transmission potential.

Figure 4b. Human health

Additional people at risk for malaria as a function of temperature, by type of transmission and parasite species.

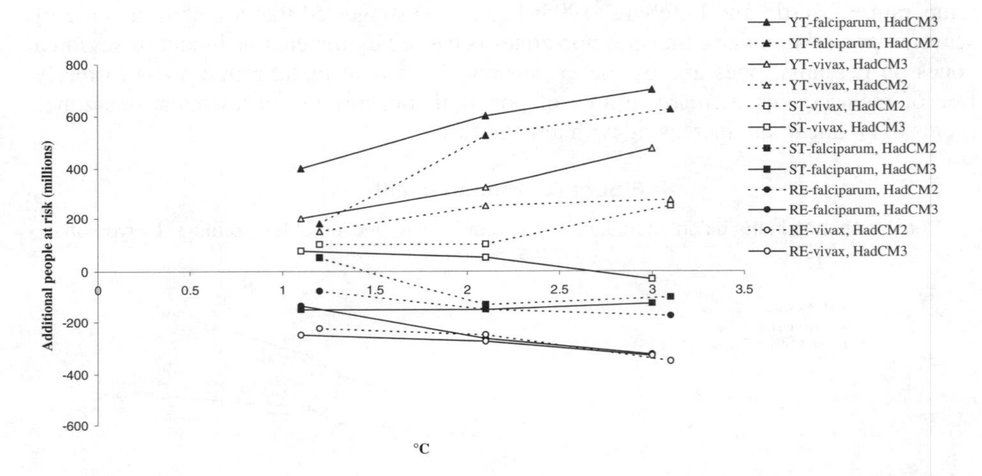

Source: Martens *et al.*, 1999.

Aggregating these various modes of transmission and types of malaria is not straightforward.[19] For instance, an increase in risk of year-round transmission is not

19 Parry *et al.* (2001) and Arnell *et al.* (2002) presented results for additional millions of people at risk of malaria for both the 2050s and the 2080s that suggest a steadily increasing trend between temperature increases of 0° and 3°C. These studies relied on a method and socio-economic assumptions that are quite similar to those of Martens *et al.* (1999). Both studies looked at the total additional population living in an area where the potential for malaria transmission exists. The two studies differed from Martens *et al.* only in how they aggregated results. Results were aggregated across different types of risk, as defined by seasonality of transmission. Total aggregate results included the populations of all areas that experience an increase in potential transmission and where the duration of the transmission season is at least one month per year. Furthermore, results were presented for only one malaria parasite,

necessarily more serious than an increase in risk of seasonal transmission. In fact, the reverse could well be true in many locations. Populations exposed to malaria year-round often develop a higher immunity than do those exposed less frequently (Gubler *et al.*, 2001). Arguably, one could simply sum the number of people at risk for malaria, regardless of endemicity or variety of parasite. Though this clearly mixes types of risk, it would provide some crude indication of how the total number of people exposed to malaria might change with climate. Doing this in Figure 2c would yield an increasing trend, suggesting that the number of people at risk of malaria over the next century does increase. This could be the case. However, such aggregation is inadvisable given that different sorts of malaria risk are likely to have different implications for actual mortality and the pitfalls in interpretation that result from aggregation.

Figure 4c. Human health

Diarrheal incidence per capita per year, shown for four different SRES scenarios

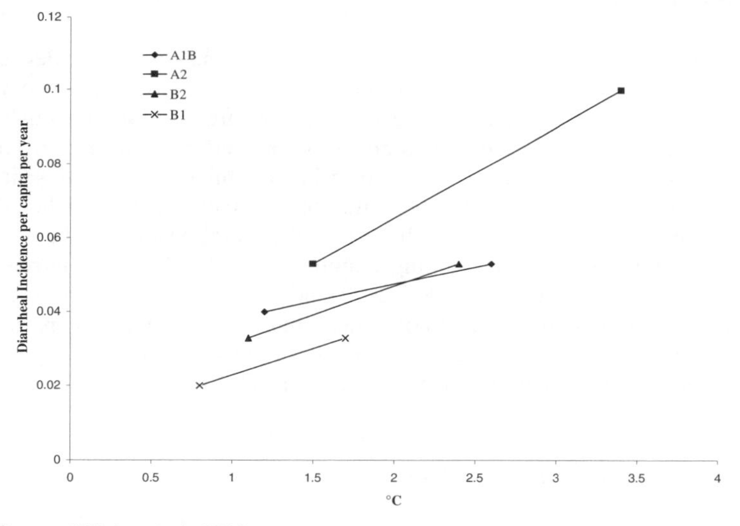

Source: Hijioka *et al.*, 2002.

Tol and Dowlatabadi (2002) took the approach of converting risk of potential transmission to mortality, which allows for aggregation across different endemicities smoothly. However, Tol and Dowlatabadi provided results of mortality as a function of only time and not temperature. They did, though, show that by the last decade of the century, global mortality from malaria is reduced to virtually zero as a result of economic growth and presumably better access to public health services. The effect of socio-economic change appears to overwhelm the negative effect of climate change alone.

---

falciparum, and much of the increase that was indicated is for what is most likely epidemic transmission in developed countries, where public health infrastructure makes it unlikely that such a risk would be realized as a significant disease burden.

The results from van Lieshout *et al.* (2004) are not shown here, but are consistent with those from the previous Martens *et al.* (1999) study. They show increases in population at risk for malaria for some socioeconomic scenarios, and decreases for other scenarios. And as with Tol and Dowlatabadi (2002), socioeconomic changes are far greater than those associated purely with climate change. The authors do report that, in general, the results suggest a northward expansion of areas with climate suitable for transmission. However, more important, the results provide some insight into why consistent trends do not emerge in this study, or in other studies of malaria. The health impacts seem to be confined to specific areas and consequently are highly dependent on assumptions regarding the spatial distribution of population and precipitation. Nonetheless, most of the additional population at risk is projected in countries that have poor current malaria control status, in the eyes of the authors.

3.4.2 Water-borne disease

Hijioka *et al.* (2002) developed a statistical model to explain the current incidence of diarrhoeal disease in 13 world regions. The model relies on two explanatory variables, water supply coverage and annual average temperature. It simultaneously accounts for the reduction in water-borne diarrhoeal incidence resulting from improvements in the water supply coverage and related sanitary conditions in developing countries (due to increasing income) and for the increase in diarrhoeal incidence resulting from the proliferation of pathogens and promotion of putrefaction due to increased temperatures in both developing countries and developed countries. They presented global results for two time slices, 2025 and 2055, for each of the four scenarios they considered. Results, as a function of temperature, are shown in Figure 4c. While there are only two data points for each scenario, these plots indicate that higher temperatures are accompanied by a higher incidence of diarrhoeal disease.

3.4.3 Heat-related and cold-related mortality

Tol (2002a) estimated the effects of climate change on both heat-related and cold-related mortality. With rising temperatures, one would expect a decrease in cold-related mortality and an increase in heat-related mortality. Tol extrapolated from a meta-analysis conducted by Martens (1998) that showed the reduction in cold-related cardiovascular deaths, the increase in heat-related cardiovascular deaths, and the change in heat-related respiratory deaths in 17 countries in the world. Tol concluded that, for the world as a whole, reduction in cold-related mortality is greater than the increase in heat-related deaths initially. He predicted reductions in mortality peak at rather moderate changes in temperature by 2050. From that point on, marginal increases in temperature result in mortality increases. His results are characterised by rather large uncertainty, but suggest that as temperatures continue to rise, reductions in cold-related mortality will be less significant while increases in heat-related mortality will dominate.

3.4.4 Main health findings

Based on our review of the literature and related analysis, we conclude that health risks are more likely to increase than decrease as GMT rises. While the results from the malaria studies we considered do not point to an unambiguous increase in risk as temperatures rise (in fact, underlying principles suggest that high temperatures might increase or decrease the survival of vectors and pathogens they transmit; see Gubler *et al.*, 2001), they do suggest that such an increase in transmission may be more likely than not. However, this may not necessarily translate to an increase in mortality or morbidity. Hijioka *et al.* (2002) also demonstrated that the threat of water-borne diseases may increase as climate changes. The limited results we examined for heat-related mortality suggest that, eventually, as temperatures rise so will total mortality. While demographic and sociological factors play a critical role in determining disease incidence (Gubler *et al.*, 2001), many of these maladies are likely to increase in low latitude countries in particular (heat stress will most likely increase in mid- and high latitudes as well). Low latitude nations have some of the highest populations in the world, tend to be less developed, and thus have more limited public health sectors. It is possible that nations in low latitudes will develop improved public health sectors, but the speed and uniformity of such development are in doubt. Taking all these considerations into account, it seems more likely that mortality and morbidity will rise than fall. We characterise the relationship between human health and climate change as one of increasing damages.

3.5 Energy

We reviewed one study, the only global study of which we are aware, that estimated the effects of climate change on the demand for global energy: the energy sector analysis of Tol's (2002b) aggregate study. Tol followed the methodology of Downing *et al.* (1996), extrapolating from a simple country-specific (United Kingdom) model that relates the energy used for heating or cooling to degree days, per capita income, and energy efficiency. Climatic change is likely to affect the consumption of energy via decreases in the demand for space heating and increases in demand for cooling. Tol, following Downing *et al.*, hypothesized that both relationships are linear. Economic impacts were derived from energy price scenarios and extrapolated to the rest of the world. Energy efficiency is assumed to increase, lessening costs. Tol analysed energy use through 2200 but did not report how temperature changes over this period, so we cannot associate a particular level of net benefits with a given temperature.

According to Tol's (2002b) best guess parameters, by 2100, benefits (reduced heating) are about 0.75% of gross domestic product (GDP) and damages (increased cooling) are approximately 0.45%. The global savings from reduced demand for heating remain below 1% of GDP through 2200. However, by the 22nd century, they begin to level off because of increased energy efficiency. For cooling, the additional amount spent rises to just above 0.6% of GDP by 2200. Thus throughout the next two centuries, net energy demand decreases. Despite the results at 2200, it is reasonable to assume that at high enough levels of temperature change, the increased spending on cooling will eventually dominate the savings from reduced expenditure on heating.

We are highly confident that global energy use will eventually rise as global mean temperature rises, but we are not certain about whether a few degrees of warming will lead to increased or decreased energy consumption. With higher temperatures, demand for heating decreases and demand for cooling increases. One can imagine that a curve relating energy demand to mean global temperature might be "U" shaped. An important question is whether we are already to the right of the low point of such a curve, in which case global energy consumption will rise with higher GMT, or whether we are still on the portion of the curve that foretells decreasing demand (left of the low point), in which case global energy consumption will first decline and then eventually rise as GMT increases. Tol's analysis suggested that we can still look forward to reductions in total consumption. However, Mendelsohn's (2001) analysis of the United States found that energy costs will increase even with an approximate 1°C increase in GMT. Since the United States consumes about one-fourth of global energy, this may be an indication that global energy demand will increase immediately as temperatures rise. Thus, based on the limited literature, we were unable to determine the effective shape of the damage relationship we face.

3.6 Terrestrial ecosystem productivity and change

Climate change could potentially affect a number of physical and biological processes on which the health and composition of terrestrial ecosystems depend. Changes in these ecosystem processes could in turn affect an equally diverse set of services on which people rely, some of which are considered elsewhere in this paper (agriculture, forestry, and biodiversity). However, a significant portion of the overall value of terrestrial ecosystems could be related to non-market sorts of goods and services or services not associated with concrete goods in any sense. Biodiversity is an example of such a good. These are difficult values to measure, and no global studies of which we are aware have attempted to quantify the impacts of climate change on terrestrial ecosystems by estimating the values of these sorts of services. Instead we focused on studies that examined the general health and productivity of terrestrial ecosystems and presumably their ability to deliver a wide range of services.

We examined two studies of the effects of climate change on terrestrial ecosystems, White *et al.* (1999) and Cramer *et al.* (2001), both of which model net ecosystem productivity (NEP), net primary productivity (NPP), and total carbon. A third study, Leemans and Eickhout (2004), looks at shifts in the extent of ecosystem types with climate change.

Figure 5 depicts the global changes in NPP and NEP as function of GMT change from White *et al.* (1999). NPP increases fairly steadily until the 2050s, or about 2°C, at which point it begins to level off. This global trend reflects an increase in NPP of northern forests in response to warming and increased atmospheric CO_2 concentrations and in some places precipitation. However, NPP decreases in southern Europe, the eastern United States, and many areas of the tropics. NEP, the difference between NPP and heterotrophic respiration, represents the net flux of carbon from between land and the atmosphere. Decreases in NEP appear after about 1.5°C of warming. The decreases in NEP were associated with the decline or death of tropical or temperate forests. Thus, White *et al.* predicted a growing terrestrial carbon sink at lower temperatures, but a collapse and reversal of this sink at higher temperatures. Leemans and Eickhout, citing a 1999 study by Cramer *et al.*, suggest that this reversal occurs somewhere between 2 and 3°C. Similarly, Cramer *et al.* (2001) indicated that the terrestrial carbon sink begins to level off by 2050 and decreases by the end of the century.

Figure 5. Terrestrial ecosystems

Change in net primary productivity and net ecosystem productivity as a function of temperature

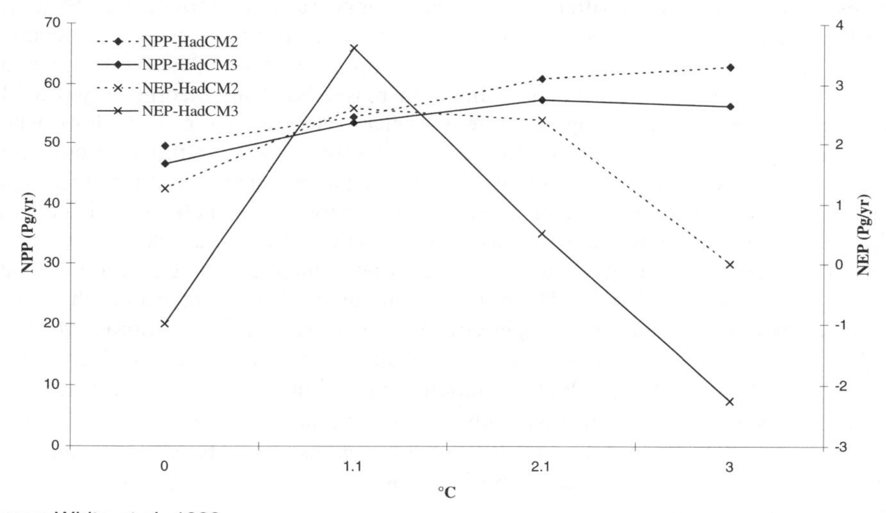

Source: White *et al.*, 1999.

Note: Results are shown as averages for the decades of the 2020s, 2050s, and 2080s.

It is reasonable to expect that the relationship between increased GMT and ecosystem productivity is parabolic. Higher atmospheric carbon dioxide concentrations will favourably affect plant growth and demand for water (although change in growth may not result in increased biomass in natural, unmanaged, systems). Higher temperatures, particularly if accompanied by increasing precipitation, could also initially be favourable for plant growth. Eventually, the increased growth will peak and then decline as the carbon dioxide fertilisation effect begins to saturate at higher CO_2 concentrations (approximately 600 to 800 ppm for C_3 plants; Rosenzweig and Hillel, 1998). Additionally, higher temperatures exponentially increase evapotranspiration, thus increasing water stress to vegetation. In summary, there are biophysical reasons to expect vegetation productivity to increase with a small rise in global mean temperature, then peak, and eventually decline. The modelling results of the White *et al.* (1999) and Cramer *et al.* (2001) studies are consistent with this hypothesis.

Leemans and Eickhout more generally addressed large-scale compositional impacts on ecosystem and landscape patterns. They looked at how climate change would affect the distribution of ecosystems and NEP over the planet, for GMT changes of 1°, 2°, and 3°C by 2100. Climate change was obtained through the standardised IPCC pattern scaling approach (Carter *et al.*, 2001).

The simulated shifts in ecosystems that Leemans and Eickhout reported increase with larger temperature increases. A 1°C warming alters approximately 10% of all ecosystems (89.6% of all ecosystems are stable). At 2°C and 3°C, 16% and 22% of all terrestrial ecosystems change, respectively. There are large differences in specific ecosystems. Even small climate changes will have substantial consequences on temperature-limited ecosystems such as tundra. And, as the authors point out, net changes in ecosystem extent often obscure the disappearance of ecosystems. However, not all changes are alike. The authors characterized changes in extent as negative, positive, or neutral, depending upon the succeeding vegetation. Positive changes are typically characterised by a shift that results in increased NEP and theoretically provides more opportunities for managing ecosystem services. Neutral changes are those where current ecosystems are replaced by new ecosystems with similar productivity characteristics but composed of different species. Negative changes are those that depict a decline in use opportunities and a release of carbon. The analysis indicated that positive and neutral NEP impacts increase with climatic warming, at least up to a point. However the authors are quick to point out, these changes are based on climatic potential, not actual dynamics. There is substantial evidence suggesting that many ecosystems cannot keep pace with rapid climate change and might deteriorate, resulting in rapid carbon loss to the atmosphere. It is possible then that the positive NEP effects the model results suggest might be substantially reduced in magnitude. In fact, Leemans and Eickhout's results indicate that with an increase of 3°C over the course of the current century (0.3°C/decade), only 30% of ecosystems might be able to adapt. The ability of species to migrate and establish themselves in new habitats in response to climate change is a critical uncertainty.

While acknowledging the possibility that their terrestrial productivity model predictions are overstated (increases in productivity might not be fully realized), we believe that their results are generally consistent with those of the other two ecosystem productivity studies we examined, at least at lower rates of climate change. Ecosystems could be negatively impacted in other important ways. In fact, Leemans and Eickhout's analysis did go further to examine multiple indicators of ecosystem impacts. Some of these indicators suggest that even small changes in climate may have other impacts on terrestrial ecosystems that are detrimental (see discussion of biodiversity that follows).

3.7 Forestry

We present temperature correlated results from one study of the impacts of climate change on global forestry, Sohngen *et al.* (2001). Other global studies of the forest sector exist (e.g., Perez-Garcia *et al.*, 2002), but do not generally present results as a function of temperature or do not evaluate the long-term economic consequences of impacts on forests.

Sohngen *et al.* estimated impacts of climate change on world timber markets. Their analysis was designed to not only capture the climate change driven ecological impacts on forest growth and distribution but also provide insight into how landowners and markets adjust and adapt to global climate change.

Sohngen *et al.* detailed changes in consumer and producer surplus under several scenarios that describe how timber species might move across landscapes in response to changing climatic conditions. Sohngen *et al.* also explored, via sensitivity analyses, the effect of higher or lower interest rates, assumptions about the ability of forests to expand, and future competition for plantation sites in the tropics. The general results were the same. Global timber supply increases and prices decline under all scenarios and assumptions. Global net surplus increases, consumers benefit because prices are lower, high latitude producers tend to lose, and low to mid-latitude producers tend to gain. Figure 6 depicts results for timber production. Global yields clearly increase over time because of two factors. First, climate change increases the annual growth of merchantable timber by increasing NPP. Second, the BIOME3 model predicts a pole-ward migration of more productive species, which tends to increase the area of these more productive species.

However, while global forest yields rise, output seems to be only loosely coupled to global temperature increases. Both the Hamburg and UIUC GCM models show comparable gains in yield at each time step, though their underlying global temperature predictions are quite different (approximately 1°C versus 3.4°C). The higher temperature scenario, UIUC, predicts slightly lower benefits than the low temperature Hamburg scenario.

We would expect the economic results for forestry to roughly track biophysical changes in terrestrial vegetation. When growth is estimated to increase, production should rise as well. If growth decreases at some point, production should too. This is also the case in agriculture. We are limited in our conclusions by a lack of forestry studies that correlate results to temperature. Furthermore, the complexities of lags resulting from decadal-long harvesting times make it difficult to draw conclusions about the impacts of rising temperature on forestry. Also, the slow dispersal times of unmanaged forest ecosystems could well limit their adaptive capacity, and reduce projected benefits. However, it does appear that everything else equal, both climate change scenarios in Sohngen *et al.* result in benefits, albeit the scenario with higher GMT has slightly lower benefits. This suggests, but does not confirm, that the relationship between GMT and global forest production is parabolic. However, without the benefit of studies that look at wider range of climate changes, we were unable to draw a more definitive conclusion.

Figure 6. Forestry

Percentage change in timber production for three 50-year time periods

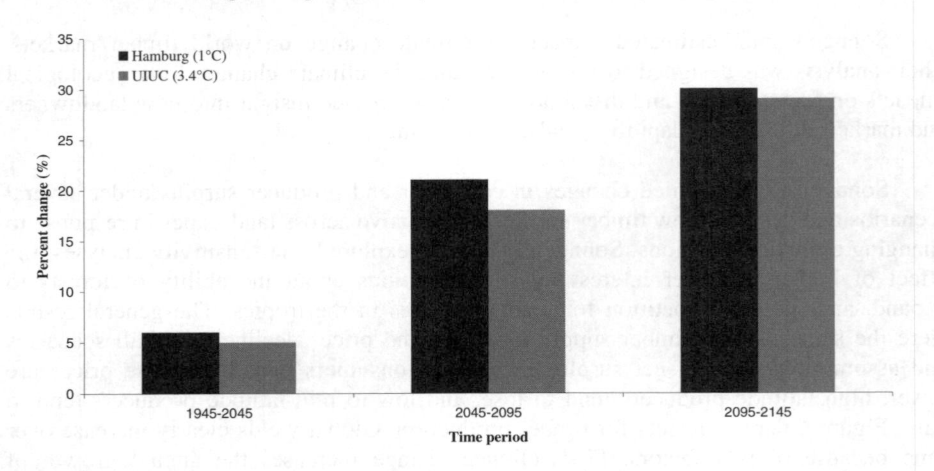

Source: Sohngen *et al.*, 2001.

Note: Figure 6 [based on Sohngen *et al.* (2001)] shows the change in timber production for three time periods and two different GCMs. Both GCMs were subjected to a doubling of CO_2, where equilibrium was assumed to occur at 2060.

3.8 Terrestrial biodiversity

We examined two studies that inform speculation regarding the impacts of climate change on global terrestrial biodiversity: Halpin (1997) and Leemans and Eickhout (2004).

Estimating the impacts of climate change on the global abundance and distribution of biodiversity is challenging. Halpin hypothesised that the survival and distribution of terrestrial plant and animal species depend on the distribution of the climates on which they depend. Specifically, he estimated the percentage of biosphere reserves that might experience a significant change in "ecoclimatic class" as well as the global average change for all terrestrial areas. A change from a current ecoclimate class to a different class was interpreted as a significant climate impact for a reserve site. The analysis predicted sites where the climatic change falls within the existing climatic range of the bioreserve and sites where the projected change exceeds the current range. It was presumed that biodiversity in reserves that have a change in climate will be threatened.

Leemans and Eickhout took a similar approach, examining ecosystem change in nature reserves. They also assume that when current vegetation disappears, it is highly unlikely that the original protection objectives of reserves can be met.

Figure 7, derived from Halpin's (1997) analysis, displays the frequency with which biosphere reserves and terrestrial areas in general experience a change in ecoclimatic class as a function of temperature. With the exception of a hitch around a 4°C change, presumably due to the difference in precipitation between GISS and GFDL, the trend is generally increasing and linear. While the GCM scenarios project major changes in the distribution of ecoclimate classes at a global scale, the more important point is that the frequency of ecoclimatic impacts on reserve areas is generally higher than the global averages. Halpin suggested a fairly straightforward explanation. The global distribution of reserves has a northern spatial bias because of the greater abundance of land mass at mid- and high northern latitudes and the fact that northern industrialised nations maintain more reserve sites. This bias coincides with the larger magnitude of climate impacts in high latitude regions projected by the GCMs that Halpin used. This produces higher rates of climate change for reserve sites than the average for terrestrial areas.

Figure 7. Terrestrial biodiversity

Percent change in ecoclimatic classes for biosphere reserves compared to global average

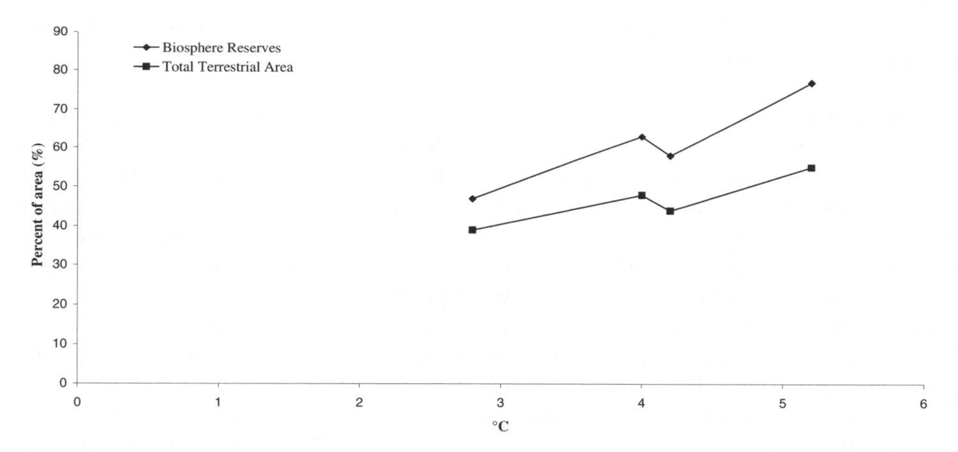

Source: Halpin, 1997.

Leemans and Eickhout conclude that ecosystem changes in nature reserves are similar to the more general patterns they report, but by definition are negative in this case, given that the chief goal of reserves is the conservation of current ecosystems. With a 3°C increase in GMT, half of all nature reserves will be incapable of upholding their original conservation objectives. In fact, negative impacts are likely to increase faster in reserves given their uneven distribution and tendency to be located in exposed or sensitive biomes, both of which reduce their inherent adaptive capacity.

It seems highly likely that larger increases in GMT this century will result in more losses of biodiversity for two reasons. Many species may be able to tolerate a limited level of change in climate, but at higher levels of change, tolerance thresholds will be exceeded. Higher GMTs also mean faster rates of change in climate, which will exceed the ability of increasing numbers of species to adapt. In addition, the threat to biodiversity from climate is much larger when considered in conjunction with the pressures of development. Habitat fragmentation and pollution, among other factors, already threaten many species. In combination with climate change, the loss could be larger (Peters and Lovejoy, 1992). We are highly confident that biodiversity will decrease with increasing temperatures; what is uncertain is whether the relationship between higher GMT and loss of biodiversity is linear or exponential.

3.9 Marine ecosystem productivity

We examined one study that analysed changes in the production of marine ecosystems due to climate change: Bopp *et al.* (2001). They investigated how climate change might affect marine primary production (production by marine plants, including phytoplankton and seaweeds). As with terrestrial ecosystem productivity, this one metric is limited and does not directly translate into fish productivity or changes in biodiversity. However, any changes in primary production would propagate up the marine food web and consequently indicate the possible effects of climate change on productivity of marine ecosystems in general.

Both biogeochemical models employed by Bopp *et al.* predicted similar responses to climate change. At $2xCO_2$ they predicted a 6% global decrease in export production (that portion of marine primary productivity that is transported below 100 m) and showed opposing changes in the high and low latitude regions. Climate induced changes in the ocean decreased export production by 20% in the low latitudes, but increased it by 30% in the high latitudes. The results in the economically important fisheries region of the equatorial Pacific indicate that export production decreased by 5 to 15%. In general, changes in production are driven by reduced nutrient supplies in the low latitudes and an increased light efficiency in the high latitudes, leading to a longer growing season there. Both changes result from increased stratification in the upper ocean. Results were not reported for lower levels of climate change, so it is not possible to determine if global export production declines with smaller increases in global mean temperature.

With only one study containing few data points, it is difficult to draw conclusions about how marine ecosystem productivity is related to increased GMT. Clearly, at some point, increasing GMT leads to reduced marine ecosystem productivity. It is reasonable to assume that further increases in GMT lead to further decreases in productivity, but we are uncertain about the relationship between GMT and marine ecosystem productivity for temperature changes less than those considered by Bopp *et al.*

3.10 Aggregate

We examined two studies that analysed the aggregate global impacts of climate change across a number of sectors and expressed results in monetary terms (as a percentage of economic output): Tol (2002a, 2002b) and Nordhaus and Boyer (2000). Other aggregate studies focus on market impacts only or are limited in geographic scope (e.g., Mendelsohn and Neumann, 1999; Mendelsohn *et al.*, 2000a, 2000b; and Mendelsohn, 2003).

Tol's (2002a and b) study considered impacts of climate change on agriculture; forestry; species, ecosystems, and landscapes; sea level rise; human health; energy consumption; and water resources. He conducted both a static analysis of the impacts of a 1°C change in global temperature on the present situation and a dynamic estimate of the potential impacts over the 2000-2200 period, taking into account the vulnerability of regions to impacts (changes in population, economies, and technology).

Tol's results showed that the impacts of climate change can be positive as well as negative, depending on the sector, region, or time period being combined. The impact on overall welfare depends on how one aggregates results. Aggregating results across regions, even when results are expressed in monetary fashion as they were in Tol's analysis, is problematic. Tol aggregated static results both as simple sums and in an equity weighted fashion, where average income by region determines the weighting factors. The simple sum results in a 2.3% increase in income globally, but risks unduly emphasizing impacts on the rich, whose marginal utility of income is apt to be less than that of the poor. The equity weighted sum reduces this figure to 0.2% of income. The picture of dynamic results was also mixed (Tol, 2002b). There are both positive and negative impacts for different regions at different points in time. Dynamic results were not aggregated globally.

Nordhaus and Boyer's (2000) aggregate analysis relied on an integrated model. The model took a willingness-to-pay (WTP) approach, estimating the insurance premiums different societies are willing to pay to prevent climate change and its associated impacts, particularly catastrophic events. Parameters were estimated based on existing studies, modification of existing results, guessing, and survey results. The Nordhaus and Boyer analysis is unique among aggregate studies in its attempted inclusion of non-market and potential catastrophic impacts as well as market impacts. The study estimated impacts in agriculture, sea level rise, other market sectors, health, non-market amenities, human settlements and ecosystems, and catastrophic events.

Nordhaus and Boyer presented aggregate damage curves for regions and by weighted summation, where weights are based on population or projected 2100 regional output. The global average of damages for a 2.5°C warming is 1.5% if weighted by output or 1.9% if weighted by 1995 population. For most countries, market impacts are small in comparison to the willingness to pay to avoid the possibility of potential catastrophic impacts. The large uncertainty associated with these WTP estimates implies that there is great uncertainty associated with the overall results.

The few studies on aggregate impacts of climate change consistently estimated that there will be damages beyond approximately 2 to 3°C of increase in GMT (Figure 8). Damages were estimated to continue increasing at higher increases in temperature. This is consistent with aggregate studies that focus on market impacts as well (e.g., Mendelsohn and Neumann, 1999; Mendelsohn *et al.*, 2000a). Disagreement among the studies concerns what happens for smaller increases in GMT. Some studies showed net benefits for a small amount of warming, while Nordhaus and Boyer showed damages at such levels. Thus, the aggregate studies did not present consistent results concerning the shape of the damages curve, but did consistently show increasing damages at higher magnitudes of climate change. For a few degrees of increase in mean global temperature, aggregate impacts appear to be uncertain.

Figure 8. Aggregate

Estimated aggregate global damages

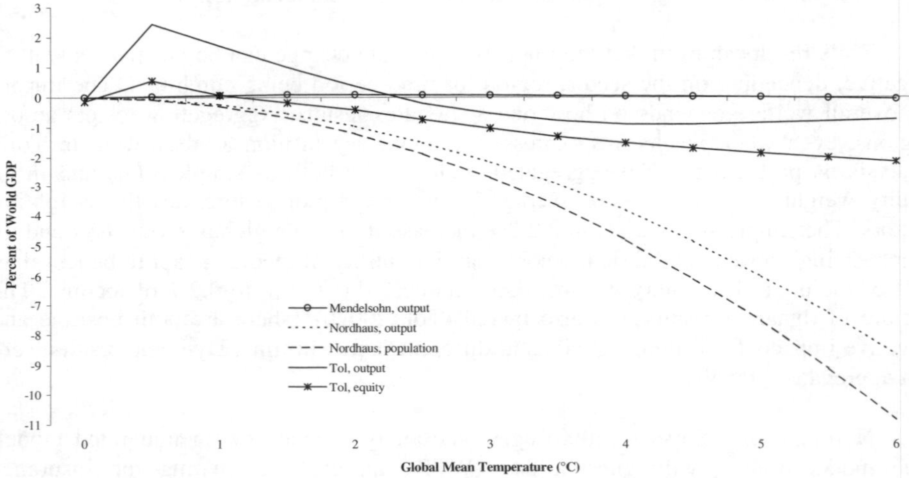

Percent of World GDP

Mendelsohn, output
Nordhaus, output
Nordhaus, population
Tol, output
Tol, equity

Global Mean Temperature (°C)

Sources: Mendelsohn and Schlesinger, 1997; Nordhaus and Boyer, 2000; Tol, 2002b.

4. Conclusions and discussion

Table 1 summarises the patterns in the studies we examined by sector. It is clear that the relationships between GMT and impacts are not consistent across sectors. Some sectors exhibit increasing adverse impacts with increasing GMT. Since the data reported in the studies are limited, we were generally unable to determine if these relationships are linear or exponential. Some sectors are characterised by a parabolic relationship between temperature and impacts, and for the others the relationship is indecipherable. The table also indicates our subjective level of confidence in our conclusions regarding the nature of the relationship between GMT and impacts in those sectors where a determination is possible.

Table 1. Summary of sectoral damage relationships with increasing temperature

Sector	Increasing adverse impacts[a]	Parabolic	Unknown	Confidence
Agriculture		X[b]		Medium/Low
Coastal	X			High
Water			X	
Health	X[c]			Medium/Low
Terrestrial ecosystem productivity		X		Medium
Forestry		X[d]		Low
Marine ecosystems	X[e]			Low
Biodiversity	X			Medium/High
Energy			X	
Aggregate			X	

Notes:

a. Increasing adverse impacts means there are adverse impacts with small increases in GMT, and the adverse impacts increase with higher GMTs. We are unable to determine whether the adverse impacts increase linearly or exponentially with GMT.

b. We believe this is parabolic, but predicting at what temperature the inflection point occurs is difficult due to uncertainty concerning adaptation and the development of new cultivars.

c. There is some uncertainty associated with this characterisation, as the results for the studies we examine are inconsistent. On balance, we believe the literature shows increasing damages for this sector.

d. We believe this is parabolic, but with only one study it is difficult to ascertain temperature relationship, so there is uncertainty about this relationship.

e. This relationship is uncertain because there is only one study on this topic.

We did not aggregate damages across sectors explicitly, given that our primary concern was to determine the general shape of damage functions and assess the consistency of results within sectors. Given this focus on deciphering trends in sectoral impacts, the magnitude of those impacts is less important in this analysis. In fact, in many of the sectors we examined, the results vary widely both within studies, from scenario to scenario or GCM to GCM, and between studies. Furthermore, given that different studies seldom use precisely the same scenarios and make precisely the same socio-economic baseline assumptions, aggregating even within a sector is fraught with difficulty. Aggregation across sectors is hindered by the fact that impacts in different sectors are expressed in different metrics.

Since the different sector studies did not demonstrate a consistent relationship over the full range of temperature increase they collectively examine, and since we did not aggregate across sectors explicitly, it is not possible to draw a definitive conclusion about whether impacts, when taken together, generally increase or take on a parabolic form over this range.

That said, one consistent pattern is that by an approximate 3 to 4°C increase in global mean temperature, all of the studies we examined, with the possible exception of those on forestry, suggest adverse impacts. It appears likely that as temperatures exceed this range, impacts in the vast majority of sectors will become increasingly adverse. Although many studies point to substantive impacts below this temperature level, there is no consistency; in some cases they are negative and in others positive.

4.1 Uncertainties

A number of important sources of unresolved uncertainty underlie this conclusion. We do not believe that these uncertainties cast significant doubt on the basic shape of damage curves we characterize. However, if resolved they might well warrant a reconsideration of our identification of 3-4°C as the point beyond which damages are adverse and increasing and would shed more light on the nature of impacts at temperatures below this range. Many of the studies we considered did not appropriately account for, or simplified, important factors that could influence our conclusion. For instance:

- The bulk of the current generation of global impact studies assumes only a change in average climate and does not address changes in climate variance. Changes in variance are plausible and have already been observed to some extent (Timmerman *et al.*, 1999; Easterling *et al.*, 2000). More impact studies are attempting to model the impacts of changes in variance at the sector level (for example in agriculture, Chen *et al.*, 2001 and Rosenzweig *et al.*, 2002), but results are too preliminary to determine how the totality of impacts in a sector would be affected.[20]

- Impact studies tend also to make simple assumptions about adaptation. It is difficult to predict exactly how affected parties will react. Smit *et al.* (1996) and West and Dowlatabadi (1999), among others, point out the complexities involved in adaptation. Adaptations in response to rapid changes in climate or changes in variance are likely to be ever more difficult to predict (Callaway, this volume). Additionally, much existing consideration of adaptation fails to account for the cost of these adaptations.

- The speed and nature of economic and technological development also raise important questions about the vulnerability of tomorrow's systems to climate change. Tol and Dowlatabadi (2002) concluded that large increases in income

---

[20] Chen *et al.* (2001) and Rosenzweig *et al.* (2002) are for the United States alone. They showed increased variability reducing the magnitude of gains in U.S. agriculture, but not necessarily resulting in net losses.

could substantially reduce the vulnerability of people in developing countries to induce malaria mortality by climate change.

- Many impact studies do not look beyond the 21st century. It is highly likely that climate will continue changing into the 22nd century and even beyond (Watson *et al.*, 2001). For systems in which there is long-term inertia, such as the climate-ocean system, long-term consequences of different levels of increase in GMT on sectors such as coastal resources may be underestimated.

- Furthermore, very few studies consider potential catastrophic changes in the climate system, such as shutdown of the ocean's thermohaline circulation, and concomitant impacts, and thus may underestimate long-term adverse impacts associated with particular increases in GMT.

- Finally, there are important linkages between many of the sectors that we considered. For example, impacts to agriculture and water resources are linked in areas where agriculture is irrigated (Hanemann, forthcoming). These linkages might result in exacerbation or, in some cases, amelioration of the impacts we report.

4.2 Recommendations for research

The resolution of many of these uncertainties may have to wait for the development of more sophisticated impact models or improved projections of climate change. However, our analysis points to several steps that might be taken now to improve the usefulness and credibility of the current generation of impact studies:

- Efforts should be made to improve methods for expressing impacts in natural sectors (i.e., terrestrial ecosystem productivity, marine ecosystems, biodiversity) in metrics that are meaningful to policy makers. Numerous studies predict how ecosystems might respond to warming. However, measures such as NEP or NPP are abstruse for policy makers, and they significantly hamper efforts at aggregating impacts across sectors. What is missing is a sense of how important these responses might be, to what extent they might be managed, and ultimately the extent to which people care. Much the same can be said for social sectors that are non-market in character (e.g., human health and amenities).

- Impacts across sectors can be compared confidently and easily only if studies share a consistent approach to the development and application of climate scenarios, socio-economic baselines, timeframes, and methods of analysis. Looking across studies that rely on different GCMs can be tricky, even when the globally averaged climate scenarios they produce are similar. Country specific temperatures and precipitations especially can vary tremendously across models. Single sector studies that rely on a single GCM are less useful than those that use a suite of models. While different GCMs can give a sense for the range of potential impacts within a single sector or in the context of a global impact model (Mendelsohn *et al.*, 2000a), cross-sector evaluation of impacts generated by different regional climate scenarios is problematic. Similarly, when different socio-economic baselines are used it can be difficult to determine to what extent the results are affected by these assumptions. Furthermore, methods of analysis must be transparent. For instance, when impacts are measured in terms of people affected, studies should provide disaggregated results that allow analysts to draw their own conclusions about net impacts, in line with the policy questions they seek to inform.

- Most existing studies are structured in a way that makes it difficult to translate results into policy insight. More studies should be designed with some explicit thought given to questions regarding mitigation or adaptation. Parry *et al.* (2001), which developed multi-sector impact assessments for several mitigation scenarios, provides a good methodological model. It also takes a consistent approach to socio-economic baselines and timeframes.

- While most of the studies we examined highlight regional results, more discussion should be devoted to the process and validity of spatial aggregation used to obtain global results. In many cases it may be reasonable that one region's gains offset another's losses (e.g., agriculture). In others, it may be better to leave results in regional form (e.g., water resources).

- Addressing sectors for which there are no global impact estimates or for which information is limited is important. Climate change impacts on tourism and recreation, and amenity values, could all involve substantial societal impacts and monetary values. In addition, there are a number of sectors for which only some impacts have been assessed or for which there are a limited number of global studies. This is particularly the case for energy and terrestrial and marine ecosystems, including terrestrial animals and fisheries. There is also limited information about impacts on developing countries in general.

- Based on this survey, there are several sensitive sectors where our understanding of the relationship between impacts and changes in GMT should be improved, most notably water resources and human health. Even small magnitudes of climate change could adversely affect many hundreds of millions of people who are afflicted with climate sensitive health impacts each year or who lack adequate and safe water supplies. On the other hand, development could substantially reduce the vulnerability of these sectors to climate change in the future (Tol *et al.*, 2004; Yohe,). To better understand the global consequences of climate change, it would help to clarify the relationship between climate change and these two sectors in particular.

ANNEX: Key features of examined studies

Table A1. Summary of global impacts studies

Sector	Metric	Driving scenario(s)	Time frame(s)	Temp. level(s)	Spatial scale/ distributional	Key factors (adaptation, baseline change, CO_2 fert.)	Change in extreme events or variance considered?	State changes or rates of change examined?	Results
Agriculture									
Rosenzweig *et al.*, 1995	No. of people at risk of hunger. Cereal production Food prices.	2xCO_2 for 3 GCMs Baseline climatology 1951-1980	2060 equilibrium	GISS4.2°C GFDL4.0°C UKMO5.2°C GISS-A 2.4°C	National;. 18 countries.	2 degrees of farm level adaptation + world economic adjustments + CO_2 effect + increasing yields + evolving baseline.	Not captured	Not captured	Figure 2
Darwin *et al.*, 1995	% change in production of agricultural commodities based on land-use change.	2x CO_2 for 4 GCMs Baseline climatology 1951-1980	2090 equilibrium	GISS 4.2°C GFDL 4.0°C UKMO 5.2°C OSU 2.8°C	Regional — major countries plus geographic regions.	Farm level adaptation, no CO_2 effect, no evolving baseline. 1990 world as benchmark.	Not captured	Not captured	Figure 2
Parry *et al.*, 1999	No. of people at risk of hunger. Food production. Food prices.	Transient GCM scenarios Baseline climatology 1961-1990	2020s, 2050s, 2080s	HadCM2 1.2-3.1°C HadCM3 1.1-3.0°C	Regional; distributional effects considered.	Farm level adaptation + world economic adjustments + increasing yields due to technology.	Not captured	Transient. Results at 2020s, 2050s, 2080s provide some insight into rates of change	Figure 2
Parry *et al.*, 2004	No. of people at risk of hunger. Food production. Food prices.	Transient GCM scenarios coupled with SRES socioeconomic scenarios	2020s, 2050s, 2080s	0.84-3.97°C	Regional; distributional effects considered.	Farm level adaptation + world economic adjustments + increasing yields due to technology as well as other SRES assumptions.	Not captured	Transient. Results at 2020s, 2050s, 2080s provide some insight into rates of change	Figure 2

Table A1. Summary of global impacts studies (cont.)

Sector	Metric	Driving scenario(s)	Time frame(s)	Temp. level(s)	Spatial scale/ distributional	Key factors (adaptation, baseline change, CO_2 fert.)	Change in extreme events or variance considered?	State changes or rates of change examined?	Results
Fischer *et al.*, 2002	Production potential, GDP of agriculture, consumption, no. of people at risk of hunger.	Various SRES scenarios for 4 GCMs	2080s	Dependent on particular scenario and GCM. Scenarios cover up to 6°C.	Regional	Farm level adaptation. No CO_2 effect.	Not captured	Not captured	Figure 2
Sea level/coastal									
Fankhauser, 1995	Direct cost (cost of protection, dryland loss and wetland loss) measured in USD	Various increments of sea level rise (SLR) relative to 1990	2100	Not applicable. SLR rise up to 200 cm	OECD Countries	Zero utility discounting, Optimal coastal protection considered but as protect or retreat dichotomy.	No change in storminess	Not captured	Figure 1
Nicholls *et al.*, 1999	No. of people at risk of coastal flooding	Transient GCM scenarios SLR relative to 1961-1990	2020s, 2050s, 2080s	HadCM2 1.2-3.1°C HadCM3 1.1-3.0°C SLR of 12, 24, and 40 cm at 2020s, 2050s, and 2080s, respectively	Selected regions and global aggregation	Evolving refer. scenario + improving flood defence standards	No changes in storminess considered. Current variance imposed on sea level rise.	Not captured	Figure 1.

Table A1. Summary of global impacts studies (cont.)

Sector	Metric	Driving scenario(s)	Time frame(s)	Temp. level(s)	Spatial scale/ distributional	Key factors (adaptation, baseline change, CO_2 fert.)	Change in extreme events or variance considered?	State changes or rates of change examined?	Results
Water									
Arnell, 1999	No. of people living in countries experiencing water stress	GCMs 4 HadCM2 scenarios 1 HadCM3 scenario Baseline climatology 1961-1990	2020s, 2050s, 2080s	HadCM2 1.2-3.1°C HadCM3 1.1-3.0°C	National Distributional differences in runoff examined at regional level	Demand unaffected by climate change in use scenarios. Use scenarios incorporated changes in populations, efficiency, irrigation, industry, etc. Land use within each catchment assumed constant. No direct CO_2 effect.	Limited treatment of 10 yr. max. monthly runoff and 10 yr. low annual total runoff.	Not captured	Figure 3.
Vörösmarty *et al.*, 2000	No. of people living under water stressed conditions	Canadian Model CGCM1 HadCM2 Baseline climatology 1961-1990	2025	Not reported	River system level	Estimates of future water efficiency incorporated, captures spatial heterogeneity by using 30' grid cells. Allowed for migration within countries to sources of water.	Not captured	Not captured	Effects of increased water demand due to population and economic growth eclipse changes due to climate. Some regional damages are masked though (e.g., Africa, South America)

Table A1. Summary of global impacts studies (cont.)

Sector	Metric	Driving scenario(s)	Time frame(s)	Temp. level(s)	Spatial scale/ distributional	Key factors (adaptation, baseline change, CO_2 fert.)	Change in extreme events or variance considered?	State changes or rates of change examined?	Results
Döll, 2002	Net irrigation requirements	ECHAM4/ OPYC3 HadCM3AA Baseline climatology 1961-1990	2020s, 2070s	Not reported	0.5° by 0.5° raster cells, regional aggregation	Allowed for changes in start of growing season, cropping patterns. Limited to 1995 irrigated areas.	Compare results to variance of past century.	Not captured	Regional analysis shows IR requirement in 12/17 world regions increases, but not more than by 10%
Alcamo *et al.*, 1997	Number of people living in water scarce conditions	ECHAM4 and GFDL	2025, 2075	Not reported	0.5° by 0.5°grid level, national aggregation	Considered three different scenarios of water use. Water use considered consumptive. Land use changes incorporated.	Limited treatment of drought cycles, but no change in variance postulated.	Not captured	Water scarcity decreases due to climate change by 2075.
Health									
Martin and Lefebvre, 1995	Area of potential transmission	$2XCO_2$ GFDLQ, GISS, OSU, UKMO, GFDL Baseline climatology 1951-1980	Not explicitly stated. Presumably between 2060 and 2090.	GISS 4.2°C GFDL 4.0°C UKMO 5.2°C OSU 2.8°C	30' by 30' grid cells	No epidemiology, no adaptation, crude water balance assumptions	Not captured	Not captured	Figure 4
Martens *et al.*, 1999	No. of people at risk of malaria	HadCM2 HadCM3 Baseline climatology 1961-1990	2020s, 2050s, 2080s	HadCM2: 1.2-3.1°C HadCM3: 1.1-3.0°C	Regional	Evolving pop. baseline, no adaptation, vector's range unchanged by climate	Not captured	Not captured	Figure 4

Table A1. Summary of global impacts studies (cont.)

Sector	Metric	Driving scenario(s)	Time frame(s)	Temp. level(s)	Spatial scale/ distributional	Key factors (adaptation, baseline change, CO_2 fert.)	Change in extreme events or variance considered?	State changes or rates of change examined?	Results
Van Lieshout *et al.*, 2004	No. of people at risk of malaria	HadCM3 and SRES socioeconomic scenarios	2020s, 2050s, 2080s	0.84-3.97°C	Regional	SRES socioeconomic assumptions. Also includes measure of future vulnerability of impacted countries.	Not captured	Not captured	Results differ from scenario to scenario. Suggest impacts confined to specific locations.
Tol and Dowlatabadi, 2002	Vector-borne mortality, malaria mortality	Var. mitigation policies applied to assumed mortality time relationship.	2100	Temp. mortality relationship hypothesised from several studies.	Regional	No evolution of control efforts, access to public health service according to income, population in regions homogenous, no land use change effects	Not captured	Not captured	Global mortality due to malaria reduced to zero by 2100 due to economic growth
Hijioka *et al.*, 2002	Diarrhoeal incidence	Temperature scenarios derived from NIES/CCSR	2025, 2055	0.8-3.4°C	Regional	Statistical model assumes that water supply coverage and temperature determine diarrhoeal incidence in future. Access to supply improves with income.	Not captured	Not captured	Figure 4

Table A1. Summary of global impacts studies (cont.)

Sector	Metric	Driving scenario(s)	Time frame(s)	Temp. level(s)	Spatial scale/ distributional	Key factors (adaptation, baseline change, CO_2 fert.)	Change in extreme events or variance considered?	State changes or rates of change examined?	Results
Terrestrial ecosystem productivity									
White *et al.*, 1999	Vegetation carbon, soil carbon, NPP, NEP	4 HadCM2 simulations, 1 HadCM3 simulation. IS92a forcing. Input to dynamic veg. model Hybrid v4.1	1860-2100	1.1°-3.1°C	Ten 200m^2 plots per GCM grid in vegetation model. Results aggregated globally	Dispersal processes not represented-potential vegetation. Land use and disturbances not modelled. No vegetation-climate feedbacks. Nitrogen deposition exogenous.	Stochastic weather generator used to create daily temperatures for input into vegetation model. No state change in variance modelled.	Not captured	Figure 5
Leemans and Eickhout, 2004	Ecosystem shifts (changes in extent and nature of change)	Integrated assessment model IMAGE implemented with SRES scenarios. IPCC pattern scaling with output from several GCMs.	2100	1, 2, and 3°C	0.5 x 0.5 grid. Results aggregated by ecosystem type.	Dispersal processes are modeled by IMAGE and land use change is incorporated.	No state change in variance.	0.1, 0.2, and 0.3°C/decade examined.	Positive productivity changes, though could be threatened by high rates of climate change.

Table A1. Summary of global impacts studies (cont.)

Sector	Metric	Driving scenario(s)	Time frame(s)	Temp. level(s)	Spatial scale/ distributional	Key factors (adaptation, baseline change, CO_2 fert.)	Change in extreme events or variance considered?	State changes or rates of change examined?	Results
Cramer *et al.*, 2001	NPP, NEP, soil C, biomass	HadCM2-SUL with IS192a forcing input to 6 DGVMs. Scenarios include 1) changing CO_2 2) changing climate with CO_2 at preindustrial 3) changes in climate and CO_2. Baseline climatology 1931-1960	2100 2199 with 100 years of unrealistic imposed equilibrium starting at 2100	Not reported as global means.	Fundamentally regional. Global coverage, but almost no results aggregated at a global level	No biogeochem. feedbacks (e.g., vegetation to climate), as DGVMs are run offline. Land use not incorporated	Variance in GCM output maintained and input to DVGMs, but no state change in variance considered.	Not captured	Terrestrial carbon sink levels off by 2050 and begins to decrease, by end of century
Forestry									
Sohngen *et al.*, 2001	% change in forest area, NPP, timber yield, welfare (USD)	2 equilibrium scenarios $2xCO_2$ Hamburg T-106 UIUC	near (1995-2045) long (2045-2145) discount rates of 3, 5 and 7% Equilibrium achieved by 2060	Hamburg 1°C UIUC 3.4°C	Regional	CO_2 fert., species migration, land use competition not modelled	Not captured	Rate of change considered to extent that dieback scenario investigated as well as regeneration.	Figure 6. Only two temperature points, results for which are similar globally. Differences are regional and reflect regional temp. differences in GCM output.

Table A1. Summary of global impacts studies (cont.)

Sector	Metric	Driving scenario(s)	Time frame(s)	Temp. level(s)	Spatial scale/ distributional	Key factors (adaptation, baseline change, CO_2 fert.)	Change in extreme events or variance considered?	State changes or rates of change examined?	Results
Perez-Garcia *et al.*, 2002	Vegetation carbon, softwood and hardwood stock, price, harvest, welfare (USD)	Transient changes in CO_2. 3 scenarios from IGCM Baseline climatology 1850-1976	Stabilisation at 2100, analysis at 2040	RRR: 745ppmv, 2.6°C HHL: 936ppmv, 3.1°C LLH: 592ppmv, 1.6°C Not reported for 2040 when economic model ends	Regional	CO_2 fertilise., revised econom. baseline including Asian Crisis, lower Russian prod., no species migration. 2 harvesting scenarios as in Perez-Garcia *et al.*, 1997	Not captured	Not captured	Need transient temperatures. Global price and harvest changes are small. Regional changes obscured though.
Marine ecosystems									
Bopp *et al.*, 2001	Change in marine export production	2 atm/ocean coupled GCMs. Constant CO_2 and 1% growth/yr until 2x after 70 years. Equilibrium for 3x and 4x CO_2. Two biogeochemical schemes.	70 years	N/A	Globally aggregated. GCM grid 2°long x1.5°lat	Biogeochemical schemes are phosphate based, lacking limitation by other nutrients (e.g., Fe, N, Si)	Not captured	Not captured	Reduction of 6% in production for $2xCO_2$ relative to control. 11% for 3x and 15% for 4x

Table A1. Summary of global impacts studies (cont.)

Sector	Metric	Driving scenario(s)	Time frame(s)	Temp. level(s)	Spatial scale/ distributional	Key factors (adaptation, baseline change, CO_2 fert.)	Change in extreme events or variance considered?	State changes or rates of change examined?	Results
Terrestrial biodiversity/species loss									
Leemans and Eickhout, 2004	Ecosystem shifts (changes in extent and nature of change)	Integrated assessment model IMAGE implemented with SRES scenarios. IPCC pattern scaling with output from several GCMs.	2100	1, 2, and 3°C	0.5 x 0.5 grid. Results aggregated by ecosystem type.	Dispersal processes are modeled by IMAGE and land use change is incorporated.	No state change in variance	0.1, 0.2. and 0.3° C/decade examined	Negative impacts likely even for low levels and low rates of climate change.
Halpin, 1997	% Change in ecoclimatic classes in biosphere reserves	$2XCO_2$ GFDL, GISS, OSU, UKMO Baseline climatology 1951-1980	Not stated explicitly. Presumably 2060-90, as with other studies using this battery of models.	GISS 4.2°C GFDL 4.0°C UKMO 5.2°C OSU 2.8°C	Global	No adaptation or baseline changes.	Not captured	Not captured	Figure 7
Energy									
Tol, 2002b	% change in GDP	FUND	2200	Not reported for dynamic study	Country specific. Extrapolated to rest of the world.	Increasing efficiency, population, income	Not captured	Not captured	Savings due to heating less than 1% GDP by 2200, costs due to cooling about 0.6% GDP.

Table A1. Summary of global impacts studies (cont.)

Sector	Metric	Driving scenario(s)	Time frame(s)	Temp. level(s)	Spatial scale/ distributional	Key factors (adaptation, baseline change, CO_2 fert.)	Change in extreme events or variance considered?	State changes or rates of change examined?	Results
Aggregate									
Tol, 2002a,b	Income	Considered both 1ºC increase and dynamic over two centuries	2200 for dynamic	1ºC for static, not reported for dynamic	9 world regions	Sector specific	Not captured	Not captured	Simple sum aggregation leads to 2.3% increase in global income
Nordhaus and Boyer, 2000	Output	Regional Integrated model of Climate and the Economy	Not reported	2.5ºC, some results extrapolated to 6ºC	13 regions	Sector specific	No captured	Catastrophic events considered	1.5% increase in output weighted by income, 1.9% weighted by populations. Figure 19.

Acknowledgements

We thank Jan Corfee Morlot and Shardul Agrawala at OECD as well as Jane Leggett at U.S. EPA for their sponsorship of this research as well as their thoughtful guidance and comments. The suggestions of two anonymous reviewers were also of great help. Christina Thomas, Erin Miles, and Shiela DeMars at Stratus Consulting provided valuable editorial and production assistance.

References

Alcamo, J., R. Leemans and G.J.J. Kreileman (1998), "Global Change Scenarios of the 21st Century", Results from the IMAGE 2.1 Model, London, Pergamon & Elseviers Science.

Alcamo, J., P. Doll, F. Kaspar and S. Siebert (1997), *Global Change and Global Scenarios of Water Use and Availability: An Application of WaterGAP 1.0.*, University of Kassel, Germany: Center for Environmental Systems Research.

Alcamo, J., G. Kreileman, M. Krol and G. Zuidema (1994), *Modeling the global society-biosphere-climate system: Part 1: Model description and testing*, *Water, Air and Soil Pollution,* **76**, 1-35.

Arnell, N.W., (1999), "Climate change and global water resources", *Global Environmental Change,* **9**, S31-S49.

Arnell, N.W., M.G.R. Cannell, M. Hulme, R.S. Kovats, J.F.B. Mitchell, R.J. Nicholls, M.L. Parry, M.T.J. Livermore and A. White (2002), "The consequences of CO_2 stabilisation for the impacts of climate change", *Climatic Change* **53**, 413-446.

Bopp, L., P. Monfray, O. Aumont, J. Dufresne, H. Le Treut, G. Madec, L. Terray and J.C. Orr (2001), "Potential impact of climate change on marine export production", *Global Biogeochemical Cycles* **15**, 81-99.

Carter, T.R., E.L. La Rovere, R.N. Jones, R. Leemans, L.O. Mearns, N. Nakícenovíc, B.A. Pittock, S.M. Semenov and J.F. Skea (2001), "Developing and applying scenarios", in *Climate Change (2001), Impacts, Adaptation and Vulnerability*, J. McCarthy, O. Canziani, N. Leary, D. Dokken and K. White (eds.). Cambridge: Cambridge University Press, 145-190.

Chen, C.C., B.A. McCarl and R.M. Adams (2001), "Economic implications of potential climate change induced ENSO frequency and strength shifts", *Climatic Change* **49**, 147-159.

Cramer, W. *et al.* (2001), "Global response of terrestrial ecosystem structure and function to CO_2 and climate change: Results from six dynamic global vegetation models", *Global Change Biology* **7**, 357-373.

Darwin, R., M. Tsigas, J. Lewandrowski and A. Raneses (1995), "World Agriculture and Climate Change: Economic Adaptations", *Agricultural Economic Report* No. 703. Washington, DC: U.S. Department of Agriculture.

Döll, P. (2002), "Impact of climate change and variability on irrigation requirements: A global perspective", *Climatic Change* **54**, 269-293.

Downing, T.E., N. Eyre, R. Greener and D. Blackwell (1996), *Projected Costs of Climate Changeover Two Reference Scenarios and Fossil Fuel Cycles*, Oxford, UK: Environmental Change Unit.

Easterling, D.R., G.A. Meehl, C. Parmesan, S.A. Changnon, T.R. Karl and L.O. Mearns (2000), "Climate extremes: Observations, modeling, and impacts", *Science* **289**, 2068-2074.

Fankhauser, S., (1995), *Valuing Climate Change: The Economics of the Greenhouse*. London, Earthscan Publications Ltd.

Fischer, G., M. Shah and H. van Velthuizen (2002), *Climate Change and Agricultural Vulnerability*, Vienna, International Institute of Applied Systems Analysis.

Gitay, H., S. Brown, W. Easterling and B. Jallow (2001), "Ecosystems and their goods and services", in *Climate Change 2001: Impacts, Adaptation, and Vulnerability*, J. McCarthy, O. Canziani, N. Leary, D. Dokken and K. White (eds.). New York, Cambridge University Press, 235-342.

Gubler D.J., P. Reiter, K.L. Ebi, W. Yap, R. Nasci and J.A. Patz (2001), "Climate variability and change in the United States: Potential impacts on vector- and rodent-borne diseases", *Environmental Health Perspectives* **109** (Suppl. 2), 223-233.

Halpin, P.N. (1997), "Global climate change and natural area protection: Management responses and research directions", *Ecological Applications* **7**, 828-843.

Hijioka, Y., K. Takahashi, Y. Matsuoka and H. Harasawa (2002), "Impact of global warming on waterborne diseases", *Journal of Japan Society on Water Environment* **25**, 647-652.

Hoozemans, F.M.J., M. Marchand and H.A. Pennekamp (1993), *Global Vulnerability Analysis*, 2nd revised edition, Rotterdam, The Netherlands: Delft Hydraulics.

Houghton, J.T., Y. Ding, D.J. Griggs, M. Noguer, P.J. van der Linden, D. Xiaosu and K. Maskell (eds.) (2001), *Climate Change 2001: The Scientific Basis*, New York, Cambridge University Press.

Leemans, R. and B. Eickhout (2004), "Another reason for concern: regional and global impacts on ecosystems for different levels of climate change", *Global Environmental Change: Special Edition on the Benefits of Climate Policy* **14**: 219-228.

Lise, W. and R.S.J. Tol (2002), "Impact of climate on tourist demand", *Climatic Change* **55**, 429-449.

Martens, P., (1999), "MIASMA: Modelling Framework for the Health Impact Assessment of Man-Induced Atmospheric Changes, ESIAM" (Electronic Series on Integrated Assessment Modelling), CD-ROM no. 2.

Martens, P., T.H. Jetten and D.A. Focks (1997), "Sensitivity of malaria, schistosomiasis and dengue to global warming", *Climatic Change*, **35**, 145-156.

Martens, P., R.S. Kovats, S. Nijhof, P. de Vries, M.T.J. Livermore, D.J. Bradley, J. Cox and A.J. McMichael (1999), "Climate change and future populations at risk of malaria". *Global Environmental Change*, **9**, S89-S107.

Martens, W.J.M. (1998), "Climate change, thermal stress and mortality changes", *Social Science and Medicine* **46**(3), 331-344.

Martin, P.H. and M.G. Lefebvre (1995), "Malaria and climate: Sensitivity of malaria potential transmission to climate", *Ambio* **24** 200-207.

McMichael, A., A. Githeko, R. Akhtar, R. Caracavallo, D. Gubler, A. Haines, R.S. Kovats, P. Martens, J. Patz and A. Sasaki (2001), "Human health", in *Climate Change 2001: Impacts, Adaptation and Vulnerability*, J. McCarthy, O. Canziani, N. Leary, D. Dokken and K. White (eds.). New York: Cambridge University Press, 451-485.

Mendelsohn, R. (2001), *Global Warming and the American Economy: A Regional Assessment of Climate Change Impacts*, Northampton, MA: Edward Elgar.

Mendelsohn, R. (2003), "Assessing the market damages from climate change", in *Global Climate Change: The Science, Economics, and Politics*, Griffin, J. (ed.). United Kingdom, Edward Elgar Publishing, 92-113.

Mendelsohn, R. and J. Neumann (eds.), (1999), *The Impacts of Climate Change on the U.S. Economy*. Cambridge, UK: Cambridge University Press.

Mendelsohn, R. and M.E. Schlesinger (1997), *Climate Response Functions*, University of Illinois: Urbana-Champaign.

Mendelsohn, R., M. Schlesinger and L. Williams (2000a), "Comparing impacts across climate models", *Integrated Assessment* **2**, 37-48.

Mendelsohn, R., W. Morrison, M. Schlesinger and N. Adronova (2000b.), "Country-specific impacts from climate change", *Climatic Change* **45**, 553-569.

Metz, B., O. Davidson, R. Swart and J. Pan (2001), *Climate Change 2001: Mitigation*, New York: Cambridge University Press.

Nicholls, R.J., F.M.J. Hoozemans and M. Marchand (1999), "Increasing flood risk and wetland losses due to global sea-level rise: Regional and global analyses", *Global Environmental Change*, **9**, S69-S87.

Nordhaus, W. and J. Boyer (2000), *Roll the DICE Again: Economic Modeling of Climate Change,* Cambridge, MA: MIT Press.

Parry, M., C. Rosenzweig, A. Iglesias, F. Fischer and M. Livermore, M., (1999), "Climate change and world food security: A new assessment", *Global Environmental Change* **9**, S51-S67.

Parry, M., N. Arnell, T. McMichael, R. Nicholls, P. Martens, S. Kovats, M. Livermore, C. Rosenzweig, A. Iglesias and G. Fischer (2001), "Millions at risk: Defining critical climate change threats and targets", *Global Environmental Change* **11**(3), 1-3.

Parry, M.L., C. Rosenzweig, A. Iglesias, M. Livermore and G. Fisher (2004), "Effects of climate change on global food production under SRES emissions and socio-economic scenarios", *Global Environmental Change* **14**, 53-67.

Perez-Garcia, J., L. Joyce, A.D. McGuire and X. Xiao (2002), "Impacts of climate change on the global forest sector", *Climatic Change*, **54**, 439-461.

Peters, R.L. and T.E. Lovejoy (1992), *Global Warming and Biological Diversity*, New Haven, CT: Yale University Press.

Rosenzweig, C. and D. Hillel (1998), *Climate Change and the Global Harvest: Potential Impacts of the Greenhouse Effect on Agriculture,* New York: Oxford University Press.

Rosenzweig, C., M. Parry and G. Fischer (1995), "World food supply", in *As Climate Changes: International Impacts and Implications*, K.M. Strzepek and J.B. Smith (eds.), Cambridge, UK, Cambridge University Press, 27-56.

Rosenzweig, C., F.N. Tubiello, R. Goldberg, E. Mills and J. Bloomfield (2002), "Increased crop damage in the US from excess precipitation under climate change", *Global Environmental Change* **12**, 197-202.

Schneider, S.H., K. Kuntz-Duriseti and C. Azar (2000), "Costing nonlinearities, surprises and irreversible events", *Pacific and Asian Journal of Energy* **10**(1), 81-106.

Smit, B., D. McNabb and J. Smithers, J. (1996), "Agricultural adaptation to climatic variation", *Climatic Change,* **33**, 7-29.

Smith, J.B., H.J. Schellnhuber, M.Q. Mirza, S. Fankhauser, R. Leemans, L. Erda, L. Ogallo, B. Pittock, R. Richels, C. Rosenzweig, U. Safriel, R.S.J. Tol, J. Weyant and G. Yohe (2001), "Vulnerability to climate change and reasons for concern: A synthesis", in *Climate Change 2001: Impacts, Adaptation, and Vulnerability*, J. McCarthy, O. Canziana, N. Leary, D. Dokken and K. White (eds.), New York, Cambridge University Press, 913-967.

Sohngen, B., R. Mendelsohn and R. Sedjo (2001), "A global model of climate change impacts on timber markets", *Journal of Agricultural and Resource Economics* **26**(2), 326-343.

Timmerman, A., J. Oberhuber, A. Bacher, M. Esch, M. Latif and E. Roeckner (1999), "Increased El Niño frequency in a climate model forced by future greenhouse warming", *Nature* **398**, 694-697.

Tol, R.S.J. (2002a), "Estimates of the damage costs of climate change, Part I: Benchmark estimates", *Environmental and Resource Economics,* **21**, pp.47-73.

Tol, R.S.J. (2002b), "Estimates of the damage costs of climate change, Part II: Dynamic estimates", *Environmental and Resource Economics,* **21**, 135-160.

Tol, R.S.J. and H. Dowlatabadi (2002), "Vector-borne diseases, development, and climate change", *Integrated Environmental Assessment* **2**, 173-181.

Tol, R., T. Downing, O. Kuik, and J. Smith (2004), "Distributional aspects of climate change impacts", *Global Environmental Change: Special Edition on the Benefits of Climate Policy* **14**: 259-272.

van Lieshout, M., R.S. Koavates, M.T.J. Livermore and P. Martens (2004), "Climate change and malaria: analysis of the SRES climate and socio-economic scenarios", *Global Environmental Change* **14**, 87-99.

Vörösmarty, C.J., P. Green, J. Salisbury and R.B. Lammers (2000), "Global water resources; vulnerability from climate change and population growth", *Science,* **289**, 284-288.

Watson, R.T. and the Core Writing Team (eds.), (2001), "Climate Change 2001: Synthesis Report", A Contribution of Working Groups I, II, and III to the Third Assessment Report of the Intergovernmental Panel on Climate Change, New York, Cambridge University Press.

West, J.J. and H. Dowlatabadi (1999), "On assessing the economic impacts of sea-level rise on developed coasts", in *Climate Change and Risk*, T.E. Downing, A.A. Olsthoorn and R.S.J. Tol (eds.), New York: Routledge, 205-220.

White, A., M.G.R. Cannell and A.D. Friend (1999), "Climate change impacts on ecosystems and the terrestrial carbon sink: A new assessment", *Global Environmental Change*, **9**, S21-S30.

Yohe, G. (2003), "Estimating benefits: Other issues concerning market impacts", OECD, Paris, (ENV/EPOC/GSP(2003)9/FINAL).

Chapter 3

INTEGRATED ASSESSMENT OF BENEFITS OF CLIMATE POLICY

by John Schellnhuber, Rachel Warren, Alex Haxeltine and Larissa Naylor,

The Tyndall Centre for Climate Change Research, United Kingdom

This paper explores how novel integrated assessment models (IAM) can provide policy relevant information through their ability to handle the level of complexity inherent in an assessment of the benefits of climate policy. We discuss how a holistic assessment of damages resulting from climate change could in principle be obtained in IAM through the systematic, state of the art collation of computer modules. We also discuss an alternative theory based approach to the definition of dangerous climate change, and finally suggest an IAM-based approach utilizing a series of (normative) damage functions constructed for each vulnerable nation, stock at risk, function or sector, each showing how damage accrues as a function of average temperature rise. We discuss how IAM could inform policy makers about choices between mitigation and adaptation, how special issues such as adaptive capacities and geological carbon sequestration options could be included in IAM, and highlight some of the co-benefits that climate policy produces in terms of air quality improvements. We conclude with recent examples of the application of IAM to assess climate policy benefits. Examples include: a "tolerable windows" approach from which policy makers can deduce the emission corridors that must be followed if environmental and economic constraints are to be met simultaneously; a coastal simulator and decision support tool that will allow assessment of various climate policies in the context of a limited geographic area; a new multi-institutional, participatory approach to IAM through a Community Integrated Assessment System (CIAS) designed to overcome the traditional mismatch between policy makers and scientists.

ISBN 92-64-10831-9
THE BENEFITS OF CLIMATE CHANGE POLICIES

CHAPTER 3. INTEGRATED ASSESSMENT OF BENEFITS OF CLIMATE POLICY

by John Schellnhuber, Rachel Warren, Alex Haxeltine and Larissa Naylor[21]

1. Introduction

Initial concerns about climate change in the 1980's highlighted the potentially enormous and damaging impacts on humans and natural ecosystems, which have iteratively been estimated scientifically by the IPCC (IPCC, 1996; IPCC, 2001). The perceived need to act resulted in a focus on estimates of mitigation costs under present socioeconomic conditions (Grubb *et al.,* 1993; Peck and Teisberg, 1995; Richels and Edmonds, 1995; Schneider and Thompson, 2000; Tol, 2000). Many studies predict high mitigation costs, which has engendered considerable doubt as to whether we can afford climate protection. Other studies predict that lower costs can be achieved through endogenous technical change and judicious application of carbon taxes, or even that costs are not high when viewed in the context of a slight delay in economic growth (Azar and Schneider, 2002; Edenhofer *et al.,* in prep). However, the widespread concern about the costs of climate policy (which are reflected in the outcome of the Kyoto process) means there is an urgent need to assess the benefits of decarbonization including: (i) avoided damages, (ii) ancillary advantages, and (iii) role models for participatory action and global partnership in facilitating holistic assessments involving both key users and our best scientists.

This is not a simple task. There is a difficulty in properly assessing these genuinely complex benefits: even factual damage costs of recent climate-related disasters like the August 2002 flood in Central Europe are hard to calculate. Further dimensions of complexity are added if: (1) mankind's potential to adapt to climate change is taken into account; (2) the climate issue is embedded in a sustainability context; (3) the assessment is carried out at multiple spatio-temporal scales; (4) the assessment allows for the use of different paradigms; (5) the assessment simulates the behaviour of various interacting actors.

---

21 The Tyndall Centre for Climate Change Research, UK. E-mail: h.j.schellnhuber@uea.ac.uk and R.Warren@uea.ac.uk.

Integrated assessment (IA) is, in principle, the only mode of scientific analysis that can cope with this complexity. The major application of IA that will be focused on here is the question of how IA models (IAM) can be used to provide a holistic assessment of damages due to climate change across different sectors and spatial scales. Such an analysis is relevant both for the case of forward simulation from climate policies to impacts, and for the investigation of stabilisation scenarios. This relates to the deeper policy question of identifying an appropriate stabilisation level that excludes "dangerous anthropogenic interference with the climate system."

The aim of this paper is to show how IAM can shed light on the benefits of climate policy given the status of current scientific understanding of relevant phenomena. We begin by proposing two methods of holistically assessing damages due to climate change: (1) the systematic collection of state of the art results, and (2) a theory-based approach to defining dangerous climate change. We continue by reviewing adaptation, ancillary and socially contingent benefits. Finally, the paper reviews three relevant examples of IA approaches to the assessment of the benefits of climate policy: (1) the ICLIPS model, (2) a coastal simulator and (3) the Community Integrated Assessment System (CIAS), in which we explain the need for a new European initiative to provide a tool for participatory integrated assessment that aims to extend mutual learning between scientists and modellers on the one hand and civil society and policymakers on the other.

2. Holistic assessment of damages due to climate change

We propose two approaches to the holistic assessment of the benefits of climate policy. By "climate policy" we mean the decision to apply economic and other instruments/incentives designed to reduce emissions of greenhouse gases, i.e. a strategic approach commonly known as "mitigation". Since these decisions are not the only drivers of future socio-economic pathways and hence emissions, analysis of climate policies needs to take into account a range of socioeconomic futures such as those encompassed by the IPCC SRES scenarios. (Population dynamics, for example, is a strong driver of emissions). For each socioeconomic future, climate policies will lead to a particular emission trajectory for greenhouse gases and thence a particular scenario for global climate change over a particular number of years. In order for a policy maker to assess the different climate policies, it is necessary to provide a holistic assessment of the damages due to climate change associated with the combination of each climate policy under different socioeconomic futures. It is this latter problem that we address here.

Two potential methods are proposed:

- A systematic, state-of-the-art integration of damages in a holistic manner across the globe for as many regions, sectors and stocks at risk covered in the IAM model.

- A theory-based approach, aimed at the more fundamental question of assisting policy makers in defining dangerous climate change.

2.1 Systematic state-of-the-art integration collection

An IAM framework eventually creates a knowledge pool large enough to generate a comprehensive, global, and yet regionally specific, impact assessment for a range of climate scenarios. This provides the basis on which consistent estimates of damages can be made. At early stages of development, incomplete assessments may be made covering key sectors or regions, particularly those where either exposure is high or the stock at risk is high or extremely valuable in economic, ecological or societal terms. Either of these situations is likely to lead to high damages. Where appropriate, such damages may also be monetised. The IAM system could also be used to implement holistic assessment schemes such as that recommended by Schneider (2004) in which five numeraires are put forward as useful indicators of climate impacts, including monetary losses, loss of life, extinction rates, and reduced quality of life.

Ideally, such a framework should also include within it estimates of the impacts resulting from changes in the frequency of extreme weather events such as that outlined by Milly *et al.* (2002). Initial work at the University of East Anglia and elsewhere is allowing IA models to address this, and the increasing intensity of such events, for the first time (Goodess *et al.,* 2003). Omission of this would lead to large underestimates in climate change damages. Schneider and Lane (this volume) also highlights the necessity to incorporate these considerations. Similarly, the framework should include within it an estimate of the impacts resulting from abrupt climate changes such as the breakdown of the thermohaline circulation (Mastrandrea and Schneider, 2001).

Such assessment requires three major challenges to be addressed:

- The relevance of multiple scales in time and space.
- The problem of uncertainty.
- The amalgamation of quantitative and qualitative information.

Taking first the issue of scaling, the problem is that no theory exists which can describe and explain dynamic behaviour at various scales of social, economic and ecological activity (Rotmans and Rothman, 2003). One approach is to use a nested hierarchy of models, linked together with appropriate boundary conditions. Such an approach has been applied in the past at the national level, as exemplified by the "Environment Explorer" decision-making tool created for the Netherlands. The system allows integration of physical, environmental and economic/institutional variables at these different scales (Engelen *et al.,* 2003). In that system, geographic information system- (GIS) based applications are used to nest models operating at local, regional and national scales. A way forward here would be to employ peer-reviewed up-scaling and down-scaling techniques such as those reviewed by Wilby and Wigley (1997) to statistically transform climate output from a general circulation model (GCM) to a finer regional scale (e.g. to 0.5 x 0.5 degrees of latitude and longitude). However, these scaling techniques often assume homogeneity and linearity, and problems exist because the range of GCMs do not produce sufficiently converged large-scale information to be processed through downscaling (IPCC, 2001).

The question of how to manage such scaling problems brings us to the second point, that of how to handle uncertainty in damage estimates, of whether that uncertainty arises from estimates of the shapes and magnitudes of dose-response curves linking local climate to local damage, and of whether it arises from the availability of a range of possible local climatic conditions matching the same level of global climate change.

Modelling system uncertainty can be addressed using Bayesian statistical techniques developed for analysing computer experiments (Craig *et al.,* 2001). In the Bayesian approach, all uncertain quantities are treated as random variables described by a joint probability distribution, and data such as field observations are used to modify beliefs about the system that is being modelled. The precise methods used depend on the complexity of the relevant model or model component. We refer to a model component as a "module". For simple modules (i.e. those with small input spaces and rapid execution times) samples can be constructed from the full probability distribution of the system variables. For more complex ones, or for entire integrated modelling systems, a linear Bayesian approach can be applied, in which uncertainty about the system is summarised by two indicators known as a mean vector and a variance matrix. In the latter approach the module is represented as a numerical "black box" of uncertain functional form. Module evaluations are employed to construct a statistical emulator which is then used to evaluate the system mean and variance. The statistical emulator can also be used off-line to provide information such as a good choice of parameter values at which next to evaluate the system, and an estimate of the benefit (e.g. in terms of the reduction in predictive uncertainty) of further evaluations. These inferential calculations can be expensive, and for very complex modules it is likely that only selected subsets of the module inputs and outputs can be considered. The approach can also be applied to modules which provide spatially varying output on a fine grid.

The results of such a robustness study could be presented to policy makers in the form of a risk management approach, which is a common method of handling decision making under uncertainty.

The question of incorporation of uncertainty into an IAM system is most easily addressed by using the novel CIAS approach outlined in Section 4, although it is applicable to other modelling systems as well. However, it is recommended the approach be complemented by intensive work on approaches to combine qualitative and quantitative assessments, such as the use of fuzzy set theory[22].

2.2 *A theory-based approach to the definition of dangerous climate change*

Although most discussions about "dangerous interference with the climate system" and "safe stabilization targets" have been plagued by concerns over intellectual and political consistency, there is no way to escape consideration of the ultimate reason for trans-national climate policy, and the post-Kyoto deliberations starting 2005 will have to face this difficult challenge. In attempting to define dangerous climate change within a formal, analytic framework designed for decision making, there are two types of thresholds that can be employed for setting long-term emissions reduction targets, namely *normative* and *systemic* ones.

Figure 1. Reasons for concern about climate change

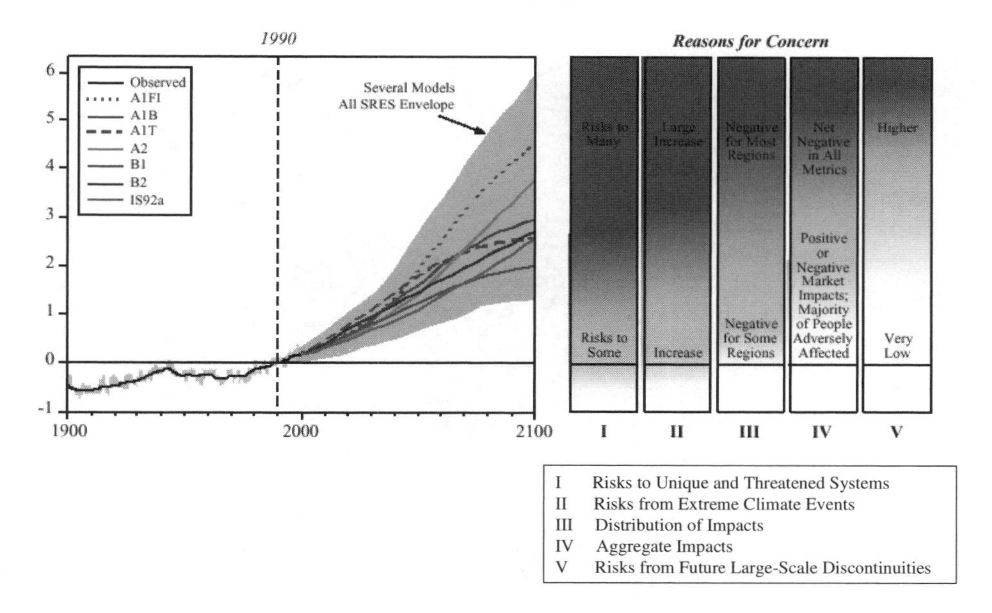

Source: taken from IPCC (2001).

In the first case (normative), multiple value judgements by pertinent actors are condensed into definitions of tolerable ranges of acceptable outcomes. In the second

22 Fuzzy set theory defines set membership as a possibility distribution. Such imprecise probabilities can provide a basis for *partial ordering of preferences*, which may be sufficient to enable policy-related decision-making, whilst better reflecting the very significant uncertainties. For further discussion, see Allen (2003).

case (systemic), major phase transitions, or points at which the behaviour and properties of the entire system suddenly alter, provide criteria for identifying completely unacceptable outcomes.

An IAM system can be used to address the question of "dangerous" anthropogenic climate excursions using the first (normative) approach, in particular by consolidating the IPCC TAR (IPCC, 2001) approach, from which Figure 1 is reproduced. However, very different threshold levels of global warming can be envisaged for different sectors of society, and for different natural ecosystems, in each region, depending on the relationship between climate changes and impacts, and upon value judgements made by those defining these normative thresholds. Numerous examples of attempts to quantify dangerous climate change exist (Parry *et al.,* 2001; O'Neill and Oppenheimer, 2002; Vaughan and Spouge, 2002). There is clearly a need to develop a more holistic approach to the problem.

Figure 2. Potentially critical elements for ecosphere operation (after Schellnhuber, 2002)

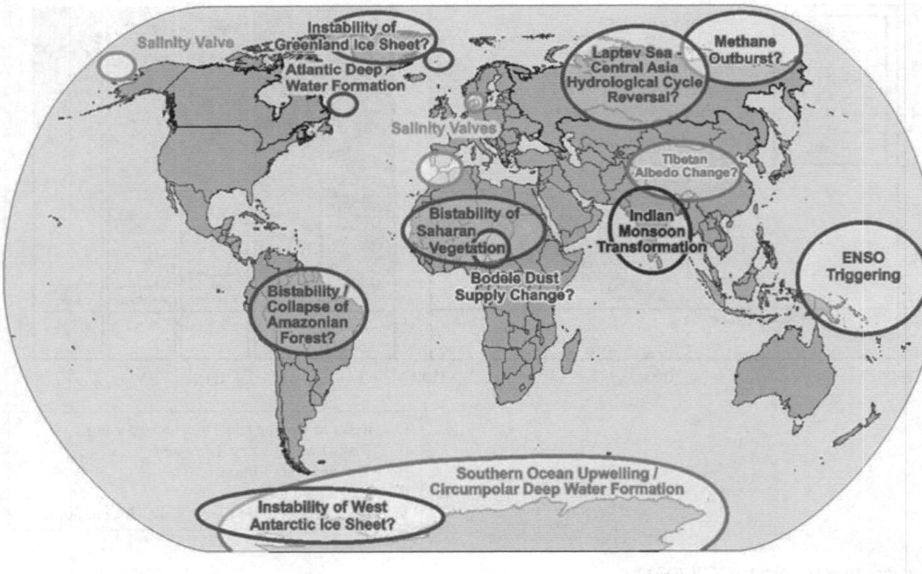

A systemic approach to the problem of dangerous climate change was used by the ICLIPS network (see below). The aim was to track down the systemic thresholds in planetary dynamics potentially transgressed in the course of unabated global warming. Figure 2 (taken from Schellnhuber, 2002) provides an overview of switch and choke elements in the Earth System that might be (de-)activated by human interference. The most prominent example of a switch element is the North Atlantic Deep Water Formation (Rahmstorf, 1995) off Greenland and Labrador, which acts as a driving force for the thermohaline circulation generating, *inter alia*, the Gulf Stream. Obviously, any human alteration of such first-order modes of operation of the eco-sphere would have severe impacts and should be avoided by all reasonable means available. It has to be emphasized though that the pertinent criticality analysis is still in its infancy and will have to be considerably advanced by Earth System modelling as part of integrated assessment systems.

Our proposal for the application of IAM to this problem is the adoption of an *intermediate* approach which combines aspects of the normative and systemic. In this, a series of (normative) damage functions may be constructed for each vulnerable nation/stock at risk/function/sector. Each damage function is shaped like the letter 'S' (or sigmoidal), showing how damage accrues as a function of average temperature rise. Damage first accrues slowly, and then rapidly rises as climate changes; and then accrues more slowly, reaching a plateau when the entire sector has been permanently damaged or lost, as shown in Figure 3.

Figure 3. Illustrative sigmoidal damage function

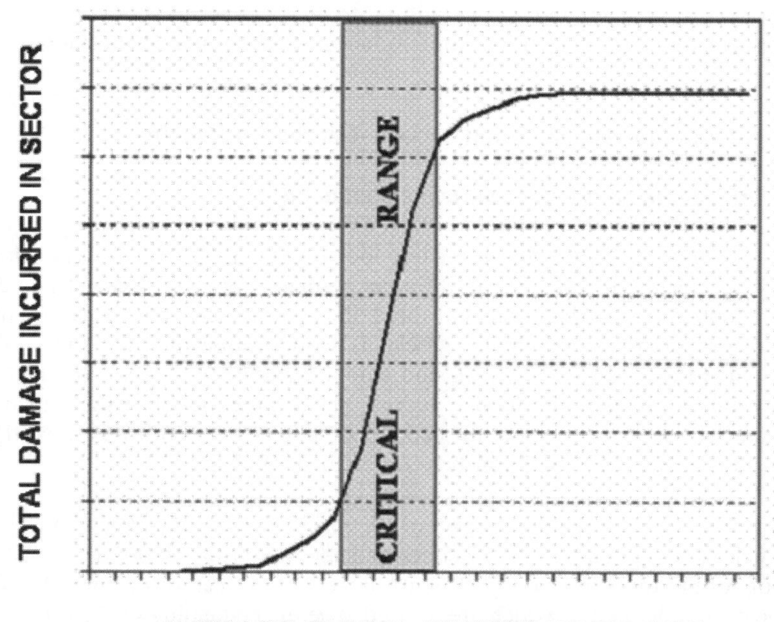

Such damage functions can be drawn from existing information held at participating institutions, from users, from the peer-reviewed literature, including reports on case studies of valuable or vulnerable sectors. For each damage function, a band of global temperatures at which climate change is considered "dangerous" for the sector in question is identified by determining the critical range for which the gradient of the sigmoidal damage function exceeds a certain threshold, as shown in the Figure. For example, for coral reefs, such a band might match a 0.5 to 1.5 degree C rise in sea surface temperature.[23] Alternatively, it might match the range of sea level rises which would cause significant additional inundation of a low-lying small island state, or the range of average global temperature changes over which significant losses in the agricultural economy of a particular area occur. Thus, most damage functions would possess a "critical range" over which dangerous effects occur (shown in the box in Figure 3). Through IAM, the spans of these critical ranges could then be plotted on a single graph, to create a probability distribution function exhibiting the number of critical ranges occupied for each average global temperature rise. Such a plot would thus determine whether there is a clustering of these critical ranges at a certain average global temperature rise. Whether or not this proves to be the case, the IAM could be used to plot how dangerous climate change accrues in various sectors as temperature rises – which gives a systemic picture. Modelling is required to indicate whether or not policy/mitigation actions will make a difference, and, if so, when, in different impact sectors/regions or regional climate systems. The policy decision would then be transformed into a new question: in which sectors/countries/stocks can we prevent a dangerous level of change, and are we willing to take the necessary actions to do so? (We have already reached dangerous climate change for the small island states).

3. Adaptation, Ancillary and Socially-Contingent Benefits

3.1 Mitigation, adaptation and sequestration policy

A major use of IAM will ultimately be to inform policy makers about the consequences, costs and benefits of different climate policy choices. Choices will include a number of combinations of three lines of attack (i) mitigation through reduction of emissions of greenhouse gases (ii) sequestration of carbon in geological formations, in forests, or in the ocean and (iii) the ability of society to adapt to the remaining climate change. IA can also address the complex issue of the effectiveness and permanence of sequestration of carbon in newly planted forests. Mitigation, adaptation and sequestration will now be referred to as M, A and S.

Mitigation is a much-studied aspect of IA (Grubb *et al.,* 1993; Peck and Teisberg, 1995; Richels and Edmonds, 1995; Schneider and Thompson, 2000; Tol, 2000). The diversity of results thus obtained is strongly linked to the nature of, and assumptions behind, the economic models which are utilised. The understanding of the robustness of IA results to the use of different modelling approaches is an important subject for future study, and one that can be taken forward by the Community Integrated Assessment System (see below). For example, different institutions may subscribe to different

[23] Global sea-surface temperature maps show that mass coral-reef bleaching episodes between 1983 and 1991 followed positive anomalies more than 1 degree C above long-term monthly averages (Goreau and Hayes, 1994), consistent with a popular model for coral bleaching (Hoegh-Guldberg, 1999).

economic theories, or different approaches to climate modelling, producing correspondingly different components for an IAM.

Although we have made suggestions as to how IAM can be used to holistically assess climate change damages under different climate and socio-economic futures, the potential for human systems to adapt to climate change is not included specifically[24]. IAM does provide a way of doing this, however. The potential to adapt may be included in each of the individual components used in the holistic assessment, or alternatively revised "damage functions under adaptation" may be considered. Owing to the inherent uncertainties of making such calculations, it is important to retain the capability to investigate the situation with and without the adaptive options included. If adaptation is to be considered holistically, then assessments of the vulnerability and adaptive capacity of all sectors/regions must be drawn upon, which is a massive task for which the data is not currently available. A practical suggestion given the state of current knowledge, and the likely impossibility of building up a comprehensive library of information on adaptive capacity, is to incorporate indicators of adaptive capacity in the IAM system through on-going work on the development of a pertinent theory of adaptive capacity.[25] The detailed treatment of adaptation would be a very difficult task, and for this reason the modelling of adaptation policies cannot realistically be incorporated in holistic assessment. Rather, the presentation of results as a risk assessment, in which levels of risk associated with different levels of investment in adaptation, such as the probability of inundation given low, medium or high investment in seawalls, as opposed to managed retreat, can provide a useful vehicle through which policymakers can view their decisions about levels of investment in mitigation, adaptation and sequestration. The problem can be presented as a risk management situation, with different combinations of policies leading to varying levels of risk (of reaching certain levels of damage that may be considered dangerous) in a diverse range of sectors.

Sequestration in newly planted forests can be studied in IA models, but only those which have a good representation of the carbon cycle and include carbon cycle feedbacks. IA models including reduced-form earth system models, such as those currently being developed under the UK GENIE project (www.genie.ac.uk), can help us to understand the permanency or otherwise of sinks in forests, and whether these sinks might be offset by other feedbacks in the climate system, such as the die-back of existing forests as a result of an inability of biomes/ecosystems to shift rapidly enough to keep up with their climate envelopes. IA can also consider the combined effects of climate change and land use change, within the context of the carbon cycle.

---

24 However, it should be noted that some policies designed to reduce emissions are also policies which could be regarded as adaptive. For example, improved building design could both reduce the energy consumed by the occupants of a building and better insulate them against an adverse and rising outside temperature.

25 Such research is ongoing at the Tyndall Centre for example (www.tyndall.ac.uk/research/theme3).

Sequestration of CO_2 in geological formations and the ocean has rarely been considered in IA models, but there is considerable potential to do so. The technology is viable: at Sleipner in the North Sea, Statoil has been removing CO_2 from natural gas since 1997 and is storing 1 million tonnes per year in a saline aquifer, in response to an off-shore carbon tax which makes it cheaper to separate and store the CO_2. Potential geological storage sites also include depleted oil and gas reservoirs, unmineable coal beds and the deep ocean. IA could incorporate the costs of the various potential pre- and post- combustion CO_2 capture techniques (e.g. direct reaction of fuel, producing hydrogen, or, amine scrubbers, for example), although there is great uncertainty as to whether these costs might decline. Realistic estimates of the geographic and technological potential of these measures need to be included, and uncertainties in these aspects are also high. Estimates of global sequestration potential also remain disparate (1500 to 11000 Gt CO_2) (Gough *et al.,* 2002). However, currently it is estimated that to collect more than 30% of the emissions from electricity generation would require a move towards use of hydrogen as fuel, derived from fossil fuels using pre-combustion decarbonisation, because CO_2 can only be captured from large installations such as power plants and industrial sources (S. Shackley, pers. comm.). Key issues for the future of the technology include not only cost but the reduction of the "energy penalty" or carbon capture (currently 9 to 34%), and the rate of leakage from storage sites. In spite of these uncertainties in current and future sequestration potentials, it is possible to use IA to scope the policy options under a range of sequestration potentials.

3.2 Ancillary benefits of climate protection

Mitigation of greenhouse gas emissions has a number of ancillary benefits. Climate change mitigation policies also reduce the concentration of many harmful air pollutants in the atmosphere that could lead, *inter alia*, to human health and eco-system benefits (Jochem and Madlener, 2003). Furthermore, if dependence on oil is reduced, other environmental impacts such as oil spills will be reduced, and significant benefits to energy security will be achieved. Davis *et al.* (2000) review the ancilliary benefits of greenhouse gas mitigation, some of which are mentioned below.

3.2.1 Ancillary benefits in terms of air pollution

Greenhouse gases share many common drivers and methods of mitigation with air pollutants such as sulphur dioxide (SO_2), nitrogen dioxide (NO_x), volatile organic compounds (VOC) and primary particulate matter (PM_{10}). Thus climate mitigation and air pollution mitigation can be achieved by employing many of the same strategies, such as switching to low-carbon or non-fossil fuels, or by increasing energy efficiency. Any process that reduces fossil fuel use will have beneficial effects in terms of mitigation of both climate change and regional scale/large scale air pollution. Carbon sequestration of exhaust gases often requires the removal of other pollutants from the waste gas stream, whilst in contrast, end-of-pipe technologies commonly used to abate air pollutants (e.g. flue gas desulphurisation for SO_2 or fabric filters for the removal of PM_{10} emissions) do not remove CO_2, and hence there may be negative synergies among these conventional pollution policies and mitigation of climate change. For example, SO_2 removal technology increases energy use for combustion facilities increasing CO_2 intensity of production processes; also in the transport area, NO_x control technologies/policies have led to an increase in N_2O.

Air pollution and climate changes also impact on a similar set of "receptors", for example human health (e.g. Kunzli *et al.,* 2000), natural and semi-natural ecosystems, agricultural practices, and the built environment (Davis *et al.,* 2000). Thus combined policy benefits accrue for human health, agriculture, and natural ecosystems, and synergistic effects between the two are also reduced.[26]

The causal commonalities between climate change and air pollution, as well as the similar nature of the stock at risk from both, create the potential for considerable synergy in policy terms. IAM is an ideal system in which to explore the complex relationships between the two subject areas, including the feedbacks resulting from the different radiative properties of air pollutants. For example, the black carbon component of primary particulate matter has a positive forcing, acting to warm the atmosphere, whilst NO_x and VOCs are the precursors of tropospheric ozone, which itself is a greenhouse gas. Conversely, secondary particulate aerosols (sulphate and nitrate) formed from SO_2 and NO_x have a negative radiative forcing acting to cool the atmosphere. Existing work in this area has already highlighted potential synergies and pitfalls (Alcamo *et al.,* 2002; T. van Harmelen *et al.*, 2002; Krupnick *et al.,* 2000).

Overall, ancilliary benefits of GHG mitigation for air pollution are large and positive (Barker and Rosendahl, 2000), whilst some air pollution policies increase, and others decrease, radiative forcing. However, this does not alter the fact that both air pollution and GHG emission can be reduced simultaneously and very cost effectively through fuel switching. In fact, it has been shown that it is cheaper to attack air pollution through fuel switching than through end of pipe technology (Sliggers and Klaasen, 1994; Sliggers, 2004.)

3.2.2 Climate policy and security

An important emerging issue is the question of how climate policy may or may not increase security. There are a number of dimensions to this issue including:

- the fact that a climate policy involving a diversification in energy sources would result in a reduction in societal and economic sensitivity to disruptions of the oil supply;

26 We are now at the 50th anniversary of the infamous London smog of 1952, where the burning of coal in the domestic sector, combined with anticyclonic conditions, led to the deaths of some 4000 people owing to the dense concentrations of SO_2 and PM.

- the potential for prevention of security problems through reduction of climate change damages.

Climate change would render parts of the world uninhabitable/agriculturally unproductive, creating potentially large movements of refugees across national borders and placing pressure on remaining food supplies. This could in turn lead to violent conflict. Environmental refugees have already been created in at least three areas. In small island states, sea level rise threatens the existence of small states. In parts of Central America, desertification and increasing demand for water is rendering land unsuitable for agriculture. Thirdly in Bangladesh, extreme weather events (cyclones, storm surges and floods) have affected 3 million people in the year 2000 alone, as a result of which emigrants are leaving the country in droves (Tanzler *et al.,* 2002). Ensuing violence has already occurred in Bangladesh where clashes between emigrating Bangladeshi and tribal people in Northern India have led to the deaths of several thousand people. Tanzler *et al.* analyse the conflict dimension of the societal and political implications of climate change in interaction with six other factors (soil erosion, hydrological cycle, water-scarcity, population growth, urbanisation, and agriculture) and conclude that there is a major gap between "the primarily natural science and economic work of the IPCC and the social science orientation of the environmental security debate", and suggests possible linkage points between the two types of research. Similarly, Brauch (2003) examines the links between environmental stress, climate-induced extreme weather events, and potential humanitarian crises, which can in turn lead to political crises. Tanzler *et al.* note that environmentally induced stress within a single country can have implications beyond its borders, with violent conflict becoming a possibility at the regional or international levels as well as the national level. Therefore, climate policy would reduce potential humanitarian crises (e.g. food supply issues) and ensuing political security risks that would otherwise be likely to develop. An IAM approach has a high potential to make contributions in this area, to indicate the potential benefits of climate policy in terms of security under different socio-economic scenarios of the future. This has to be placed in the context of recent security developments and the potential for terrorist attacks to energy installations and other infrastructure which is of increasing importance.

3.2.3 Avoiding considerable costs of fossil fuel transports like tanker accident damages

Climate mitigation that reduces dependence on fossil fuels for energy would also avoid the costs associated with the transport of fossil fuels about the globe. For example oil pollution incidents such as the recent Prestige disaster close to the North Iberian coast would be avoided. This has caused a marine ecological disaster on an unprecedented scale, the costs of which are likely to run in to the billions. Over the past 20 years, the total damage costs of such incidents run into tens of billions. For example, the use of the contingent valuation method following the Exxon Valdez incident in

Alaska found damage costs of USD 10 billion[27] (Marine Conservation Society, personal communication).

4. Some recent examples: use of integrated assessment to assess climate policy benefits

In this section we select three recent examples to demonstrate the application of IAM to the benefits issue:

a. the ICLIPS project of the Potsdam Institute for Climate Impact Research;

b. the "Coastal Simulator" being pioneered at the Tyndall Centre; and

c. the novel Community Integrated Assessment System, also being pioneered at the Tyndall Centre.

a) The ICLIPS project at the Potsdam Institute

An integrated modelling system was assembled a few years ago to address the issue of a definition of dangerous climate change. The ICLIPS project, initiated and coordinated by the Potsdam Institute in 1995, gave birth to an international and interdisciplinary network of eminent experts on relevant climate-change aspects for an inverse analysis of climate management strategies using the "Tolerable Windows Approach" (Toth *et al.,* 2002; Leimbach, 2000; Schneider and Toth, 2003). The basic idea was to identify/define *intolerable* global (or regional) impacts of anthropogenic global warming and to calculate, via causal-backwards modelling, the admissible GHG emissions corridors for different macro-actors (like the Annex-I countries) over the next centuries. The ICLIPS community model calculates many crucial variables, such as ecosystem impacts, regional costs of mitigation measures and timing of emissions reductions. It acts in a semi-coproductive way already: generating and sharing a common pool of model source codes and data sets to provide the crucial building blocks for (inverse) integrated assessment.

A recent publication emerging from the ICLIPS project (Toth *et al.*, 2002) illustrates the use of normative thresholds described in Section 3 above. The ICLIPS integrated assessment model was used to assess the influence of variations in three normative guardrails on the existence and shape of necessary emission corridors. The three guardrails are: first, an *impact guardrail* that indicates the percentage of ecosystems worldwide (agricultural areas excluded) that would undergo a change in

[27] In court, Exxon managed to argue that these be reduced to USD 4 billion.

biome-type;[28] second, an *economic guardrail* that limits the loss in consumption due to emissions reductions (compared to the high-emissions reference path) in any region at any time; third, *a timing constraint* that excludes effective emissions reductions before a specified date.

An emission corridor is an envelope drawn on a graph of emissions against time, encompassing all the possible emission pathways which satisfy the normative guardrails. Note however, that any arbitrary path within the envelope is not necessarily a permitted path. Figure 4 presents, as an example, the admissible emissions corridors for an illustrative impact guardrail that prohibits biome changes of more than 35% worldwide (i.e. at least 65% of the world's ecosystems are to be preserved). Whilst maintaining this impact guardrail, the economic guardrail is varied from a loss in regional consumption of between 0.3 and 3%, or the timing constraint is varied so that emission reductions start between 2005 and 2035.

Since 65% preservation would constitute a rather unambitious ecosystem protection target, impact guardrails of 30% and 25% biome transformation were also studied. The 30% limit results in a drastically narrower emissions corridor, whilst no corridor exists for the 25% limit. In other words, a 75% protection criterion is impossible to achieve by reducing CO_2 emissions alone if willingness to pay is limited to 3% of regional consumption. Future studies will address how much flexibility would be provided by reducing also the other greenhouse gases.

The single emission pathway shown in Figure 4 is the result of an ICLIPS welfare optimisation calculation, subject to the constraints of the guardrails. This follows the baseline emissions initially and then switches to a path of accelerating reduction once autonomous and learning-by-doing types of technological development make mitigation efforts less expensive.

b) The "Coastal Simulator" project at the Tyndall Centre

The "Coastal Simulator" is being designed as an innovative decision-support tool which will allow assessment of various climate response policies in the context of limited geographic area (the North Norfolk coast, UK). This will be made possible through the integration of climate change scenarios and policy response options with information on sediment transport, biodiversity, sea defences and socio-economic activities. Typical policy responses considered include managed retreat and sea wall construction. For each coastal future generated by the model, the probable effects on biodiversity, sediment dynamics, landform characteristics and socio-economics will be estimated. The outputs of the model will be linked to a Geographic Information Systems (GIS) framework that will display biodiversity data and socio-economic information,

---

[28] The study by Leemans and Eickhout (2004) also links the percentage of ecosystems that will be preserved to the magnitude of temperature changes over the coming century using the IMAGE integrated assessment model.

allowing detailed case studies to be visualised using cutting-edge virtual reality GIS techniques. This provides a sophisticated and attractive method of replying to user demands and of displaying results.

Figure 4. Variation of the socio-economic constraints

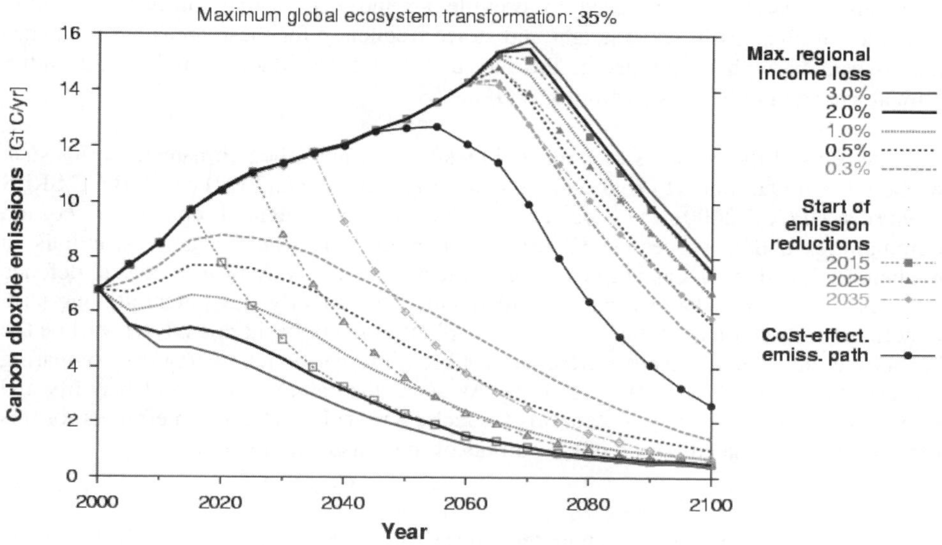

Source: after Toth *et al.* (2002).

Note: Admissible corridors for energy-related CO_2 emissions for different levels of regional income loss if at least 65 % of the world's ecosystems are to be preserved under climate change. The figure shows emission corridors for variations of a regional mitigation cost constraint (from 0.3% to 3% loss of consumption without timing restrictions) and for variations of a timing constraint (from a start date of 2005 to 2035 for a regional mitigation cost constraint of 2%). The diagram shows eight pairs of lines representing emission corridors and one particular emission pathway. The outermost envelope of solid lines indicates the widest emission corridor, i.e. that a wide range of emission trajectories satisfy a 3% loss in consumption and a 65% preservation criterion, if emission reductions begin now. If society would tolerate only lower losses in consumption, the corridor narrows (other solid, dotted and dashed lines) and the maximum possible annual emission rate which is reached during the period decreases. If society delays emission reductions, the corridor also narrows (squares, triangles, diamonds). For comparison, the (middle) line of black dots show the optimal emission reduction path (i.e. that which maximizes global utility) whilst meeting a constraint to preserve 65% of ecosystems.

The model is being developed using a fuzzy-logic expert-system approach where the probabilities of a range of different coastal futures can be estimated for a given section of coast. This approach is preferable to a more mechanistic, black-box model design as both qualitative and quantitative data can readily be incorporated into the simulator while the model can be continually informed and modified by user involvement. Moreover, this technique allows the tool to be flexible as probabilities of future change can be readily updated as climate model estimates are refined. These qualities of a fuzzy-logic approach allow the model to be readily used in a policy-relevant context, where the trade-offs and benefits of different coastal futures can be evaluated. The long-term aim for the simulator will be a quantitative modelling approach where expert opinions will be gradually confirmed, refuted or replaced by quantitative models, as our scientific understanding improves. For example, the results of recent research will be used to provide an index of vulnerability[29] generating estimates of changes in wave height and storm frequency for these areas. The changes are associated with variations in the North Atlantic Oscillation[30] index and future climate change predictions (Tsimplis, 2004).

Aspects of the coastal simulator will make use of published climate scenarios such as the UK Climate Impact Programme scenarios (Hulme *et al.,* 2002) or the IPCC SRES scenarios (IPCC, 2000) to generate estimates of the potential change in coastal communities and/or processes. Hence, much of the direction for the research is in evaluating the adaptive capacity of systems (ecosystems, governance or coastal defence policies) to climate change, rather than quantifying directly the potential long-term benefits of mitigation activities. However, implicit in many of these analyses will be the capacity to analyse the relative effects of different mitigation levels (e.g. by comparing the effects of an A1 with B2 scenario). As such, the relative costs and benefits and effects of different mitigation scenarios (or lack thereof) in different timeframes for this region can be evaluated as one of the outputs of the coastal simulator.

c) The Community Integrated Assessment System: a multi-institutional, participatory approach to IA

In this section we highlight the need for a new European initiative to provide a tool for participatory integrated assessment that overcomes the inherent limits of existing IA systems, and also allows the study of the robustness of the output of the IAM to the use of, for example, different economic theories or climate modelling schemes.

---

29 Coastal communities are more vulnerable to climate change than inland communities because, in addition to changes in meteorological parameters, they are also affected by changes in oceanic parameters, especially increases in sea level and wave heights. Both direct effects (for example changes in coastal erosion, storm surges and water temperature) and indirect effects (like reductions in fishing stocks and in the number of days suitable for fishing) will have physical and socio-economic impacts on coastal communities.

30 The NAO is the dominant mode of winter climate variability in the North Atlantic region ranging from central North America to Europe and much into Northern Asia. The NAO is a large scale seesaw in atmospheric mass between the subtropical high and the polar low. The corresponding index varies from year to year, but also exhibits a tendency to remain in one phase for intervals lasting several years.

There is a traditional mismatch between questions raised by policy makers and society and information available from scientists. It has often been pointed out that involvement of non-scientists in integrated assessment is necessary in order to ensure the relevance and later acceptance of the results of analytical modelling (Hordijk, 1991); and a variety of participatory modelling approaches have been used in attempts to resolve this mismatch (Rotmans, 1998; van Asselt and Rijkens-Komp, 2002). Hence we propose a tool for participatory integrated assessment that aims to extend mutual learning between scientists and modellers on the one hand and civil society and policymakers on the other.

Consider now the inherent limits of existing knowledge production systems based on IA (e.g. Alcamo, 1984; Dowlatabadi and Morgan, 1993; Dowlatabadi, 1995; Hulme *et al.,* 1995; Kainuma *et al.,* 2003; Matsuoka *et al.,* 1995; Morgan and Dowlatabadi, 1996; Plambeck *et al.,* 1997; Prinn *et al.,* 1999; Rotmans, 1990; Rotmans *et al.,* 1994). These are (i) the impossibility of building monolithic all-purpose models that assemble a comprehensive array of components at a single site, and (ii) the impossibility of providing responses to policy-relevant questions with a turn-around time that is acceptable to policy-makers.

To overcome these obstacles, the Tyndall Centre and the Potsdam Institute are together pioneering a novel approach to integrated assessment in which a set of participating institutions build a Community Integrated Assessment System (CIAS). CIAS consists of an IAM and a participatory process with stakeholders. This requires that the CIAS IAM be distributed across the participating institutions that provide components and data. Each institution contributes at least one component, known as a module, to the system. Examples of such modules include a model of the world economy, or a simple model of the earth's climate system. We refer to these model components as "modules" because they become components of the IAM. We say that a module is "used" when it forms an internal component of the IAM.

These components must then be linked together in a modular fashion. This modularity results in a highly flexible integrated modelling system.

Its principal advantages over traditional systems are:

a. one can investigate the robustness of the output of the IAMs to the use of different theories concerning the same discipline. Consider how different institutions will frequently study the same discipline, such as economics, but will subscribe to slightly, or vastly, differing theories concerning that discipline. Therefore, they will each produce very different modules for that discipline, since the modules are written based upon the theory which is studied at the institution in question. The CIAS system's flexibility and modularity allows one to easily study an important aspect of uncertainties in the output of key IAM results, such as cost benefit analysis. This is possible because one can study the robustness of these results to the different theories which exist about, for example, economics. The flexible structure means that it is relatively easy to remove a single module from the IA model, and substitute it with another from another institution. For example, an economic module produced by one institution can be substituted for an economic module produced by a different institution. In this way the effect of using different theories or "paradigms" can be studied. The system thus encourages a pluralistic approach. Contrast this with the situation that exists where each institution builds its own IAM. Since the two IAMs will differ in all their components, and not just one, it is then very difficult to understand precisely why the outputs of (for example, cost-benefit analysis) will differ between the IAMs;

b. similarly, the system allows exploration of robustness with respect to the use of modules of the same discipline (e.g. climate) of differing complexity;

c. it allows the assembly of a large number of components addressing climate impacts in sectors and regions; and

d. the flexible nature of the system allows us to more readily tune it to address stakeholders' evolving needs.

Figure 5 shows our vision of how the Community Integrated Assessment System forms a bridge between the knowledge space, i.e., into which the scientific community – and users – contribute their expertise, and the problem space, the societal and institutional context within which decisions are made. The overall result is the co-production of knowledge relevant to policy and decision-making, such that the integrated modelling system and other relevant knowledge can be combined to solve real world climate-change related problems. For further details see Warren (2002).

It is important to note the iterative nature of the interaction between the knowledge and problem spaces: hence we represent the system as a wheel, rotated by the successive actions of operators, called demander, surveyor, composer and responder, respectively. We now explain the role of these four.

We begin with the demander, who first identifies pertinent policy-relevant research questions to be addressed by the scientific community. To do so, the potential users of the CIAS output need to be identified and the demander then co-produces these research questions with the users, who are the ultimate "problem-owners". These questions are then supplied to the modelling team, which comprises the surveyor and the composer. The surveyor identifies the scientists who have relevant information, and obtains from them the components necessary to address the questions. At this stage, the surveyor can make a rapid assessment as to whether the question has the potential to be answered by the scientific community. If so, the wheel proceeds to the composer, who combines the components selected by the surveyor into an integrated assessment modelling system: a modular, flexible and multi-institutional Community Integrated Assessment Model. The composer must now ensure that the control flow matches the question being asked by the demander. Again, at this stage there is further assessment of whether the question can actually be answered, and to what degree of accuracy: the aforementioned Bayesian approach is important here. Finally, the responder determines the most appropriate form of feedback of the results, and communicates them to the users and to society as whole.

Figure 5. Community Integrated Assessment System

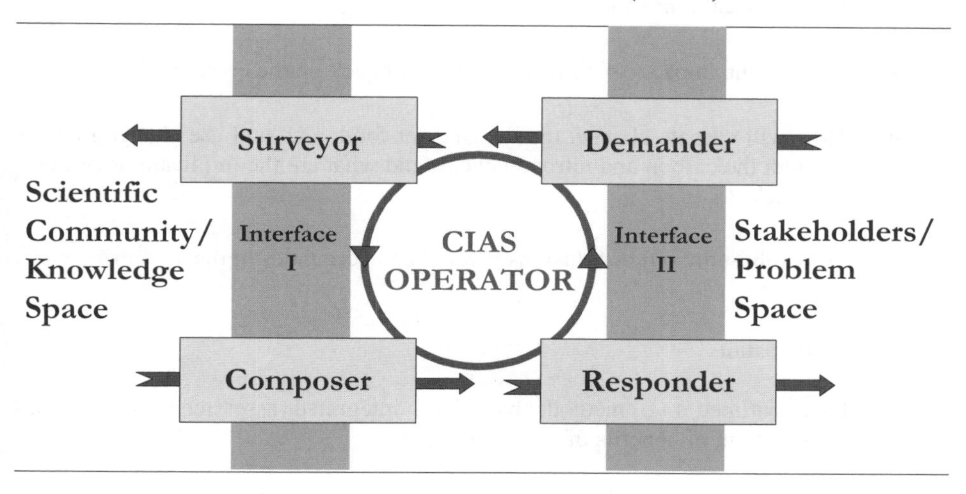

A core group of institutions (including the Tyndall Centre and PIK) are involved in the initial phase of this project, with the expectation that many more will participate in due course. Operators of the system will consist of expert groups from a range of European research institutes. So far our modelling team has accumulated a range of modules for potential use in the CIAS system, together with information from the literature and completed projects. Some of these have been assembled into a simple prototype model. Model design will evolve over time, being co-produced by the modelling team in reaction to the demander's requirements. This co-evolution is being facilitated by an Interactive Integrated Assessment Process (IIAP), which comprises (1) a scoping study of user needs for research on climate change, involving a series of some 60 structured interviews with key potential users across industry, the public sector and NGOs in the UK and (2) the development of a conceptual framework of the problem space through an initial mapping of the climate change policy network with which the research must interact. (Turnpenny *et al.*, in prep).

Ultimately, an advanced CIAS system could be applied to answer a wide range of policy questions, e.g.:

- What are the costs and benefits of different levels and timescales of stabilisation of greenhouse gas concentrations in the atmosphere? More specifically, which economic pathways lead to which stabilisation scenarios and what are the implications for climate change damage and for adaptation?

- What are the consequences of various divisions of investment between mitigation, adaptation and sequestration?

- What is the efficacy of the application of carbon taxes and investment in technological innovation?

- How do the impacts of climate change feed back on the economy?

- How will climate change, the demand for food, and land use change interact through the carbon and nitrogen cycles and what are the implications of these feedbacks?

- How robust are our conclusions given the uncertainties in the system?

5. Conclusions

We have outlined two methods by which integrated assessment models can address the assessment of benefits of climate policy:

a. The community might pool its resources to provide policy makers with holistic estimates of the (non-monetised) climate-change damages associated with different climate futures. Where appropriate, monetisation might be carried out and linked back to economic models to analyse cost-benefit issues. We recommend the use of Bayesian uncertainty analysis in combination with peer-reviewed up- and down-scaling approaches to address the problems or accuracy and scaling.

b. A theory-based approach utilising damage functions might be applied. Damages could be summarised using, for example, the set of numeraires proposed by Schneider (2004). Stabilisation scenarios could be analysed in a similar way. Such approaches can help define "dangerous climate change", through the assimilation of "thresholds" in the non-linear damage functions. Such functions could also be modified to reflect the adaptive capacity of different regions or sectors or society.

We have shown three examples of the application of integrated assessment models to assess the benefits of climate policy, taken from projects under way at the Tyndall Centre for Climate Change and the Potsdam Institute for Climate Impacts Research. Our first example is the ICLIPS impacts tool, based on the identification of "tolerable windows" of climate change-induced damages and economic costs. Our second example describes how a coastal simulator can be applied to solve climate policy problems at the local scale. Finally we have in our third example pointed the way forward for integrated assessment of the benefits of climate policy, proposing a novel Community Integrated Assessment System based on a modular, flexible and multi-institutional design. This approach can help overcome the traditional mismatch between science and policy. Further, it will allow us to distil robust modelling conclusions in the face of uncertainties, and to understand the robustness of conclusions to the use of different modelling paradigms originating at different institutions.

Acknowledgements

Thanks are due to Jan Corfee-Morlot and Hans Martin Fuessel for their helpful suggestions on an earlier draft of this paper.

References

Alcamo, J. (ed.), (1984), "IMAGE 2.0: Integrated Modeling of Global Climate Change", Kluwer Academic Publishers, London.

Alcamo, J., P. Mayerhofer, R. Guardans, T. van Harmelen, J. van Minnen, J. Onigkeit, M. Posch and B. de Vries (2002), "An integrated assessment of regional air pollution and climate change in Europe: findings of the AIRCLIM project", *Environmental Science and Policy*, **5**, 257-272.

Allen, M. (2003), "Climate forecasting: possible or probable?" *Nature,* **425**, 242 pp.

Azar, C., and S.H. Schneider (2002), "Are the Economic costs of Stabilising the Atmosphere Prohibitive?" *Ecological Economics*, **42**, 73-80.

Barker, T. and K.E. Rosendahl (2000), "Ancillary Benefits of GHG Mitigation in Europe: SO_2, NO_x and PM_{10} reductions from policies to meet Kyoto targets using the E3ME model and EXTERNE valuations", Ancillary Benefits and Costs of Greenhouse Gas Mitigation, Proceedings of an IPCC Co-Sponsored Workshop, March, 2000, OECD, Paris.

Brauch, H.G. (2003), "Climate Change, Environmental Stress and Conflict", AFES Press Report for the Federal Ministry for the Environment, Nature Conservation and Nuclear Safety.

Craig, P.S., M. Goldstein, J.C. Rougier, A.H. Seheult (2001), "Bayesian forecasting for Complex Systems using Computer Simulators", *Journal of the American Statistical Association*, **96**, 717-729.

Davis, D.L., A. Krupnick and G. McGlynn (2000), "Ancilliary Benefits and Costs of Greenhouse Gas Mitigation: an overview", in OCED *et al.*, Proceedings of an Expert Workshop, Paris, pp. 9-49.

Dowlatabadi, H. (1995), "Integrated Assessment Models of Climate Change - An Incomplete Overview", *Energy Policy*, **23**, 289-296.

Dowlatabadi, H. and M.G. Morgan (1993), "A Model Framework for Integrated Studies of the Climate Problem", *Energy Policy,* **21**, 209-221.

Edenhofer, O, N. Bauer, E. Kriegler, "The Impact of Technological Change on Climate Protection and Welfare. Insights from the MIND model" submitted to Ecological Economics.

Engelen, G., I. Uljee and R. White (2003), "Environment Explorer: Spatial Support System for the Integrated Assessment of Socio-economic and Environmental Policies in the Netherlands", in Proceedings of the 1st biennial meeting of the International Environmental Modeling and Software Society, 24-27 June 2003, Vol. 3, 109-114.

Goodess, C.M., C. Hanson, M. Hulme and T.J. Osborn (2003), "Representing climate and extreme weather events in integrated assessment models: a review of existing methods and options for development", in press.

Goreau, T.J. and R.L. Hayes (1994), "Coral bleaching and ocean 'hotspots", *Ambio*, **23**, 176-180.

Gough, C., S. Shackley and M. Cannell (2002), "Evaluating the Options for Carbon Sequestration", Tyndall Technical Report No. 2, 108 pp., www.tyndall.ac.uk.

Grubb, M., J. Edmunds, P. ten Brink and M. Morrison (1993), "The costs of limiting fossil fuel CO2 emissions: a survey and analysis", *Annual review of energy and the environment*, **18**, 397-478.

Hoegh-Guldberg, O. (1999), "Climate change, coral bleaching and the future of the world's coral reefs", *Marine and Freshwater Research*, **50**, 839–866.

Hordijk, L. (1991), "Use of the RAINS Model in Acid Rain Negotiations in Europe", *Environmental Science and Technology,* 25, No. 4, 596-603.

Hulme. M., J.G. Jenkins, X. Lu, J.R. Turnpenny, T.D. Mitchell, R.G Jones, J. Lowe, J.M. Murphy, D. Hassell, P. Boorman, R. McDonald and S. Hill (2002), "Climate Change Scenarios for the United Kingdom: The UKCIP02 Scientific Report", Tyndall Centre for Climate Change Research, School of Environmental Sciences, Norwich, UK., 120 pp.

Hulme, M., Raper, S.C.B. and Wigley, T.M.L. (1995), "An Integrated Framework to Address Climate Change (ESCAPE), and Further Developments of the Global and Regional Climate Modules", Energy Policy 23, 347-355.

IPCC (1996), "Climate Change 1995 – Economic and Social Dimensions of Climate Change", Contribution of Working Group III to the Second Assessment Report of the Intergovernmental Panel on Climate Change, Cambridge University Press, UK, 448 pp.

IPCC (2000), *Special Report on Emission Scenarios*, Cambridge University Press, UK, 570 pp.

IPCC (2001), "Climate Change 2001 – Impacts, Adaptation and Vulnerability", Contribution of Working Group II to the Third Assessment Report of the Intergovernmental Panel on Climate Change, Cambridge University Press, UK, 1032 pp.

Jochem, E., and R. Madlener (2003), "Ancillary Benefit Issues of Mitigation: Innovation, Technological Leapfrogging, Employment, and Sustainable Development" in OECD *et al.*, *Ancilliary Benefits and Costs of Greenhouse Gas Mitigation*, Proceedings of an Expert Workshop, Paris.

Kainuma, M., Y. Matsuoka and T. Morita (2003), (eds.), "Climate Policy Assessment: Asia Pacific Modelling", Springer-Verlag, Tokyo. 402 pp.

Krupnick, A., D. Burtraw and A. Markanyda (2000), "The Ancilliary Benefits and Costs of Climate Change Mitigation: A Conceptual Framework" in OECD *et al.*, *Ancilliary Benefits and Costs of Greenhouse Gas Mitigation*, Proceedings of an Expert Workshop, Paris. http://www.oecd.org/oecd/pages/home/displaygeneral/0,3380,EN-document-517-nodirectorate-no-21-5396-8,00.html.

Kunzli, N., R. Kaiser, S. Medina, M. Studnicka, O. Chanel, P. Filliger *et al.* (2000), "Public Health Impact of Outdoor and Traffic-Related Air Pollution : a European Assessment", Lancet 356, 795-801.

Leemans, R. and B. Eickhout (2004), "Another reason for concern: regional and global impacts on ecosystems for different levels of climate change", *Global Environmental Change: Special Edition on the Benefits of Climate Policy* **14**: 219-228.

Leimbach, M. (2000), "ICLIPS – Integrated Assessment of Climate Protection Strategies: Political and Economic Contributions", Research Report 296 41 815, Potsdam Institute for Climate Research, Germany.

Marine Conservation Society (2003), info@mcsuk.org and www.mcsuk.org.

Mastrandrea, M.D. and S.H. Schneider. (2001), "Integrated Assessment of Abrupt Climatic Changes", *Climate Policy*, **1**, 433-449.

Matsuoka, Y., M. Kainuma and T. Morita (1995), "Scenario Analysis of Global Warming Using the Asian Pacific Integrated Model (AIM)" *Energy Policy*, **23**, 357-371.

Milly, P.C.D., R.T. Wetherald, K.A. Dunne and T.L. Delworth (2002), "Increasing Risk of Great Floods in a Changing Climate", *Nature*, **415,** 514-517.

Morgan, M.G. and H. Dowlatabadi (1996), "Learning from Integrated Assessment of Climate Change", *Climatic Change*, **34**, 337-368.

O'Neill, B.C. and M. Oppenheimer M. (2002), "Dangerous climate impacts and the Kyoto Protocol", *Science*, **296**, 1971-1972.

Parry, M. *et al.* (2001), "Millions at risk: defining critical climate change threats and targets", *Global Environmental Change*, **11**, 1-3.

Peck, S.C. and T.J. Teisberg (1995), "International CO_2 Emissions control: An Analysis Using CETA", *Energy Policy*, **23**, 297-308.

Plambeck, E.L., C. Hope and J. Anderson (1997), "The PAGE95 Model: Integrating the Science and Economics of Global Warming", *Energy Economics*, **19**, 77-101.

Prinn, R., H. Jacoby, A. Sokolov, C. Wang, X. Xiao, Z. Yang, R. Eckhaus, P. Stone, D. Ellerman, J. Melillo, J. Fitzmaurice, D. Kicklighter, G. Holian, and Y. Liu, (1999), "Integrated Global System Model for Climate Policy Assessment: Feedbacks and Sensitivity Studies", *Climatic Change*, **41**, 469-546.

Rahmstorf, S. (1995), "Bifurcations of the Atlantic Thermohaline Circulation in response to changes in the hydrological cycle", *Nature*, **378**, 145-149.

Richels, R. and J. Edmonds (1995), "The Economics of Stabilizing Atmospheric CO_2 Concentrations", *Energy Policy*, **23**, 373-378.

Rotmans, J. (1990), "IMAGE: An Integrated Model to Assess the Greenhouse Effect", Kluwer, The Netherlands.

Rotmans, J., M. Hulme, and T.E. Downing (1994), "Climate Change Implications for Europe: An application of the ESCAPE Model", *Global Environ. Change 4*, 97-124.

Rotmans, J. (1998), "Methods for IA: The Challenges and Opportunities Ahead", Environmental Modelling and Assessment 3, no.2, 155-179

Rotmans, J. and D. Rothman (eds.), (2003), "Scaling Issues in Integrated Assessment", Swets and Zeitlinger, The Netherlands.

Schellnhuber, H.-J (2002), "Coping with Earth-System Complexity and Irregularity" in W. Steffen, J. Jager and C. Bradshaw (eds.), *Challenges of a Changing Earth*, Springer, 2002.

Schneider, S.H. (2003), "Abrupt Non-Linear Climate Change, Irreversibility and Surprise" in OECD *et al.*, *Ancilliary Benefits and Costs of Greenhouse Gas Mitigation*, Proceedings of an Expert Workshop, Paris.

Schneider, S.H. and S.L. Thompson (2000), "A Simple Climate Model Used in Economic Studies of Global Change" in S.J. Decanio, R.B. Howarth, A.H. Sanstad, S.H. Schneider and S.L. Thompson (eds.), New Directions in the Economics and Integrated Assessment of Global Climate Change, PEW Centre on Global Climate Change, pp. 59-79.

Schneider, S.H., and F.L. Toth, (2003), (eds.). "Integrated Assessment of Climate Protection Strategies", *Climatic Change*, **56**, Nos. 1-2 . Special Issue, January I-II.

Schneider, S.H. (2004), "Abrupt non-linear climate change, irreversibility and surprise", *Global Environmental Change: Special Edition on the Benefits of Climate Policy* **14**: 245-258.

Shackley, S. (pers. comm. in 2002) Co-Manager of Tyndall Research Theme Decarbonising Modern Societies, UMIST, Manchester, UK.

Sliggers, J, and G. Klaassen (1994), "Cost Sharing for the Abatement of Acidification in Europe", *European Environment*, **4**, 1.

Sliggers, J. (2004), "The need for more integrated policy for air quality, acidification and climate change: reactive nitrogen links them all", *Environmental Science & Policy*, **7**, 47-58.

Tanzler, D., A. Carius and S. Oberthur (2002), "Climate Change and Conflict Prevention Report on behalf of the German Federal Ministry for the Environment, Nature Conservation and Nuclear Safety", Prepared for Special Event at the 16th Meeting of the Subsidiary Bodies to the UNFCCC, Bonn, 10 June 2002.

Tol, R.S.J. (2000), "The Marginal Costs of Greenhouse Gas Emissions", *The Energy Journal*, **20**, 61-81.

Toth, F.L., T. Bruckner, H.-M. Fussel, M. Leimbach, G. Petschel-Held and H.-J. Schellnhuber (2002), "Exploring Options for Global Climate Policy: A New Analytical Framework", *Environment*, **44**, 22 -34.

Tsimplis, M. (2004), "Towards a Vulnerability Assessment for the UK Coastline", www.tyndall.ac.uk/research/theme4/final_reports/it1_15.pdf.

Turnpenny, J.R.; A. Haxeltine, T. O'Riordan (in press), "A scoping study of user needs for integrated assessment of climate change in the UK context: Part 1 of the development of an interactive integrated assessment process", *Integrated Assessment.*

van Asselt Marjolein, B.A, and N. Rijkens-Komp (2002), "A look in the mirror: reflection on participation in Integrated Assessment from a methodological perspective", *Global Environmental Change*, 12, 167 – 184.

van Harmelen, T., J. Bakker, B. de Vries, D. van Vuuren, M. den Elzen and P. Mayerhofer (2002), "Long-term reductions in costs of controlling regional air pollution in Europe due to climate policy", *Environmental Science and Policy*, **5**, 349-365.

Vaughan, D.G. and J.R. Spouge (2002), "Risk estimation of collapse of the West Antarctic Ice Sheet", *Climatic Change*, **52**, 65-91.

Warren, R. (2002), "A Blueprint for Integrated Assessment," Tyndall Centre Technical Report 1, 20 pp. www.tyndall.ac.uk.

Wilby, R.L. and T.M.L. Wigley (1997), "Downscaling general circulation output: a review of methods and limitations", *Progress in Physical Geography*, **21**, 530-548.

Chapter 4

THE BENEFITS AND COSTS OF ADAPTING TO CLIMATE VARIABILITY AND CLIMATE CHANGE

by John M. Callaway,

UNEP-RISØ Centre, Denmark

This chapter presents a framework for defining and estimating the benefits and costs of adapting to climate change under uncertainty. Adjustment to climate variability is linked to adjustment to climate change in a economic framework, in which lack of reliable information about climate change constrains decision makers from adjusting fixed and quasi-fixed factors, while still allowing them to adapt to climate change as if it were climate variability, by adjusting variable inputs in the short run. The chapter illustrates the role of reliable climate information in allowing decision makers to re-plan capital investments and shift from "partial" to "full" adjustment to climate change both in theoretical terms and through the use of a numerical example of river basin planning. It also shows how decision makers can quantify the costs of being too cautious or not cautious enough in capacity planning in an ex ante, ex post framework. The chapter concludes with a discussion of the determinants of adaptation capacity. It suggests that the features which limit the ability to partially adjust to climate change are much more prevalent in developing than developed countries and that sound development policies focusing on structural and capital improvements in developing countries will increase their adaptive capacity, even if the objective of these policies is not related directly to climate change.

ISBN 92-64-10831-9
THE BENEFITS OF CLIMATE CHANGE POLICIES

CHAPTER 4. THE BENEFITS AND COSTS OF ADAPTING TO CLIMATE VARIABILITY AND CLIMATE CHANGE[31]

by John M. Callaway[32]

1. Objectives and policy implications

This paper has two main objectives. The first is to extend an earlier framework for evaluating the benefits and costs of adaptation developed by Fankhauser (1997) and Callaway *et al.* (1999) so that it can address the following issue: how to link adjustments to both climate variability and climate change in a way that not only describes how these adjustments may actually be occurring, but also provides a long-run planning paradigm for adjusting to climate change? The second objective of the paper is to show how this framework might be operationalised using a hypothetical example from a river basin.

The paper is written from the perspective of an economist. I assume that individuals, as economic agents – whether as consumers, investors, managers, or workers – act rationally. They pursue certain objectives with the technology and information available to them, subject to the constraints they face in terms of the resources at their disposal and the environment in which they function. Some of the comments I make are normative in this regard, but they are not intended to be prescriptive. Policy makers are concerned about the economic impacts and the costs of their decisions and also about how people behave in the face of policies that change the structure of economic incentives, whether these consequences are intended or latent. However, economics is just one aspect of policy-making and has to be balanced against other considerations. I am not suggesting that policy-making should be, or is, based entirely on the principles of neo-classical economics, just that economics does have something to add to policy decisions. Finally, while I am operating from the perspective

---

31 This paper is a substantially revised version of one that was originally written for an OECD workshop, The Benefits of Climate Policy: Improving Information for Policy Makers (Callaway, 2003). The first several sections of that paper dealt with a variety of policy issues related to adaptation and mitigation. However, these parts of the paper have been revised and will be published elsewhere (Callaway, 2004).

32 Ph.D., UNEP-RISØ Centre, PO 49, 4000 Roskilde, Denmark.

that economic agents are rational, the adjustments they make both to climate variability and climate change may well be motivated by objectives other than economic efficiency. In the example given, I have tried to show that the framework is broad enough to encompass this.

This chapter has several main policy implications, both descriptively and normatively, for natural resource sectors. A crucial descriptive element of the framework presented, here, is that adaptation to climate change is partially dependant on our ability to either detect it or take action based on the reliability of climate forecasts. Yet most planners and managers in natural resource industries find themselves in a situation where they cannot detect climate change occurring at local scales in ways that are meaningful to them and they do not believe climate forecasts are reliable enough to use for most planning (long-run) and some management (short-run) decisions. In this situation, climate change looks like, and is treated as, part of the existing climate variability. Adjustments to climate change can be made, but these adjustments are "partial" in the sense that planners and managers will only make short-run adjustments to climate change, which is perceived as climate variability.

The extent of partial adjustment to climate change depends on two factors. The first, is the flexibility or adaptive capacity that we have to deal with climate variability, for example, built into our current infrastructure, management practices and more generally "our ways of doing things". The second is the extent of the "new" climate variability that we experience (but are unable to detect) in relation to the historical climate variability for which the capital stock, management practices, and other determinants of adaptive capacity were originally designed. By contrast, full adjustment to climate change requires a long-run perspective in which the capital stock can also be varied to increase the long-run capacity to adapt to a different climate.

The framework and the river basin planning example link short-run and long-run decisions to describe how natural resource planners and managers are currently adapting, partially, through short-run decisions to climate change they cannot detect. They also provide a normative framework (i.e., economic efficiency) for making long-run planning decisions under uncertainty. One way to address this descriptive situation in a normative way is through the use of *ex-ante, ex-post* planning, where planners can estimate the net benefits of adjusting capital stocks on an *ex-ante* basis over a wide range of climate forecasts and then balance these against the *ex-post* cost consequences of being too cautious or taking too much precaution on an *ex-ante* basis. Interestingly enough, the net costs of being too cautious and doing nothing beyond partial adjustment represent the *ex-post* net benefits of full adjustment, if the climate does change as expected.

The framework also suggests that from a macro-economic perspective the capacity to make both partial and full adjustments in natural resource industries are determined by a number of factors. These include: resource scarcity, the relative mobility of the factors of production between sectors and regions, the degree of differentiation and specialization within the location and space economy, the technological substitution possibilities among inputs and outputs, the breadth of markets, the degree of integration between domestic and international markets, and the extent of market distortions created by domestic policies. These are important limiting factors in most developing countries. Development policies that address these more general structural problems will increase the capacity of natural resource sectors to partially and fully adjust to climate change. Moreover, the implementation of these "no regrets" policies carries with it no risk of over- or under-protecting from climate change damages, since they are beneficial with or without climate change.

2. A framework for estimating the benefits and costs of adaptation

2.1 The original framework

In our earlier paper, Callaway *et al.* (1999) outlined a conceptual framework for estimating the benefits and costs of adaptation that can be applied at almost any scale. This framework was largely based on earlier work by Fankhauser (1997).

Box 1. Defining adaptation in this framework

In this chapter the term adaptation is used broadly in this chapter to refer to any adjustments that individuals, singly and collectively (in whatever organizational framework), make *autonomously* or policy makers undertake *strategically* to avoid (or benefit from) the direct and indirect effects of climate change. Individuals and organisations can do this *autonomously* in the sense that economic, social and political institutions with which they interact provide incentives for them to adjust "automatically" to climate change, without political intervention. Or, they can adapt *strategically* through decisions made by governments. Stated in this way, the line between autonomous and strategic adaptation is not easy to draw in all cases, since, conceptually, governments do not usually operate outside the underlying incentive systems that have developed to guide behaviour. In the case of a market economy, this distinction is fairly easy to make based on differences between the private objectives of individuals, firms and factor owners and the social objectives of governments and the divergence of these objectives due to market failures. In this type of political economy, private decision makers respond to whatever incentives they face, whether these are market or public policy driven. As such, private decision makers can be expected to respond autonomously to strategic climate change policies, but would not take the same actions on their own, except for altruistic reasons.[33]

Callaway *et al.* (1999) outlined a conceptual framework for estimating the benefits and costs of adaptation that can be applied at almost any scale. The core ideas in the

---

[33] Maybe this distinction holds true for all societies, but I would not use it generally, since there may well be workable incentive systems in some societies where private and collective welfare are better aligned than in market or socialist economies in developed countries.

conceptual framework can be explained with the use of Table 1.[34] The top left box describes a situation in which society is adapted to the existing climate, C_0, through adaptive behaviour A_0. This is sometimes referred to as the Base Case. The lower right box represents a situation where society is adapted to a change in climate from C_0 to C_1 that has changed over time through behaviour A_1. The top right box describes a situation in which society behaves as if the climate was not changing, and is adapted to the existing climate through behaviour A_0, but not the altered climate, C_1. The bottom left box represents a case in which society decides to behave as if the climate had changed by altering its behaviour to A_1, when in fact the climate has not changed, C_0.

Table 1. Alternative Adaptation Scenarios for Estimating Adaptation Costs and Benefits

Adaptation Type	Existing Climate (C_0)	Altered Climate (C_1)
Adaptation to existing climate (A_0)	Existing climate. Society is adapted to existing climate: (C_0, A_0), or Base Case	Altered climate. Society is adapted to existing climate: (C_1, A_0).
Adaptation to altered climate (A_1)	Existing climate. Society is adapted to altered climate: (C_0, A_1).	Altered climate. Society is adapted to altered climate: (C_1, A_1).

Source: Modified from Fankhauser (1997).

Alternatively, we can look at this table in a new way, from the perspective of *ex-ante, ex-post* planning, in which there are the same two, *ex-post* climate states and two sets of actions that can be taken, *ex-ante*, namely: do nothing additional (A_0) or adapt to climate change (A_1). In this framework, the top left box represents a situation in which "the planner" does not believe climate will change, *ex-ante*, does nothing additional, and that expectation turns out to be true. The top right box describes the situation where the planner also does not believe climate will change, does nothing additional, but the climate does change. The lower left box shows the case where the planner believes the climate will change *ex-ante*, adapts to climate change, but the climate does not change. Finally, in the lower right box, we have the situation where the planner also believes climate will change, *ex-ante*, adapts to climate change, and the climate does change. So, the top right and the lower left boxes are representative of situations in which planners make mistakes because their *ex-ante* expectations about climate change turn out to be false, *ex-post*

In the paper by Callaway *et al.* (1999), this table was used to define the various benefits and costs associated with climate change and adaptation. The last two

34 It should be noted that this framework was originally developed in the context of a market economy, but in this paper I will refer more generally to behaviour in any kind of political economy.

definitions were added in Callaway (2004) to address the issue of making planning "mistakes" in an *ex-ante ex-post* framework. These are:

- *Climate change damages* - the net cost to society if climate changes and society does not adapt to it.

- *Net benefits of adaptation* - the value of adaptation benefits minus adaptation costs. This is the difference between:

  – *Adaptation benefits* - the value of the climate change damages avoided by adaptation actions.

  – *Adaptation costs* - the value of the resources society uses to adapt to climate change.

- *Imposed damages of climate change* - the net climate change damages that are not avoided by adaptation. This is the difference between climate change damages and net adaptation benefits.

- *The cost of precaution* - the net cost to society of adapting to climate change, based on *ex-ante* expectations, when climate does not change, *ex-post*. This is the cost of adjusting to climate change when it does not occur.

- *The cost of caution* – the net cost to society of not adapting to climate change, based on *ex-ante* expectations, when climate does change, *ex-post*. This is the cost of not adjusting to climate when it does occur.

The various benefits and costs associated with adapting (or not adapting) to climate change can be calculated by defining net social welfare, however it is measured, as W(A,C) and then using Table 1 as a guide. *Climate change damages* can be measured by the welfare loss that occurs from moving from moving from the Base Case (top left box) to a "state" where the climate changes but society does not adapt to it (top right box), or $W(C_1, A_0) - W(C_0, A_0)$. However, these damages can be reduced and social welfare can be improved by adapting to climate change through a movement from the top right box to the lower right box in Table 1. Thus, *the net benefits of adaptation* = $W(C_1, A_1) - W(C_1, A_0)$. The net "cost" of climate change, taking into account adaptation can then be measured by the climate change damages that are not reduced by adaptation, defined here as *the imposed damages of climate change*. This is calculated as the difference between net social welfare in the lower right box, and the net social welfare in the top left box or $W(C_1, A_1) - W(C_0, A_0)$. For the case of *ex-ante, ex-post* planning (the last two definitions), *the cost of precaution* is the difference between net social welfare in the bottom left box and net social welfare in the top left box, or $W(C_0, A_1) - W(C_0, A_0)$, while *the cost of caution* is the difference in net social welfare between the top right box and the lower right box, or $W(C_1, A_0) - W(C_1, A_1)$, which interestingly enough turns out to be just the negative value of the net benefits of adaptation. Thus, making the mistake of acting too cautiously can be overcome by adapting to climate change, but if you act according to the precautionary principle and climate is not changing, society ends up with an overcapacity to adjust to climate change and this is costly.

Callaway *et al.* (1999) illustrated the original welfare measures in the market for a single good produced in a climate sensitive industry. In Figure 1, I have added the *ex-ante, ex-post* measures to the original ones. The effects of climate change on the production of a good in this market are illustrated by shifts in the supply curve for the good S(A,C). The market supply curves for the good depend on the extent of the adjustment to climate change (through changes in A)[35] and on the climate. Thus, each supply curve corresponds to one of the four scenario boxes in Table 1. The line D indicates the market demand curve for the good. Presumably, the market demand for a good or an input is often influenced directly by weather and climate, but showing these changes complicates the graphic analysis. Therefore in this Figure and others in the chapter, we assume for the sake of simplicity that D does not respond to climate change.

There is economic logic in the way the four market supply curves are arranged in Figure 1. The market supply curve, $S(A_0,C_0)$ lies below all the rest if we assume that climate change has the effect of making resources scarce and thus raising production costs. The supply curve $S(A_1,C_0)$ must lie above $S(A_0,C_0)$, since adaptation is no longer optimal for the current climate. The supply curve $S(A_1,C_1)$ must also lie above $S(A_0,C_0)$, since adaptation cannot completely eliminate the scarcity effects of climate change.[36] Finally, the supply curve $S(A_0,C_1)$ must lie above $S(A_1,C_1)$, since the adaptation to climate change under the former is not complete.

Given these supply curves, the negative sum of the bold rectangular areas A+B+C equals the loss in consumer and producer surplus associated with *climate change damages. The imposed damages of climate change* are represented by the loss in consumer and producer surplus indicated by the negative sum of the areas B+C, while the positive value of the area A represents the positive *net benefits of adaptation*. In the *ex-ante, ex-post* framework, the negative value of the area A represents *the costs of caution*, while the negative value of the area C represents *the cost of precaution*.

Figure 1 also illustrates the changes in welfare associated with these different definitions. However one can use the same conceptual approach to characterise the physical damages of climate change in terms of climate change damages and the residual damages of climate change. In the same way, the difference between the two,

---

35 As we will later see, the extent of adjustment to climate change depends on the climate.

36 The relationship between $S(A_1, C_0)$ and $S(A_1, C_1)$, as shown in Fig. 1, implies that adapting to climate change that does occur is more costly than adapting to climate change that does not occur; however, this is an empirical issue, strictly speaking, since long-run investments made to adapt to climate change may preclude or limit adjustments in variable inputs that would have occurred in the absence of climate change and adaptation.

which are the net adaptation benefits can be characterised in terms of the physical climate change damages avoided by adaptation measures. The inputs used to "create" and implement adaptation measures can in some sense be treated as the physical analogue to the real resource costs of adaptation, although not all inputs, such as information, are easy to characterise in physical terms, even if they have a real cost. The point is that this framework could be applied using many different damage metrics.

Figure 1. Illustration of adaptation in a goods market (old framework)

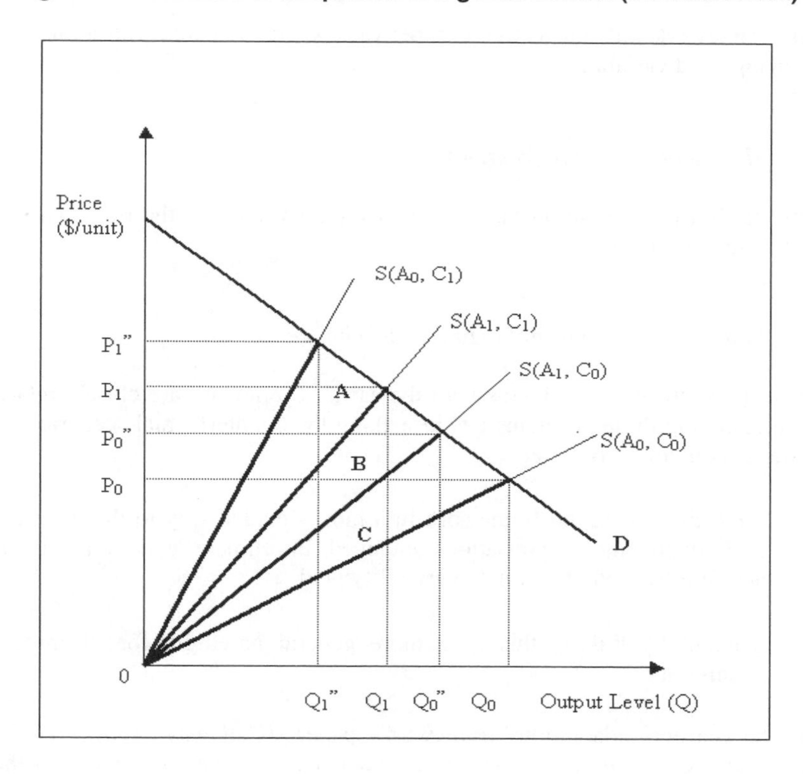

In passing we should note that this framework answers the objections raised by Darwin and Tol (2001) about the use of opportunity cost measures to capture the benefits of adaptation. In our framework (Callaway *et al.*, 1999), adaptation cost is not a measure of avoided climate change damages, it is strictly a measure (monetary or physical) of the real resources used to avoid climate change damages.

A number of problems with this framework were presented in Callaway *et al.* (1999) and others have emerged since it was published. Most notable among these are:

- The link between adaptation to climate variability and climate change was missing, making the framework both deterministic and incomplete.

- The counterfactual scenario, (C_1, A_0), as stated, could not be observed, and did not seem to make empirical sense.

- More generally, the framework did not relate this process of adaptation to planning under uncertainty, both through planning models and how this process could be integrated into assessment models at the sectoral and project levels.

The framework did not distinguish between short- and long-run adjustments to climate change and variability.

2.2 *Modifications to the framework*

This section follows up on these problems and shows how the framework can be modified to address them.

2.2.1 *Linking climate variability to climate change*

Most if not all of the problems with the earlier framework are closely related and the key to resolving these problems lies in making the counterfactual scenario (C_1, A_0), believable. This involves two steps:

- First, by linking the framework in a more detailed way to the way in which individuals and organisations engaged in climate and weather-sensitive activities respond to climate variability; and

- Second, by linking this to a more general paradigm for planning under uncertainty.

In the earlier adaptation framework paper (Callaway *et al.*, 1999), the counterfactual scenario in which producers did not adjust their input use to the new climate was necessary to isolate the adjustment that individuals and organisations made to climate change. In support of this construct, we know there are historical examples of places that were settled due to lack of sufficient information about the true climate and then abandoned when random (perhaps even "persistent") periods of above-average rainfall eventually gave way to the dominant, drier climate pattern. We also know that no matter how much information we have about climate and climate variability, "you don't need to be smart get out of the rain". Thus, experience and common sense tell us that, while it is indeed possible to be poorly-adapted to the "true" climate without sufficient information, one does not need a lot of information to make some adjustments to climate variability, whereas other adjustments require a little more information and, thus, take longer.

At this point, it may be useful to better relate adjustments to climate variability with those to climate change. Smit (1993) and Carter *et al.* (1994) and Smit *et al.* (1996, 2000) suggested that farmers are always "optimally" adjusted to climate variability, given the information available to all of them formally and informally, but that at any given time farmers also face constraints, imposed by past investment decisions, to adjust to climate variability. This does not mean that they can always guess the weather with precision, but rather: a) given their knowledge about the joint distribution of meteorological variables, individuals and organisations build a certain amount of flexibility into their activities in order to be able to cope with the weather variability they and others have faced in the past, and b) this flexibility may enable individuals in regions with a great deal of weather variability to better cope with climate change than those living in regions where there is very little weather variability, *even if they do not have information that the climate has in fact changed.*

The point Smit (1993) and Smit *et al.* (1996, 2000) make about the importance of information in adapting to weather variability helps us to link the strongly counter-factual adjustment case used for measuring adaptation benefits to the actual behaviour of individuals under uncertainty, both in terms of their adjustment to climate variability and climate change.

When people, singly or collectively, make short- or long-term decisions that are weather or climate sensitive, they take the information available to them from the joint distribution of weather and/or climate variables into account, both in planning and implementation. The type of information that is relevant to them – weather or climate – depends on the time horizon associated with a specific action and the flexibility they have in adjusting their behaviour. These time horizons vary widely. For example, information about the weather next week is crucial at harvest time for farmers, as is the day's wind speed and wave height for a coastal fisherman. On the other hand, when individuals make investment decisions they usually take a long-term perspective, because investment expenditures in land and capital goods and other tangible assets are long-lived and often "lumpy" and it is costly to replace them once the investment decision is made. An investment decision to plant a particular type of forest species for harvest requires a planning horizon that spans the rotation length for the species and the products for which it will be harvested. The useful life of a water supply reservoir for irrigation water may be 50 years or more. Thus, investment decisions are more likely made on the basis of climate, as opposed to weather information.

The flexibility that individuals and organisations have to adjust to climate variability is partly related to the above distinction between the short- and long-run, but it also is related to environmental and technological features that characterise the structure of production activities in the household, in firms and organisations more generally. When a crop is damaged by hail, for example, the damaged plants cannot be brought back to life. In the same way, once an ear of corn has tasselled, there is no substitute for sunlight and persistent rains after this stage of development will reduce yields. There is little the farmer can do about this, except cut his economic losses and hope for better weather in the next crop season.

A formal definition of this link between adaptation to climate variability on the one hand and adaptation to climate change on the other is therefore important to estimation of the costs and benefits of adaptation (see Section A.1.1).

The starting point for making this link is to recognise the difference between weather and climate and between climate variability and climate change. Managers and planners in natural resource industries regard the *weather* in terms of the observed outcomes of meteorological variables that are partially random and partially deterministic in nature. The underlying process that drives the weather is composed of a system of deterministic functions and parameters that generate the deterministic part of the weather and "random" errors that are associated with the random factors that influence weather and unexplained errors. Even if they are not, the outcomes of meteorological variables often appear to be random. So for their own purposes managers and planners in natural resource industries often characterise *climate* in terms of the parameters of the partial and joint distribution functions of the meteorological variables they are interested in. They use historical observations of the meteorological variables to estimate the parameters of these functions. Accordingly, *climate variability* refers to the parameters of these distributions that reflect variations from mean values and the shapes of the distributions, while *climate change* refers to changes in the actual parameters of these distributions.

To illustrate how decision-makers in the private and public sectors deal with the randomness of climate and climate variability in their capacity as resource planners or managers we introduce the idea of the production function (See Section A.1.1). The production function in a climate-sensitive sector, industry, or even more broadly "an activity" is a way to characterise the relationship between the outputs of an activity and the inputs used to "create" that output, including meteorological variables. Once it is constructed and its parameters are either estimated or calibrated, planners and managers can use a production function to simulate both how existing climate variability and climate change can change output levels.

The concept of the production function is general. It is not only applicable in traditional production activities, such as in the agriculture, forestry, fisheries and energy sectors, but can also be used to characterise the relationship between the services provided by a wide variety of household and commercial activities. For example, the concept is broad enough to characterise the role of land and sea level rise to the output commercial enterprises that are located in low-lying coastal areas. The concept of the production function is also broad enough to apply, not only to activities that are associated with the production of goods and services that are sold in markets, but also to household production and the production of "non-market" goods and services that are not sold in markets (Freeman, 1994). Perhaps the most important example of non-market production in the climate change arena is the production of ecosystem services (Toman, 1997).

Throughout this chapter, climate and the weather are treated as exogenous: the planner/manager cannot change the climate. This assumption is restrictive for making global decisions, as well as in countries that emit sizeable fractions of global GHG emissions. In these cases, emissions and climate are either completely or partially under the control of decision makers. However, only a handful of nations produce a sizeable enough fraction of global GHG emissions to significantly influence their own climates and weather. Moreover, at the sectoral and activity levels, where most adaptation occurs, decision makers have no control over the climate or weather they face. Thus, from here on, it must be understood that the analysis contained in this chapter is based on the notion of "partially optimal" adaptation, taking the level of GHG mitigation as exogenous.

To show how the climate and weather influence production and how changes in climate and weather affect the decisions of planners and managers in climate sensitive sectors and industries, the idea of *ex-ante, ex-post* planning is again useful. The *ex-ante* problem that confronts planners is how to select the amounts of quasi-fixed factors that are optimal for the climate they face. This is a long-run decision since, once the investment in quasi-fixed factors is made, they become fixed factors. The *ex-post* problem of managers is a short-run problem: how to adjust their variable inputs to cope with the weather, given that the amounts of the fixed factors have already been determined.[37] An *ex-ante, ex-post* planning problem combines both types of decisions and this approach is commonly used in planning new electrical generating capacity, where a given amount of capacity must be able to cope with weather-driven peak demands.

2.2.2 *Making the framework stochastic*

The *ex-ante, ex-post* approach to planning is amenable both to autonomous adaptation and to strategic adaptation planning decisions. The *ex-ante* part of the model makes sense from the perspective of either type of adaptation decision since there is no restriction on the objectives to be followed or the constraints imposed. The *ex-post* part of the model takes into account autonomous adaptation that occurs in response to the *ex-ante* decision regarding investment in quasi-fixed factors. Again, there is no reason why the autonomous adaptation that is characterised in *ex-post* decisions should be economic, or market driven. Autonomous adaptation is simply a general term used to describe how people respond on their own to political, social and economic incentive systems and rules, regardless of the objectives of this system. Nor does this approach necessarily involve the use of mathematical models. A farmer in a developing country, whose only objective is to feed his family, has to plan on the number of large animals he needs for food and to perform work under a variable climate and then has to live with these decisions through good times and bad (up to a point).

Annex 1 (Section A.1.2) contains a simple example of the first-stage of a two-stage, dynamic stochastic programming model to maximise the expected physical yield of a single crop for a single climate state. In this model, the physical crop yield depends

37 For a variety of reasons, it is often too costly or physically difficult to adjust quasi-factors to unplanned events.

on variable inputs that can be changed only in the short-run, a single quasi-fixed factor that can only be changed in the long-run, and the random outcomes of a meteorological variable, which are simulated by Monte Carlo methods from a known distribution. The results from this exercise demonstrate that the optimal value of the quasi-fixed factor depends on climate, while the variables factors depend on the simulated expected values of the weather, and the value of the quasi-fixed factor, once the quasi-fixed factor is determined. This emphasis the importance of climate, and not weather, in the selection of the quasi-fixed factor, and the importance of the selection of the quasi-fixed factor on the *ex-post* choice of variable inputs in the second-stage of the analysis. Obviously, if the climate was changing and the possibilities for climate change were not represented in the simulated values of the meteorological variable, the selection of both the quasi-fixed and variables factors would not be optimal. This is analogous to a situation, where decisions about the capital stock were made under one climate and the selection of variable inputs was made under a different climate, which no one could detect.

2.2.3 Linking the counterfactual case and damage/benefit definition to adjustments to climate variability and climate change

Now, ordinarily the second *ex-post*, stage of this planning problem would involve solving a maximization problem to determine the optimal values of the variable inputs, holding the capital stock constant, but with revised forecasts for the meteorological variables based on actual weather conditions. But, for the purposes of looking at how individuals adjust to climate change, I want to structure the *ex-post* stage slightly differently to suggest how managers and planners in natural resource industries currently are acting and what the consequences of this are.

In Annex 1 (Section A.1.3), a second-stage evaluation of the crop yield model is presented, where the climate has changed, but individuals can't detect this change and are unable or unwilling to change their *ex-ante* decisions about the size of the capital stock based on climate change information from global and regional climate models. As such, this model illustrates the path from partial to full adjustment. In this case, how decision makers will react to this climate change in their selection of variable inputs depends on two factors:

- Whether or not climate changes can be detected or predicted with enough reliability that decision makers are willing to act (i.e., make investment decisions) on the new information; and

- How much flexibility exists, both in economic terms (i.e., short-run vs. long-run), and physical/environmental terms to adjust the variable inputs and quasi-fixed factors. Presumably the amount of flexibility is a function of pre-existing adaptation to current climate variability and the "overlap" between climate change existing climate variability.

If the information from either type of source is reliable, then decision makers will adjust the capital stock and institutional arrangements under their control consistent with the results of planning models that characterise this climate change. One can call this process "full adjustment"[38] and it represents the fullest kind of adaptation that can occur with reliable information about climate change.

The earlier framework decomposed the transition between optimal adjustments for two climates into two partial steps, with the counterfactual case of no adjustment to climate change, lying in between full adjustment process. Now, it can be seen that by linking adjustment to climate variability to adjustment to climate change the transitional process becomes both more complicated and more realistic. The process of full-adjustment between two climate states can be decomposed into three parts, based on how complete the adjustment is. They are:

1. Pure effect of climate change (a purely physical response, with no human adjustment possible): This, first part of the adjustment involves what I have elsewhere (Callaway and Ringius, 2002) termed "the pure climate effect", in which the output responds *only* to the change in the weather (under the new climate). The quasi-fixed factors are not adjusted in this case because the climate change has not been detected or forecast with sufficient reliability to allow re-planning. The variable inputs are not adjusted due to the restrictive nature of the production technology or just the sheer inability to alter the effects of "nature", as in the case of severe hailstorms. Can this pure effect be modelled? I will deal with this question later on, in connection with measuring the benefits of partially adjusting to climate change.

2. Partial adjustment to climate change (limited adjustment to perceived climate variability): The next part involves adjusting the variable input to be optimal for the change in weather (under the new climate), holding the fixed factors at their optimal values under the old climate. Callaway and Ringius (2002) have termed this as a "partial adjustment" to climate change. The adjustment is considered partial because the quasi-fixed factors are not fully adjusted to climate change and this also limits the adjustment of the variable inputs. However, since decision makers experience the change in weather, they are still able to make some adjustment, even if the quasi-fixed factors are not adjusted. This type of adjustment can also be observed in many different settings, and what is important about it, is that decisions makers do not need to have reliable information about climate change to make these adjustments. Rather, they are responding to perceived climate variability.

3. Full adjustment to climate change (adjustment to climate change given reliable – not perfect – information about climate change): The next part involves adjusting the variable input to be optimal for the change in weather (under the new climate), holding the fixed factors at their optimal values under the old climate.

---

[38] If a known non-stationary process generates climate, then these "full adjustments" will be gradual and perhaps discontinuous over time, but the process of adjustment can nevertheless be referred to as a full adjustment process.

Callaway and Ringius (2002) have termed this as a "partial adjustment" to climate change. The adjustment is considered partial because the quasi-fixed factors are not fully adjusted to climate change and this also limits the adjustment of the variable inputs. However, since decision makers experience the change in weather, they are still able to make some adjustment, even if the quasi-fixed factors are not adjusted. This type of adjustment can also be observed in many different settings, and what is important about it, is that decisions makers do not need to have reliable information about climate change to make these adjustments. Rather, they are responding to perceived climate variability.

2.2.4 *Measuring climate change damages and benefits of adaptation*

Corresponding to these three types of adjustment (or non-adjustment), there are also comparable measures of damages and benefits that are consistent with the definitions in the earlier framework. Since the objective function of the problem, used here, is measured in terms of yields, the damages associated with climate change and the benefits of avoiding climate change damages are also measured in terms of yield changes. However, as I indicated earlier the same approach could be used to measure welfare changes in monetary or other terms. The formal definitions for these measures are given in Annex 1 (Section A.1.4). Conceptually, there are five measures, as follows:

1. Maximum climate change damages: These are the climate change damages associated with the "pure climate change effect", where even adjustment of variables inputs in the short-run is impossible. The "pure effect of climate change" is difficult to model, except at the level of an individual economic activity and individual weather events. While it can be observed in relevant cases (such as crop damage from severe hail storms), it is best used as a theoretical (and somewhat artificial) construct to define the highest possible damages that can be caused by climate change. This can be thought of as a pure measure of vulnerability to climate change, without taking into account any form of human adjustment to climate change. Presumably, this can be altered in the long-run by adopting production functions that allow meteorological inputs to play a less important role in the production function of an activity. For example, the production function of green-house crops would be less sensitive to changes in the natural environment than field crops. Of course, this is an extreme example. Nonetheless, it illustrates the point.

2. Imposed damages of climate change after partial adjustment (to climate variability)= Climate change damages prior to full adjustment: This expression measures the changes in physical or monetised climate change damages, taking into account the partial adjustments that are made in response to perceived climate variability, but not climate change (since it is not detected). It has an advantage over our (Callaway et al., 1999) previous definition of the imposed damages of climate change in that the first term is derived directly from an observable adjustment that individuals and organisations make to climate variability when they do not have reliable enough information to detect or plan on the climate changing. More importantly, being able to measure this adjustment allows us to decompose the imposed damages of climate change into two parts: one due to partial adjustment and one due to full adjustment (next definition). Conceptually, this expression represents the residual damages that are left over after partial adjustment, that is: without explicitly planning for climate change and acting on these plans.

3. *Imposed damages of climate change = Climate change damages after both partial and full adjustment:* This measure represents the damages from climate change after both partial and full adjustment are taken into account, using and acting on available information about climate change.

4. *Net benefits of partially adjusting to climate variability:* Conceptually, this is the physical or monetised value of the damages avoided associated with moving from the "worst case" in which no human adjustment occurs, either to climate variability or climate change, to one in which partial adjustment to perceived climate variability takes place (in the absence of reliable information about climate change.

5. *Net benefits of fully adapting to climate change:* It represents the damages avoided, once partial adjustment to climate variability has taken place, as a result of fully adjusting the quasi-fixed factors used in the production function to expected changes in climate.

One of the questions raised earlier was whether or not it was possible to simulate the pure effect of climate change in existing models and so measure the net benefits associated with this adjustment. The answer is that it may be technically possible to do this using physical models, for example the crop yield models used by Parry *et al.* (1999), Rosenzweig *et al.* (1995) and Rosenzweig and Parry (1994). Such models allow one to simulate changes in crop yields using meteorological data from a climate change scenario while holding the application of managed inputs at the values prevailing for the existing climate. In fact this is what both sets of authors did to simulate climate change, initially, and then they "imposed" adaptation scenarios on top of this, by varying management. But, in general, the concept of "no adjustment" makes practical sense only at the farm level and the pure effect of climate change is probably too limiting a concept for practical use in sector and national models.

Figure 2 illustrates the changes made in the framework as applied to welfare changes.[39] This figure characterises the damage and benefit measures in the market for a good in the same way as Figure 1, in terms of the maximisation of consumer and producer surplus in a market. If we think of the vertical axis as measuring marginal output, instead of output price, the diagram fits the case of benefits measures conceptually, leaving out other exogenous variables that influence production. Of

---

39 To avoid over-complication, I have dropped the supply curve $S(X_0, K_1, M_1)$ that would characterise the situation in which the quasi-fixed factors were adjusted, *ex-ante*, to climate change changes that did not occur, *ex-post*, leaving it possible only to partially adjust the variable inputs to be optimal for the current climate.

course the marginal and optimal output would be different under the two different objectives (physical costs and benefits vs. economic costs and benefits).

But now that changes in quasi-fixed factors have been added to the analysis, I need to add a word of caution. Typically, when long-run adjustments in market supply take place, they are in response to demand shifts. However, this case is somewhat different because even without increases in demand, detection or anticipation of climate change by producers may create incentives to adjust quasi-fixed factors. This and later figures isolate that effect without consideration of normal long-run adjustments to demand increases. Taking both factors into account is, of course, important in determining the effect that climate change and adjustment to it will have on the long-run supply curve of an industry, but a complete graphic analysis of both the effects of climate change and demand growth on producer supply adjustments would obscure the main point.[40] Demand growth and anticipation or detection of climate change may well interact in real life. In cases where fixed or quasi-fixed factors are "lumpy", the anticipation or detection of climate change may not change the marginal benefits of adaptation sufficiently to justify long-run adjustments. However, increased demand for a good may alter that picture in the long-run and make it profitable to adjust to both influences i.e. demand and climate changes.

In Figure 2, the top supply curve has been added to reflect the pure effect of climate change. Net welfare will always be lower in this case than in the others, because it is more highly constrained. The welfare loss represented by the areas C+A+B is equal to *the maximum climate change damages*. The welfare gain from partial adjustment, represented by a shift to the next lowest supply curve, and measured by the area A, represents *the net benefits of partially adjusting to climate variability* when these changes are treated as existing climate variability. Therefore, the welfare loss characterised by the area, B+C, which is left after partial adjustment to climate variability takes place, is equal to the *imposed damages of climate change damages after partial adjustment*. The welfare gain from full adjustment, alone, involving a shift from the partial adjustment supply curve to the full adjustment supply curve, just lower down, and measured by the area B, represents *the net benefits of fully adapting to climate change*, once the quasi fixed factors have been adjusted using information about climate change. Thus the area C represents *the imposed damages of climate change*, once all adjustments have taken place.

---

40 The long-run supply curve for a constant-cost industry is perfectly elastic (horizontal). However, if both increases in demand and climate change impacts are time-dependent, then the long-run supply curve of an industry that adjusts to those impacts will slope upward, as in the increasing-cost industry case, assuming climate change impacts shift the short-run supply curve to the left, as shown in Figures 1 and 2.

Figure 2. Illustration of adaptation in a goods market (new framework

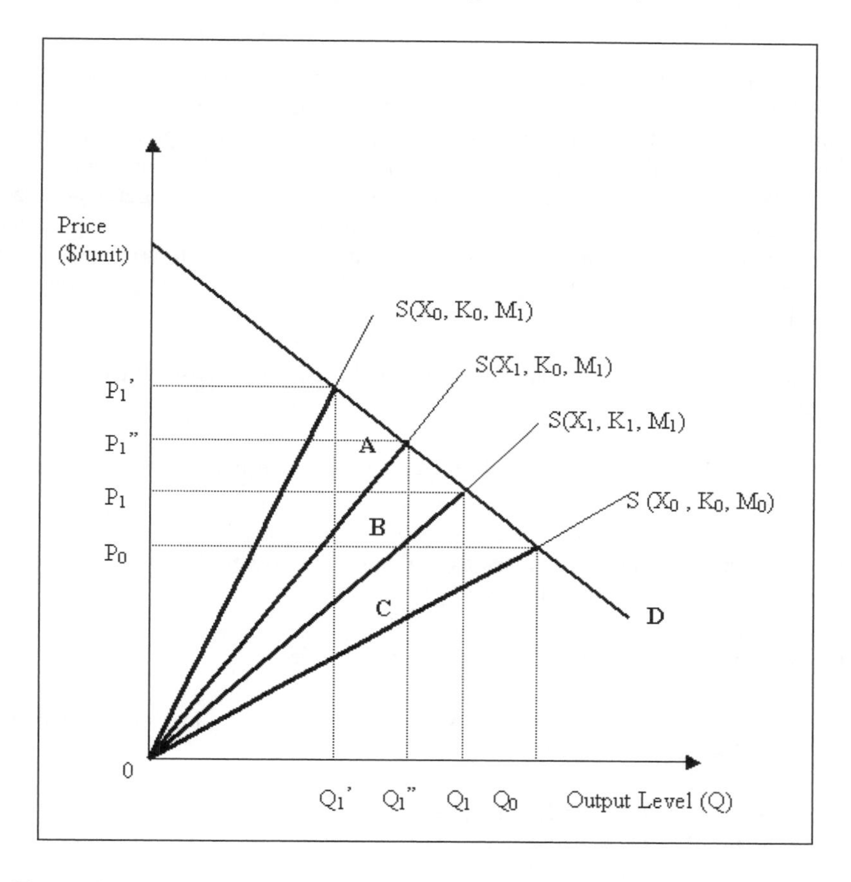

To see how these concepts can be applied to planning under uncertainty and the assessment of the benefits and costs of adaptation options at the sectoral level and projects at the local level, the section below introduces a "made up" example of *ex-ante, ex-post* planning.

3. An example

3.1 The model

This example involves the use of an *ex-ante, ex-post* planning model to determine the optimal capacity of a small reservoir under *ex-ante* uncertainty about climate. (See Annex 1, Section A.1.5 for the mathematical representation of this model).

The model can be used in two different ways: either as an *ex-ante* planning model, or as an assessment model, used *ex-ante or ex-post* to estimate climate change damages, the imposed damages of climate change and the benefits (or costs) associated with adapting (or not adapting) to climate change.

The basin model is not as complicated as others that can be found in the literature.[41] It consists of a single runoff source that discharges into a potential reservoir site. Immediately downstream there is an aggregate diversion point for irrigated agriculture and below that another aggregate diversion point for municipal use. The river then discharges into the sea. The demand for water, by both aggregate uses, is modelled seasonally, over four water seasons, with the use of inverse water demand functions, where the marginal returns to water are a function of consumptive use. Initially, there is no reservoir and, by law, at least 75% of the annual runoff must go to agricultural use.

Planners want to select among four options, based on the discounted net present value of the benefits of:

- Option 1: Doing nothing and leaving the situation unchanged (i.e., the status quo).

- Option 2: Adopting markets to allocate water, seasonally, between agricultural and municipal uses.

- Option 3: Building an optimally sized reservoir.

- Option 4: Building an optimally sized reservoir and adopting water markets (per 2 above).

In this hypothetical river basin, there are two prevailing types of climate patterns: one in which runoff occurs in the late winter and early spring and another in which runoff is reduced by about 10% and occurs in late fall and early winter. Each of these climate states is characterised in the model by seasonal runoff distributions. In the model these distributions are "mixed" using probability combinations pr(state 1, state 2), where p(state 1) + p(state 2) = 1. Just two climate states are used in the example (1=current climate and 2=climate change), but this was done only to keep the analysis simple. We solved this model for a range of probability combinations, but only show the results for a selected number of them.

The model developed to look at this problem is a "price-endogenous spatial equilibrium model, patterned after those developed in Hurd *et al.* (1999), with two important differences. First it is stochastic in that seasonal runoff is generated randomly according to distribution functions for the two different runoff distributions. Second, the model jointly optimises reservoir capacity, water storage, allocation and consumption

---

41 For example see those developed by Hurd *et al.*(1999) for the EPRI study, *The Impacts of Climate Change on the US Economy* (Mendelsohn and Neumann, 1999).

simultaneously under the two states of nature, based on the expected net present value of the willingness of users to pay for water less operational costs, less the capital cost associated with reservoir capacity. Thus, the reservoir capacity depends, *ex-ante* both on climate variability associated with each climate state plus the probability of occurrence of each of these states. Once the optimal reservoir storage capacity is determined, *ex-ante*, the management of the reservoir is also constrained *ex-post*, by capacity and the resulting values for the endogenous variables in all periods depend on the fixed reservoir capacity. However, both long-run and short-run decisions are explicitly modelled with perfect foresight,[42] but with risk

To keep the example focused on the benefits and costs of adaptation, only the welfare and reservoir storage capacity estimates for the various cases are shown. This is somewhat regrettable, because changes in seasonal reservoir storage and water allocation to users differ quite a lot under the different climate probability combinations and institutional rules.

The methodology used to address this planning problem was to solve the planning problem to obtain optimal values for the endogenous variables over a number of probability combinations, pr(1, 0), ..., pr(0, 1) for each of the four project options.[43] This set of runs simulated optimal adjustment to climate change, since the storage capacity was allowed to adjust freely. This information was used to calculate the development benefits of the four project options, assuming the climate did not change, and the value of the imposed damages of climate change after partial and full adjustment for each option, including doing nothing.

Partial adjustment solutions were also obtained for each of the probability combinations and option cases. These simulations were conducted by taking the value of optimal reservoir capacity for Options 3 and 4[44] for a given climate probability combination and holding it constant in all the runs for the remaining probability combinations in which the climate and runoff were different. This set of runs was designed, in a planning capacity mode, to simulate a situation in which planners did not have reliable enough information about climate change to use in their planning and wanted to explore the expected benefits and costs of each option under different climates. From an assessment perspective, it represents a situation in which water managers and users are only partially adjusted to climate change through their adaptation to existing climate variability.

---

42 One could also introduce adaptive expectations about runoff into the model, but I have not done this.

43 GAMS (1998) was used to solve the planning problem.

44 There is no partial adjustment solution for the status quo or option 2, since no reservoir is built in these two cases.

Table 2 shows the welfare, project benefit and capacity results of the first set of runs, for the probability combinations pr(1, 0), pr(.9, .1), pr(.25, .75) and pr(0,1) and their associated values of expected annual runoff.

Table 2. Expected Net Welfare and Optimal Reservoir Capacity for Project Options Under Alternative States of Nature – Full Adjustment

Project Options	States of Nature Probability Combinations			
	Pr(1, 0)	Pr(.9, .1)	Pr(.25, .75)	Pr(0, 1)
	Expected Annual Runoff x 10^3 m^3			
	9210	9100	8380	8100
	Expected Net Welfare x USD 10^6			
1. Status quo	85.557	82.330	61.351	53.282
2. Market only	118.119	112.965	79.467	66.582
3. Dam only	96.313	93.983	79.612	74.476
4. Market + dam	119.389	116.663	106.454	102.945
	Project Net Benefits in Relation to Status Quo x USD 10^6			
2. Market only	32.562	30.636	18.116	13.300
3. Dam only	10.756	11.653	18.261	21.194
4. Market + dam	33.831	34.334	45.104	49.663
	Optimal Capacity (10^3 m^3)			
1. Status quo	---	---	---	---
2. Market only	---	---	---	---
3. Dam only	3732	3946	5143	5740
4. Market + dam	775	2531	4514	4754

For each option, the value of net welfare falls as climate shifts, over the indicated probability combinations, from pr(1, 0) to pr(0, 1). Under any single expected climate, the net benefits of Option 4 (instituting water markets and building a reservoir) dominate all other options. But note that the difference between the net benefits of Option 4 and Option 2 (instituting water markets, only) under the climate pr(1, 0) is quite small, but increases fairly dramatically as the probability of winter and spring runoff falls and the probability of fall and winter runoff increases. As this occurs, the importance of reservoir storage capacity in both Options 3 (building a reservoir, only) and 4 in adjusting to climate change becomes more and more important relative to water markets. This can be seen, not only by comparing the net benefits of these three options relative to optimal storage capacity increases across the various probability combinations, but also by noting that when we compare the optimal storage capacity estimates for Options 3 and 4, water markets become a poorer and poorer substitute for reservoir capacity as the climate shifts from pr(1, 0) in the direction of pr(0, 1).

Tables 3 and 4 show the expected net welfare results of the partial adjustment runs for Options 3 and 4, in which capacity is fixed at its optimal value as indicated in the first column of the table and the climate probability combinations are varied over the rows. The diagonal values in both tables are just the optimal expected net welfare values from Table 2.

Table 3. Expected Net Welfare for Project Option 3 (Dam only) Holding Reservoir Capacity Fixed and Varying Climate – Partial Adjustment

Reservoir Capacity		States of Nature Probability Combinations			
		Pr(1, 0)	Pr(.9, .1)	Pr(.25, .75)	Pr(0, 1)
		Expected Net Welfare x USD 10^6			
Pr(1, 0):	3732	***96.313***	93.962	78.678	72.800
Pr(,9, .1):	3946	96.292	***93.983***	78.976	73.204
Pr(.25, .75):	5143	95.347	93.249	***79.612***	74.367
Pr(0, 1):	5740	94.647	92.630	79.518	***74.476***

Table 4. Expected Net Welfare for Project Option 4 (Dam + Water Markets) Holding Reservoir Capacity Fixed and Varying Climate – Partial Adjustment

Reservoir Capacity		States of Nature Probability Combinations			
		Pr(1, 0)	Pr(.9, .1)	Pr(.25, .75)	Pr(0, 1)
		Expected Net Welfare x USD 10^6			
Pr(1, 0):	775	***119.389***	115.465	89.960	80.150
Pr(.9, .1):	2531	118.697	***116.663***	103.444	98.360
Pr(.25, .75):	4514	117.102	115.682	***106.454***	102.905
Pr(0, 1):	4754	116.856	115.465	106.423	***102.945***

The results in these two tables can be used in a variety ways. First, each diagonal row entry (in italics) represents the maximum expected net welfare that can be achieved when the reservoir capacity is fixed at the level in column 1. Each off-diagonal column entry, on the other hand, shows how changes in reservoir capacity affect expected net welfare given the climate probability combinations for that column. All the off-diagonal elements in any given column must be smaller in value than the diagonal element for that column, indicating partial adjustment.[45] Conceptually, it shows that when the long-run reservoir capacity is not adjusted as climate changes, the resulting short-run behaviour of water managers and users is also not optimal for the altered climate.

The information in these two tables can be used to calculate the loss in expected net welfare as a result of acting too cautiously (the cost of caution) or not cautiously enough (the cost of precaution).[46] An example from Option 3 can be used to illustrate this. Suppose we currently experience a climate characterised by pr(.9, .1) and we

---

[45] This is due, mathematically, to the LeChatelier principle.

[46] The costs of precaution and caution can also be couched in terms of Type I and Type II errors. A Type I error is a "false-positive" assumption – in this case, it means that decision makers assume, *ex ante*, that the climate is changing, and adjust to it, but in fact climate change does not occur, *ex post*. A Type II error is a "false-negative" situation – assuming, *ex ante*, that climate is not changing, and making no adjustments to it, when in fact the climate changes, *ex post*.

assume that the null hypothesis (no climate change) is true. So, we build a reservoir that is optimally sized for that expectation. However, let's suppose, *ex-ante*, that the climate does change to a state characterised by the probability combination pr(.25, .75). In other words, we accepted the null hypothesis, but it turned out to be false. In this case, we have been too cautious.. What is the cost of this error? Since the climate has changed, *ex-post*, net welfare is reduced from USD 93.982 million to USD 78.976 million, a loss of USD 15.007 million. This is basically the cost of not being in full adjustment. Had we correctly rejected the null hypothesis and optimally sized the reservoir for the probability combination pr(.25, .75), then net welfare would have fallen by less, from USD 93.982 million to USD 79.612 million, or a loss of USD 14.371 million. Thus, the cost of being too cautious (i.e. not adjusting to climate change) is equal to USD 15.007 million – USD 14.371 million = USD 636,000.

Now let's say we had rejected the hypothesis that the climate was not changing and we did build the optimally sized reservoir, based on the *ex-ante* expectation of a climate change associated with the probability combination pr(.25, .75), but the climate did not change, *ex post*. In that case we have not acted cautiously enough, an error of precaution. Had we not optimally sized the reservoir for this expectation, net welfare would have been equal to USD 93.982 million, but since we did build the larger reservoir net welfare is reduced to USD 93.249 million. The difference between the two is the cost precaution: USD 93.249 million – USD 93.982 million = USD 734,000.

By assessing all of these errors, *ex ante*, for each option one can construct a "regrets" matrix in which the top diagonal elements present the costs of being too cautious and the bottom elements capture the costs of being too precautious (not being cautious enough) for each row action taken, *ex ante*. Thus, even if one is not sure about the probabilities of the various states of nature, one can still examine the value of the regrets one will experience from making either kind of error. This kind of information can be extremely useful in planning under uncertainty.[47]

In an adaptation assessment framework, the information in the last three tables can be used to estimate the damages due to climate change and the benefits of adjusting to climate change, both partially and fully. Unfortunately, it was not possible to estimate the *maximum climate change damages* in this example or the *net benefits of partially adapting to climate variability*. The pure effect of climate change could not be simulated using the model here because the water demand functions were not derived, explicitly, from production functions.[48] These estimates are contained in Tables 5

---

47 To save space, the regrets matrix for options 3 and 4 has been omitted from the text. However, if calculates these, then it can be seen that the cost of caution is higher than the cost of precaution for all *ex ante* choices for both options, and the gap between the two is much larger in absolute and relative terms for option 4.

48 An effort was made to simulate this by holding water use constant and existing reservoir capacity constant, but it resulted in many infeasible solutions. These infeasibilities are, of course, instructive of the strain water users and managers would face if they could not adapt to climate variability, which they can.

(Option 3) and 6 (Option 4). These tables show only the partial results for climate that is changing in just one direction. They reflect an increase in the joint probabilities that runoff will occur earlier in the water year rather than later, in other words: the probability of late winter and early spring runoff decreases, while the probability of late fall and early winter runoff increases.

Table 5. Estimates of Climate Change Damages, Imposed Damages of Climate Change and Net Benefits of Adaptation for Project Option 3

Damage-Benefit Categories For the Different Probability Combinations	States of Nature Probability Combinations		
	Pr(.9, .1)	Pr(.25, .75)	Pr(0, 1)
Pr(1, 0)	**Expected Welfare Estimates x USD 10^6**		
Climate change damages[1]	-2.351	-17.635	-23.514
Imposed damages of climate change[2]	-2.331	-16.701	-21.838
Net benefits of adaptation[3]	0.021	0.934	1.676
Pr(.9, .1)			
Climate change damages		-15.007	-20.778
Imposed damages of climate change		-14.371	-19.507
Net benefits of adaptation		0.636	1.271
Pr(.25, .75)			
Climate change damages			-5.245
Imposed damages of climate change			-5.137
Net benefits of adaptation			0.108

Notes:

1. With partial adjustment.
2. With both partial and full adjustment.
3. Due to full adjustment.

The rows in Tables 5 and 6 present the estimates expected for the different damage-benefit categories by the four *ex-ante* probability combinations. The columns in the two tables represent the expected *ex-post* values for the different probability combinations. Thus, if planners assumed the subjective probability combination Pr(1,0) in designing the project, but the probability combination that actually occurred (the true probability combination) was Pr(.9,1), then the expected value of climate change damages would be 2.351 million dollars, whereas this would increase to 17.635 million dollars if the true probability combination was Pr(.25, .75), and would increase still further to 23.514 million dollars if the true probability combination was Pr(0, 1).

These two tables contain a lot of information relating to vulnerability and adaptation. Going back to the previous discussion about errors of caution and precaution, one of the most important conceptual points about these two tables is that climate change damages that are avoided by full adjustment (i.e., the *net benefits of adaptation*) are also the cost associated with acting too cautiously in this example (with the sign reversed). The concluding section tries to draw this all together in some suggestions about how the various measures in these tables might be conceptually related to vulnerability and adaptive capacity. However before doing so, a few points are relevant.

First, the adaptation benefits associated with the various changes in climate tend to be much larger in absolute and relative terms for Option 4 than for Option 3. What this suggests is a strong interaction between water markets and reservoir storage in adjusting to climate change. The net adaptation benefits of being able to adjust storage capacity in the presence of water markets is substantially greater than the net benefits of adjusting storage capacity when there are no water markets. This is understandable due to the highly constrained allocation rules associated with Option 3. However, note that Option 4 is more vulnerable to climate change than Option 3 in the sense that the climate change damages associated with Option 4, are greater both in absolute terms and relative to the full adjustment net welfare estimates shown in the diagonal elements of Tables 4 and 5. These differences narrow somewhat as the reference climate shifts toward Pr(0, 1) – that is a situation where climate change is expected. This is surprising in the sense that intuition and the difference in net welfare levels under full adjustment for the two options tells us that combining water markets with a reservoir is economically more efficient than just building the reservoir. But this is explainable in a more general conceptual framework of vulnerability and adaptation, to be taken up at the end of the chapter.

Table 6. Estimates of Climate Change Damages, Imposed Damages of Climate Change and Net Benefits of Adaptation for Project Option 4

Damage-Benefit Categories for the Different Probability Combinations	**States of Nature Probability Combinations**		
	Pr(.9, .1)	Pr(.25, .75)	Pr(0, 1)
Pr(1, 0)	**Expected Welfare Estimates**		
Climate change damages[1]	-3.924	-29.429	-39.238
Imposed damages of climate change[2]	-2.725	-12.934	-16.443
Net benefits of adaptation[3]	1.199	16.495	22.795
Pr(.9, .1)			
Climate change damages		-13.219	-18.303
Imposed damages of climate change		-10.209	-13.718
Net benefits of adaptation		3.010	4.585
Pr(.25, .75)			
Climate change damages			-3.549
Imposed damages of climate change			-3.509
Net benefits of adaptation			0.040

Notes:

1. With partial adjustment.
2. With both partial and full adjustment.
3. Due to full adjustment.

3.2 Adding no regrets

The final issue in this example is the estimation of climate change damages, the imposed damages of climate change and the net benefits of adaptation for "no regrets" projects. This is explored in the context of Options 3 and 4.

It seems perfectly acceptable to view Options 3 and 4 as development projects. The net benefits of these two options are shown in the middle panels of Table 2 for various climate combinations. As already shown, these projects can also reduce climate change damages, and this is not taken into account in these estimates. To find out what the adaptation benefits of a development project are, we have to ask the question for each option: "what would climate change damages be if we built the reservoir without taking into account the possibility of climate change during the life of the reservoir?" The answer to this question lies in the results in Tables 5 and 6. For example, if the historical data suggested the existing climate was characterised by pr(.9, .1) and the climate was in fact changing so that it would be more like pr(.25, .75), the climate change damages associated with Option 3, taking into account only partial adjustment, would still be USD 15.007 million and the net adaptation benefits of building a larger reservoir USD 0.636 million. For Option 4, the corresponding damage and net benefit values would be USD 13.219 million and USD 3.01 million. Thus, the general definitions of various climate change benefits and costs apply equally to all types of projects, no matter what the objective.

While the net adaptation benefits may be a good measure for capturing the contribution of a development project to a climate-related objective, it is important to distinguish between no-regrets projects (that are subject to making errors of caution and precaution) and those that are not. Options 3 and 4 always improve social welfare compared to the status quo. However, they both face the risk that the optimal reservoir storage capacity levels will be too small (or too large) if climate does (or does not) change in the way we expected. The costs of these planning "errors" for Option 3 under these same probability combinations (but not for Option 3) have been calculated and are shown above. While many "no regrets" projects may be better than other alternatives under any climate, they still may face "climate regret" if the physical investment is climate sensitive. So, it seems reasonable to make a distinction between "no regrets" options that face potential climate regrets and those that do not.[49]

Option 2 is important from this perspective in that it represents an anticipatory, no regrets action that has no climate risk associated with it. This point is illustrated for Option 2 in Figure 3, involving the same shift in the climate probability combinations from pr(.9, .1) to pr(.25, .75). In this figure the demand curve, D, is the derived aggregate demand for water, W, and P is the marginal value of water in the basin. As in the previous figures, the demand curve is portrayed so that it is not affected by climate change, to keep the graphic analysis as uncomplicated as possible. Initially there are no

---

49 I have not heard that distinction being made in the literature about no regrets or anticipatory adaptation (Smith and Lenhart, 1996), but I think it may be implied.

water markets and climate is characterised by M_0. The aggregate water supply curve in the basin, corresponding to this, is S [W_0(No Mkt), M_0]. After water markets are adopted and the climate changes, the relevant aggregate water supply curve is S[W_1(Mkt), M_1]. The expected net welfare loss between these two is represented by the area A. These are the net project benefits of Option 2, taking into account climate change and the change in water allocation rules. It is the smallest welfare loss that can be achieved, relative to the no action cases.

The figure is presented to compare Option 2 with the next best alternative, the status quo. The welfare loss due to climate change under the status quo option is measured by the areas between the supply curves, S[W_0(No Mkt), M_0] and S[W_1(No Mkt), M_1], or A+C. If water markets are instituted, under either climate, the welfare change associated with doing this is positive. Under the initial probability combination, pr(.9, .1), the resulting expected welfare gain associated with the supply curve, S[W_0(Mkt), M_0], is measured by the area B. If climate has changed the relevant supply curve is S[W_1(Mkt), M_1], and the resulting expected welfare gain is measured by the area C. In other words, water managers and users are always better off no matter what the climate is. Thus, there is no climate regret associated with this option, as welfare is always improved by adopting water markets.

Figure 3. Illustration of adaptation for a no "climate regrets" project

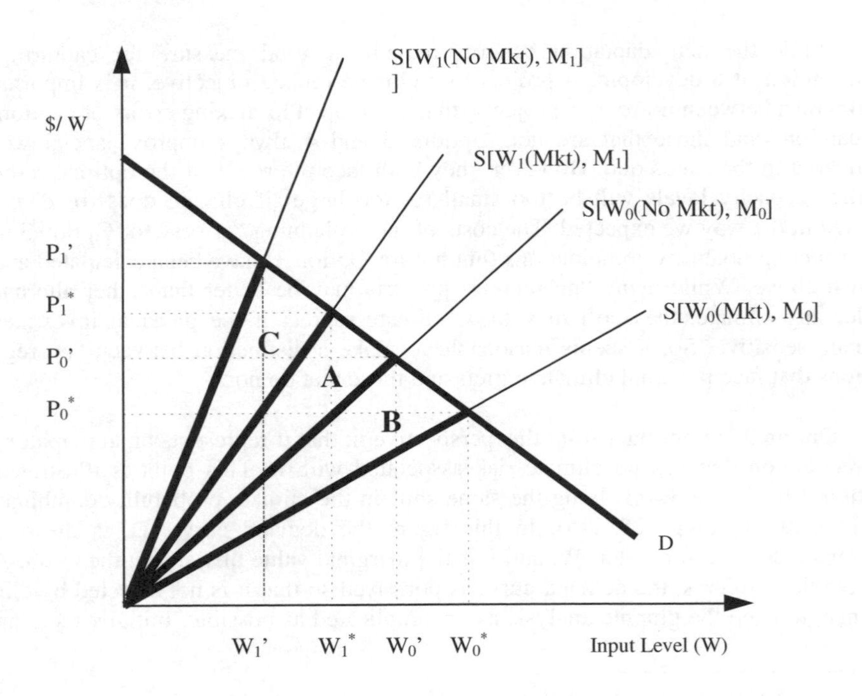

So, what are the net adaptation benefits of Option 2, if water markets are adopted and *ex-post* we learn that climate has changed? By adopting water markets sooner rather than later, water managers and users will avoid the potential expected welfare loss measured by the area C. As such, the area C represents the adaptation benefits of instituting water markets as a form of insurance against climate change, which water managers and users may not even be able to detect and can only predict with error. In short, adapting water markets will always dominate over not adapting them, no matter how much information we have or don't have about climate change and doing this will avoid potential climate change damages that water managers and users can not detect or predict with certainty.

In conclusion, it is important in selecting various options for anticipatory or proactive adaptation to separate out those that have climate regrets associated with them and those that do not, and concentrate on those that face no climate regret. At the same time, it is important to quantify the expected value of the net adaptation benefits associated with these options to give an idea of how much better off they will make society, if climate is changing and we are not able to detect it yet. The job of assessing the adaptation benefits and costs of policies that have potential climate regrets is not a technically complicated task. Even though we may not have very much reliable information about climate change to undertake resource planning at small geographic scales, this should not prevent us from looking at the expected adaptation benefits of these projects under alternative *ex-ante, ex-post* subjective probabilities to better assist us to understand the costs of acting too cautiously or not cautiously enough. There is ample precedent for doing this in water resources planning when there is risk associated with using historical stream flow records in planning storage capacity. Adding climate change risk is conceptually no different.

3.3 Assessment modeling

One of the points made in Callaway *et al.* (1999) was that there were no sector level studies available that measured the net benefits of adaptation or climate change damages in developed or developing countries. This remains almost as true today as it was then. Since the appearance of Mendelsohn and Neumann (1999) this and several other studies (Mendelsohn *et al.,* 1994; Yohe *et al.,* 1996; Fankhauser, 1994) have concentrated on measuring the imposed damages of climate change, only. This measure, as we have indicated, places a monetary value on the impacts of climate change, taking into account the fact that adaptation takes place. More recently, Darwin and Tol (2001) constructed the proper paradigm for separating out the benefits and costs of adaptation, but were unable to relate the imposed damages of climate change to adaptation benefits and costs in a consistent accounting framework because they used different models to calculate adaptation costs and the imposed damages of climate change.

More recently, Mendelsohn *et al.* (2000) and Mendelsohn and Dinar (1999) have suggested that the net benefits of adaptation can be "parsed" out by comparing their earlier estimates of the imposed damages of climate change in the US agricultural sector

(Mendelsohn *et al.* (1994) using the "Ricardian"[50] approach with so called "agronomic"[51] studies of climate change damages by Adams *et al.* (1993, 1999) for the US agricultural sector. The difference between the two, he suggests, will yield an estimate of the net benefits of adaptation. The problem with this argument, as Darwin (1999) has pointed out, is that the "agronomic" studies have been conducted using price-endogenous sector models that may, in fact, include more adaptation possibilities than does the Ricardian method. Thus, according to Darwin, both sets of studies have actually presented estimates of the same measure, the imposed damages of climate change. Since the studies use vastly different databases for the US agricultural sector, it is probably these differences and not adaptation that explain the differences in the results of the two sets of studies (Hanemann, 2000).

So, how the economic research community can best estimate climate change damages, net adaptation benefits and the imposed damages of climate change remains an open question. The proposal outlined here is based on implementing a conceptual framework that links adaptation to climate change to adaptation to climate change variability. This proposal opts for partial equilibrium price-endogenous sector models (McCarl and Spreen, 1980), at the national or global (i.e., world trade) level, due to a basic question about the ability to capture "everything" in the context of a global or national CGE models. This preference is based on the data and computational problems associated with embedding a model detailed enough to depict adjustments to climate variability and climate change in a climate-sensitive sector model within a detailed CGE framework.

The "ideal"approach would involve:

- Explicit linkages between outputs and input use and exogenous meteorological/environmental variables, such that changes in these exogenous variables:
  - Directly influence output;
  - Endogenously influence input and output prices; and as such

---

50 The "Ricardian" approach involves developing a regression relationship between net farm income in a cross-sectional, time series database and meteorological variables. The imposed damages of climate change are simulated by evaluating the regression equation with different values for the meteorological variables in a Base Case and in a Climate Change Scenario. This valuation function is also better known as a "hedonic price" function and it was initially developed to explain the impacts of environmental externalities on human welfare through changes in housing/land prices.

51 So called because Mendelsohn did not believe that these studies allowed crop prices to reflect the pure effects of climate change. A true "agronomic" model would hold crop prices constant at Base Case levels and changes in net farm income would only reflect the impacts of the pure effect of climate change. If that had been true in the studies cited by Mendelsohn, his point would have been theoretically correct.

- Endogenously influence input and output adjustment to both climate variability and climate change.

- The ability to include management, technological, and infrastructure options for adaptation to both climate variability and climate change and allow these to be selected endogenously in the model, based on marginal benefits and costs.

- The ability to simulate, and distinguish between, short-run adjustments of variable inputs and long-run adjustments to quasi-fixed inputs.

- A stochastic approach to modelling both climate variability and climate change by:

  - Simulating inputs of meteorological/environmental variables stochastically, and

  - Relating climate change to climate variability by explicit changes in the distributions of (or the probabilities of occurrence of specific values) these variables.

  - Linkages to the rest-of-the world through trade, partially, through excess demand (or supply) functions or fully.

This may seem like a tall order, especially on the climate scenario side, given that there is not very much information about changes in climate variability associated with climate change, and emission scenario modelers are reluctant to attach subjective probabilities to the emissions scenarios that drive GCMs. However this should not let hold us back, especially if, as so many climate experts tell us, GCM results should not be treated as forecasts. In that regard, it is also tempting to believe that we need a lot of meteorological information in climate forecasts, if only because state of the art impact models have voracious appetites for data. Yet, the source of much of the meteorological information that already goes into these models for climate runs is highly "massaged" in the first place. Thus, we should not become too focused on forecasts of individual meteorological variables and the need to fit the partial and joint distributions of a large number of meteorological variables in constructing climate scenarios.

Instead, it is important to concentrate on developing climate scenarios, or climate states, that depend on only a few "random" variables in the historical record. These should be varied based not only on the weak links we have to global and regional GCM models, but more importantly based on the existing historical record of climate variability, closer to the extremes than the middle. Good examples of this would be to shift the intensity and timing of Asian monsoons or African rainy seasons to coincide with those events that occur more rarely and to which Asian and African resource managers and governments are presumably not that well adapted (even though they do occur). By doing this we can gain, not only insights into the autonomous and strategic benefits and costs of investments required to better adapt to more infrequent, but very important, weather patterns, but also a better appreciation of the role which short-run adjustments to climate variability already play, and will play, in adapting to these changes.

One of the features that are not easy to incorporate in large-scale assessment models is climate variability. This is a problem, computationally, because of the need to simultaneously simulate climate variability and market adjustment to it in a number of regions. However, this is possible at smaller geographic assessment scales, for example in river basins, and in planning models for individual projects. Thus, how to relate adjustments to existing climate variability to adjustments to climate change in large scale assessment model represents an important area of further research and development.[52]

4. Conclusion: vulnerability, adaptation and economic development

The paper has tried to illustrate how adaptation to climate variability can play an important role in adapting to climate change. However, adaptation to climate variability and climate change are affected by many of the same factors. This final section links the adaptation framework developed in the previous section to two questions:

- What makes a region vulnerable to climate change, and
- How does this relate to the capacity of a region to adapt to climate change?

In addition, these issues are related here to more general sustainable development issues and policies in developing countries.

One of the issues that natural scientists and economists have been debating, implicitly, is how to relate the physical effects of climate change to the adjustments that individuals and organisations make to avoid these effects. The previous section showed that adjusting to climate change occurs along a continuum from no, or extremely limited adjustment, as a result of the pure effect of climate change, to partial adjustment. Partial adjustment occurs when the climate is changing but it is perceived as climate variability, and so only short-run adjustments are made, to full adjustment where individuals and organisations adjust quasi-fixed factors based on reliable information about climate change.

The question is, do we define vulnerability to climate change before or after these adjustments take place. Or, put in terms of the definitions presented in this chapter, do

52 Such an effort is underway, I believe, under the title *Close-Coupling of Ecosystem and Economic Models: Adaptation of Central U.S. Agriculture to Climate Change* at Montana State University.

we define vulnerability to climate change in terms of the pure effect of climate change, where climate change damages are at a maximum, or in terms of the imposed damages of climate change, after partial and full adjustment have taken place? I think this question is best approached by looking at three extreme cases, where:

1. The maximum climate change damages are very large in relation to the existing climate and the imposed damages of climate change are small in relation to the maximum climate change damages (implying relatively large net adaptation benefits),

2. The maximum climate change damages are very large in relation to the existing climate and the imposed damages of climate change are also very large in relation to the maximum climate change damages (implying relatively small net adaptation benefits), and

3. The maximum climate change damages are quite small relative to the existing climate and the imposed damages of climate change are large in relation to the maximum climate change damages (implying relatively small net adaptation benefits).

Each of these cases is perfectly plausible. In each case, it is assumed that society is well adapted to the existing climate. Therefore, Case 1 represents a situation where the climate is very different, after climate change, but the capacity to adjust partially and fully to climate change is large. This is to be contrasted with Case 2, where climate change is large, but the capacity to partially and fully adjust to it is limited. Case 3 represents a situation where not much adaptation needs to take place, because the climate does not change a great deal.

These three cases focus attention on the importance of adaptation adjustments relative to climate change damages. Breaking down adaptation to climate change into its two parts, partial adjustment and full adjustment, provides further insights. As indicated previously, partial adjustment, in itself, creates net benefits measured by the reduction in maximum climate change damages. These benefits are due to the already built in capacity to adjust to existing climate variability, without making new investments in quasi-fixed factors. To measure these net benefits explicitly one has to be able to simulate the pure effects of climate change. If this cannot be done, as in the case of the example in this chapter, then there is no way to measure the contribution of partial adjustment to adaptation to climate change. This point illustrates the practical importance of the capability to simulate the pure effects of climate change in sectoral assessment models. Full adjustment to climate change gives rise to further adjustments that individuals and organisations make to adapt to climate change by adjusting quasi-fixed factors consistent with the altered climate. As indicated previously, full adjustment generally will not occur unless information about climate change is reliable enough for individuals and organisations to risk incurring relatively large investment costs. The example in this chapter illustrated this point for measures where society faced potential climate regret as a result of making the wrong decision regarding reservoir capacity.

Thus, it seems reasonable to suggest that adaptation capacity is also composed of two parts: the capacity to adapt to climate change due to the ability to adjust to existing variability and the capacity to adapt to climate change by making long-run investments. The effects of both of these are measurable in terms of avoided damages, both physical and economic,[53] and by characterising the actions that individuals and organisations are able to take to avoid climate change damages. One way to measure the contribution of adaptation capacity to avoiding climate change damages totally, or in its parts, is by the ratio(s) of avoided damages to maximum climate change damages.

Going back to our original question, we are still left with the following conceptual possibilities for quantitatively evaluating the relationship between vulnerability and adaptation. These are: to measure vulnerability independent of adaptation through various measures of climate change damages, or to measure vulnerability after we net out adaptation through partial and full adjustment through measures of the imposed damages of climate change. In the first case, it is possible to create both physical and economic indices of vulnerability by the ratio of the appropriate measure of climate change damages to a base case measure. From a welfare accounting perspective this index could be represented by the ratio of the maximum climate change damages to base case welfare. In the second case, vulnerability can also be measured in relative terms through the ratio of the imposed damages of climate change to a base case measure, which could be characterised in welfare accounting terms by the ratio of the imposed damages of climate change to base case welfare.

In a recent unpublished paper, Antle *et al.* (2002) have arrived independently at similar conclusions. Combining their definitions with the terms in Callaway *et al.* (1999) and this paper, one can identify the following indices:

- Relative climate variability (without adaptation) = Climate change damages/Base Case Welfare.

- *Relative climate variability (with adaptation)* = Imposed damages of climate change/Base Case Welfare.

- *Relative adaptive gain* = Net benefits of adaptation/Base Case Welfare.

Much of the literature on climate change impacts and adaptation to climate change has tended to view the issues of vulnerability and adaptation, fundamentally, as resource management issues (See for example: Mendelsohn and Neumann, 1999; Downing *et al.*, 2001; McCarthy *et al.,* 2001). The IPCC's *Third Assessment Report* (2001b) investigated the links between sustainable development and adaptation capacity, but in doing so failed to relate it in specific terms to the underlying structural features in the macro-economies of developing countries that make it difficult for them to adjust to environmental and other "shocks." They also left out any reference to the growing body

[53] Through both Marshallian surplus and Hicksian variation measures and through changes in sectoral income accounts.

of literature on sustainable development using computable general equilibrium models. This is perhaps reflective of the fact that sustainable development as a concept and sustainable development financing as a practice have tended to focus on resource and environmental management and on human and social capital development, as opposed to infrastructure development and domestic structural adjustment policies (in the macro-economic sense). While this perspective makes a great deal of sense when looking at partial adjustments to climate change, it misses the point somewhat when it comes to full adjustment. And thus, the emphasis on adaptation, as a resource management issue is especially overstated when it comes to assessing vulnerability and adaptation in developing countries.

There are at least two reasons that more attention needs to be given to the relationship of adaptation to climate change to infrastructure development and structural adjustment policies. First, the ability to partially adjust to climate change in developing countries is primarily limited by structural factors that, in many cases, can only be adjusted in the long-run. While it is often true that resource managers in some developing countries have trouble responding to climate variability, in the short-run, due to poor resource management practices, this may not be the key problem. Rather, it may be due to the fact that the economic systems in which they operate are often highly constrained in structural terms. This limits the ability of resource managers to respond to environmental shocks of any kind, whether this comes from global climate change or hoards of locusts. These limiting factors, which I have taken loosely from Winters *et al.* (1998), include briefly:

- Highly constrained resource mobility, nationally and internationally.
- Lack of differentiation and specialisation in natural resource sectors that are dominated by household production.
- Limited technological possibilities for output and input substitution.
- High prices of inputs that can be substituted for, or counteract the effects of, environmental inputs affected by climate change, relative to their marginal productivity in natural resource sectors.
- Thin domestic product markets in natural resource sectors due to poor distribution and marketing systems.
- Lack of integration in international product markets, compounded by developed country subsidy programs.

Second, if one looks over the list of factors, above, that limit the adaptive capacity of individual producers, sectors, or groups of sectors in a nation or region, it can be seen quite easily that these factors are fundamentally related to economic development issues and not resource management issues. The nexus between the capacity of individuals, singly or collectively, to adapt to climate change and their stage of economic development is critical and, until recently, has been neglected in much of the literature on adaptation to climate change. That is to say: policies that are effective in modernising the economy of a developing country, by increasing the productivity of natural resource sectors, making output and input substitutions more elastic, better integrating domestic markets into the national and international economy, and reducing both economic and non-economic constraints on resource mobility are also good climate change policies both in the long-run and the short-run. This goes for long-run adjustments in technology, infrastructure and institutions that face climate regrets and those that do not. If the climate were changing very rapidly in time scales measured by decades, the almost total focus on resource management might be warranted. But the time scale of climate change and its possible impacts are in fact quite close to the time scales we ordinarily think of in terms of economic development – perhaps half a century or more.

Annex 1: Modelling the costs and benefits of adaptation

This technical annex is divided into four sections. Section A.1.1 focuses on defining the differences between weather and climate and climate variability and how this is linked to the production for a firm or industry whose outputs are influenced by climate variability and climate change. Section A.1.2 presents a model to characterise the adjustment to climate variability, and Section A.1.3 extends this model to characterise adjustments to both climate variability and climate change. Section A.1.4 shows how the new benefit and cost definitions are derived consistent with the model in Section A.1.3. Section A.1.5 presents the planning model used in the river basin example.

A.1.1 Climate variability, climate change and the production function

For the purposes of modelling the costs and benefits of adaptation, it is useful to formally define weather, climate and climate variability. To understand these differences, we can define the weather as the observed values of the vector of meteorological variables M_{kt}, for k=1,…,K variables over t=1,…, N time periods. The weather is generated by a climate process, whereby the observed values of the weather variables are approximated by:

$$M_{kt} = f(M_{kt-1},\ldots,M_{kt-N}; D_t,\ldots,D_{t-N}) + e_{kt} \qquad \text{for all k and t,} \qquad \text{EQ.1.1}$$

where D represents a vector of exogenous "driver" variables, some of which may be random, and the e_{kt} are random error terms that captures the unexplained variation in the observed values of the meteorological variables. While the generating process is part deterministic through the process f( ) and part random, as captured in the error terms, it is common practice to focus on the randomness in the observed outcomes of M, and to use historical data to fit a distribution function to M that allows us to characterise the joint distribution of M_t in the following way:

$$M_t \sim \Phi_M(\upsilon_M, \sigma_M^2, \Omega_M, \Theta_M), \qquad \text{EQ.1.2}$$

where υ_M, σ_M^2, and Ω_M are, respectively, the means and variances of the partial distributions of the individual random variables in M, and Θ_M represents a vector of the higher order moments of the distribution of M. In this framework, climate is characterised by the distribution parameters of M, while the observed (or predicted) weather is characterised by the observed (or predicted) values of the various meteorological variables that comprise the joint distribution of M. For example at time t this would be: M_{1t}, M_{2t}, … , M_{Kt}. Climate variability, which is a part of climate, is characterised by the variances of the partial distributions of meteorological variables, the co-variances between the meteorological variables and the higher-order moments of the partial and joint distributions of the meteorological variables, such as skewness and kurtosis. Finally, climate change is characterised by changes in any of the moments of this distribution.

Now, to show how decision-makers in the private and public sectors deal with the randomness of climate and climate variability in their capacity as resource planners or managers we introduce the idea of the production function. The production function in a climate-sensitive sector, industry, or even more broadly "an activity" is a way to characterise the relationship between the outputs of an activity and the inputs used to "create" that output, including meteorological variables. Here, I will use a very simple form of a production function, where there is just one output:

$$Q_{st} = f(X_{st}, K_{st}, M_{st}), \qquad \text{EQ.2}$$

where the subscripts s and t represent, respectively, the state of nature (or climate) and t represents time. Q stands for the single output and X and K are, respectively, vectors of variable inputs (X) and fixed or quasi-fixed factors (K), while M is a vector of random meteorological variables, as characterised in EQ.1 above. Note that a single value for any of these meteorological variables, M_{ist}, is a weather value, for example the observed or predicted precipitation in a particular month.

A.1.2 Making the framework stochastic

This model illustrates the principle of *ex-ante, ex-post* planning for climate variability in a mathematical framework, just to be concise. And, since the approach is general, I will illustrate it first through a non-economic model of yield maximisation, first for a single arbitrary climate state s, over $\tau = 1, \ldots, T$ random trials of weather observations (or predictions) for each climate. The expected values of the objective function and the other random variables in all of the models in this chapter are approximate, based on averaging the realisations of random variables over a number of climate trials using Monte Carlo simulation. The first, *ex-ante*, stage of the problem involves selecting the *ex-ante* values of K and, less importantly $X_{s\tau}$ that will:

$$\text{Maximise yield: } Z = \sum_{\tau=1}^{\Delta} (1/\Delta) * f(X_{s\tau}, K_s, M_{s\tau}), \qquad \text{EQ.3.1}$$

where:

$$M_{s\tau} \sim \Phi(M_s, \tau) \qquad \text{for } \tau = 1,\ldots,\Lambda \qquad \text{EQ.3.2}$$

where Φ (M_s, τ) is shorthand used to characterise the joint distribution of the random weather values for $M_{s\tau}$ over T random trials for each climate. In this model Z in EQ.3.1 is the expected value of yield that is optimal over all T trials, while EQ.3.2 defines the values of the weather variables that are generated in each trial. In this model the objective function is being maximised over a number of different trials for a single climate state. As such, the model only deals with climate variability associated with a single climate state, and there is only one optimal *ex-ante* value for the quasi-fixed factor(s), K_s^*, but there is an *ex-ante* optimal value, $X^*_{s\tau}$, for each τ for the variable input(s).

An optimal *ex-ante* solution to the yield maximisation problem requires that the partial derivatives of Z with respect to K_s^* and $X_{s\tau}$ be equal to zero. Assuming for simplicity that the model solutions for these variables can be written in closed form, the *ex-ante* (long-run) solution equations for the vectors of quasi-fixed factors, variable inputs and maximum yields are:

$$K_s^* = k_s^*(M_s), \qquad \text{EQ.4}$$

$$X_{s\tau}^* = x_s^*[k_s^*(M_s), M_{s\tau})] = x_s^*(M_s, M_{s\tau}), \text{ and} \qquad \text{EQ.5}$$

$$Q_{s\tau}^* = f^*[x_s^*(M_s, M_{s\tau}), k_s^*(M_s), M_{s\tau}] = q_s^*(M_s, M_{s\tau}). \qquad \text{EQ.6}$$

The important result here that the *ex-ante* values of the optimal quasi-fixed factors depend on climate, which can be characterised by $\Phi(M_s, \tau)$, while the *ex-ante* values of the variable inputs and output depend on both climate, through K, and the simulated, *expected* values of weather values. This means that when K is fixed, as it is in the second stage of this problem (which we do not show here), the values of Q and X will depend on K and on the values of the actual weather variables.

A.1.3 Linking the counterfactual case and damage/benefit definition to adjustments to climate variability and climate change

The purpose of this model is to find the optimal values of the variable input, given that climate is changing, from s = 0 in the initial *ex-ante* planning stage and s = 1 in the *ex-post* planning stage, but individuals can't detect this change and are unable or unwilling to change their *ex-ante* decisions about the size of the capital stock based on climate change information from global and regional climate models. Using this approach, the second-stage planning problem used to illustrate partial adjustment is to

$$\text{Maximise Yield} = \sum_{\tau} (1/\Delta) * f(X_{1\tau}, K_0^*, M_{s1}), \qquad \text{EQ. 7}$$

where $M_{1\tau} \sim D_1(M_1, \tau) \neq M_{0\tau} \sim D_0(M_0, \tau)$.

This model can be used to characterise full adjustment to climate change, which occurs when the capital stock is adjusted for climate change and the variable inputs are adjusted for the new climate variability. It can also be used to characterise a series of steps from "no adjustment" to climate change to full adjustment. If climate change can be detected or forecasted with a reliable degree of confidence, then decision makers will adjust the capital stock and institutional arrangements under their control consistent with the results of planning models that characterise this climate change. One can call this process "full adjustment" and it represents the fullest kind of adaptation that can occur with reliable information about climate change. I will characterise the "change path" of relevant variables by arrows from state 0 to state 1. Furthermore, I will limit this part of the analysis to describing only the change path for the yield variable; however, the underlying economic logic applies to the variable inputs, as well. The change path in the short-run values of the yield variable[54] associated with this process of full adjustment can be written as:

$$Q_{0\tau}^* = f\left(X_{0\tau}^*, K_0^*, M_{0\tau}\right) \longrightarrow Q_{1\tau}^* = f\left(X_{1\tau}^*, K_1^*, M_{1\tau}\right), \qquad \text{EQ.8}$$

where

$$Q_{0\tau}^* = f\left(X_{0\tau}^*, K_0^*, M_{0\tau}\right) = f^*[x_0^*(\overline{K}_0^*, M_{0\tau}), \overline{K}_0^*, M_{0\tau}] \text{ and}$$

$$Q_{1\tau}^* = f\left(X_{1\tau}^*, K_1^*, M_{1\tau}\right) = f^*[x_1^*(\overline{K}_1^*, M_{1\tau}), \overline{K}_1^*, M_{1\tau}].$$

In the earlier framework I helped to develop, we looked at the transition between these two sets of optimal adjustments in terms of just two partial steps, with the counterfactual case of no adjustment to climate change, lying in between full adjustment

54 We can also show similar paths for the optimal expected value of the objective function.

process shown in EQ.8. Now, it can be seen that by linking adjustment to climate variability to adjustment to climate change the transitional process becomes both more complicated and, I think, more realistic.

If climate change is not detected or the information about predicted changes is too unreliable for planning purpose, this puts decision makers effectively in a short-run situation where the quasi-fixed factors are fixed *and* they respond to changes in weather by varying their variable inputs.

The change in the value of output in these circumstances can be decomposed into three parts, as follows:

1. *Pure effect of climate change (a purely physical response, with no human adjustment possible):*

$$Q^*_{0\tau} = f\left(X^*_{0\tau}, K^*_0, M_{0\tau}\right) \longrightarrow Q'_{1\tau} = f'\left(X^*_{0\tau}, K^*_0, M_{1\tau}\right), \quad \text{EQ.9}$$

where $Q'_{1\tau} = f'\left(X^*_{0\tau}, K^*_0, M_{1\tau}\right) = f'[x'_0(\overline{K}^*_0, M_0), \overline{K}^*_0, M_{1\tau}]$.

2. *Partial adjustment to climate change (limited adjustment to perceived climate variability):*

$$Q'_1 = f'\left(X^*_0, K^*_0, M_{1\tau}\right) \longrightarrow Q''_{1\tau} = f''\left(X''_{1\tau}, K^*_0, M_{1\tau}\right), \quad \text{EQ.10}$$

3. *Full adjustment to climate change (adjustment to climate change given reliable – not perfect – information about climate change):*

$$Q''_{1\tau} = f''\left(X''_{1\tau}, K^*_0, M_{1\tau}\right) \longrightarrow Q^*_{1\tau} = f\left(X^*_{1\tau}, K^*_1, M_{1\tau}\right) \quad \text{EQ.11}$$

A.1.4 Measuring climate change damages and benefits of adaptation

The decomposition of the adjustment paths that occur between one climate state and another gives rise to comparable set of measures of damages and benefits that are consistent with the definitions in the earlier framework. Since the objective function of the problem, used here, is measured in terms of yields, the damages associated with climate change and the benefits of avoiding climate change damages are also measured in terms of yield changes. However, as I indicated earlier the same approach could be used to measure welfare changes in monetary or other terms.

1. Maximum climate change damages:

$$f'\left(X_{0\tau}^{*}, K_{0}^{*}, M_{1\tau}\right) - f\left(X_{0\tau}^{*}, K_{0}^{*}, M_{0\tau}\right). \quad \text{EQ.12}$$

2. Imposed damages of climate change after partial adjustment (to climate variability)= Climate change damages prior to full adjustment:

$$f''\left(X_{1\tau}^{''}, K_{0}^{*}, M_{1\tau}\right) - f\left(X_{0\tau}^{*}, K_{0}^{*}, M_{0\tau}\right). \quad \text{EQ.13}$$

3. Imposed damages of climate change = Climate change damages after both partial and full adjustment:

$$f\left(X_{1\tau}^{*}, K_{1}^{*}, M_{1\tau}\right) - f\left(X_{0\tau}^{*}, K_{0}^{*}, M_{0\tau}\right) \quad \text{EQ.14}$$

4. Net benefits of partially adjusting to climate variability:

$$f''\left(X_{1\tau}^{''}, K_{0}^{*}, M_{1\tau}\right) - f'\left(X_{0}^{*}, K_{0}^{*}, M_{1\tau}\right) \quad \text{EQ.15}$$

5. Net benefits of fully adapting to climate change:

$$f\left(X_{1\tau}^{*}, K_{1}^{*}, M_{1\tau}\right) - f''\left(X_{1\tau}^{''}, K_{0}^{*}, M_{1\tau}\right) \quad \text{EQ.16}$$

A.1.5 Model for the example

This example involves the use of an *ex-ante, ex-post* planning model to determine the optimal capacity of a small reservoir under *ex-ante* uncertainty about climate. The model can be used in two different ways: either as an *ex-ante* planning model, or as an assessment model, used *ex-ante* or *ex-post* to estimate climate change damages, the imposed damages of climate change and the benefits (or costs) associated with adapting (or not adapting) to climate change.

The model can be expressed as:

Maximize the expected value of: EQ.17

$$Z = \sum_{\tau}^{\Lambda}(1/\Lambda) * \{\sum_{s}^{S} pr_{s} * \sum_{t}^{T}\sum_{j}^{J}\sum_{i}^{N}(1+\sigma)^{-t}[\Theta_{sij} * (c_{sij}W_{si\tau jt}) - .5 * \Psi_{sij} * (c_{sij}W_{si\tau jt})^{2} - p_{si} * W_{si\tau jt}]\} - (\gamma * K + .5 * \delta * K^{2})$$

Subject to:

Reservoir Storage Balance: EQ.18

$$S_{s\tau jt} - S_{sjt} + REL_{s\tau jt} = RO_{s\tau jt} \overset{d}{=} \Phi_{sj}(\upsilon_{sj}, \sigma_{sj}^2, \Omega_{sj}, \Theta_{sj})$$ for all s, τ, j and t

Diversion Balance for Agriculture: EQ.19

$$W_{s,ag,\tau jt} + F_{s,ag,\tau jt} = REL_{s\tau jt}$$ for all s, τ, j and t

Diversion Balance for Municipal: EQ.20

$$r_{s,ag,\tau jt} W_{s,ag,\tau jt} + F_{s,ag,\tau jt} = W_{s,muni,\tau jt} + F_{s,muni,\tau jt}$$ for all s, τ, j and t

Flow to Sea Balance: EQ.21

$$r_{s,muni,\tau jt} W_{s,muni,\tau jt} + F_{s,muni,\tau jt} = OUT_{s\tau jt}$$ for all s, τ, j and t

Reservoir Capacity Constraints: EQ.22

$$S_{s\tau jt} = 0$$ for all s, j and t

$$S_{s\tau jt} \leq K$$ for all s, τ, j and t

No Market Constraint: EQ.23

$$\sum_{j}^{J} W_{s,ag,\tau jt} = .75 * REL_{s\tau jt}$$ for all s, τ and t

Initial and Terminal Condition: EQ.24

$$S_{s,0,0} = S_{sJT}$$ for all τ

Where:

Indexes (sets):

τ = 1,…, Λ: random trial periods (Λ is also used to scale the objective function)

s = 1, …, S: states of nature

t = 1, …, T: years

j = 1, …, J: water seasons

i = 1, …, N: activities (agriculture, municipal)

Endogenous Variables:

Z = Expected net present value of the sum of producer and consumer surplus less capital costs

W = Water diversion

S = Reservoir storage

K = Reservoir storage capacity

REL = Reservoir releases

F = Flow in river remaining after diversion

OUT = Outflows to sea

Parameters and Exogenous Variables:

pr_s = The probability of occurrence of state s: pr(winter+spring runoff, fall+winter runoff)

σ = Opportunity cost of capital

Θ = Intercept of inverse water demand function

Ψ = Slope of inverse water demand function

γ = Intercept of storage cost function

δ = Slope of storage cost function

p = Variable cost of delivering water

c = Consumptive use fraction of diverted water

r = Return flow fraction of diverted water

RO = Runoff into the reservoir

And W, S, K, REL, F, OUT ≥ 0.

The objective function, Z, (EQ.17) represents the expected net present value of consumers and producers surplus associated with the operation of the reservoir and the delivery and consumption of water minus the capital cost of constructing the reservoir. The reservoir balances (EQ.18) maintain the intertemporal continuity between storage, releases, and runoff into the reservoir. Seasonal runoff is random and varies by water season and the climate state. Water is released from the reservoir and flows downstream to satisfy agricultural water demands and municipal water demands that are farther downstream. The two diversion balances (EQ.19 and 20) take into account that water, which is not diverted from the river, remains in the channel and then is combined with the return flows from use, before reaching the next diversion activity. The flow to sea balance (EQ.21) adds the water in the river channel that is not diverted by municipal water users and combines it with the return flow from municipal uses to account for the flow that reaches the sea.

The last two sets of constraints are used to formulate different problems (or cases) to look at institutional changes in water allocation rules and the interaction between these rules and storage capacity. There are two reservoir capacity constraints (EQ.22). One restricts water storage in each period to the capacity of the reservoir, for cases where reservoir storage is assumed. The other sets storage to zero in each period for cases where no reservoir is assumed and run-of-river conditions prevail. The No market constraint (EQ.24) requires that 75% of all the water released in each year be allocated to agricultural diversions[55]. The single initial-terminal constraint (EQ.24) requires that terminal storage and initial storage values are equal. These are determined endogenously, along with the values for seasonal diversions, seasonal reservoir, seasonal reservoir releases, seasonal in stream flows at each diversion, seasonal outflows to the sea and reservoir storage capacity.

References

Adams, R.M., R.A. Fleming, B. McCarl and C. Rosenzweig (1993), "A Reassessment of the Economic Effects of Global Climate Change on US Agriculture," *Climatic Change*, **30**, 147-167.

Adams, R.M., B. McCarl, K. Seegerson, C. Rosenzweig, K. Bryant, B. Bixon, R. Cooner, R. Evenson, and D. Ojima (1999), "The Economic Effects of Climate Change on US Agriculture" in Mendelsohn, R. and J. Neumann (eds.) *The Impacts of Climate Change on the US Economy*. Cambridge University Press, Cambridge, England.

Antle J.M., S.M. Capalbo, E.T. Elliott and K.H. Paustian (2002), "Adaptation, Spatial Heterogeneity, and the Vulnerability of Agricultural Systems to Climate Change: an Integrated Assessment Approach", www.climate.montana.edu/pdf/MTcc.pdf.

Callaway, J.M (2004), "Adaptation Benefits and Costs: Are they important in the Global Policy Picture and How can we Estimate them?" Forthcoming, *Global Environmental Change*, September 2004.

Callaway, J.M (2003), "Adaptation Benefits and Costs – Measurement and Policy Issues", Environment Directorate, Environment Policy Committee, OECD. ENV/EPOC/GSP(2003)10/FINAL. Paris, France.

[55] Note that consumptive use is less than the amount of water diverted, due to evaporative losses and return flows. Also total diversions in a year do not necessarily equal total releases due to the presence of return flows.

Callaway, J.M. and L. Ringius (2002), "Optimal Adaptation: a Framework. Completion Report Submitted to the Danish Energy Ministry", UNEP Collaborating Centre on Energy and Environment, Risø National Laboratory, Roskilde, Denmark.

Callaway, J.M., L. Ringius, and L. Ness (1999), "Adaptation Costs: A Framework and Methods" in Christensen, J. and J. Sathaye (eds.), Mitigation and Adaptation Cost Assessment Concepts, Methods and Appropriate Use. UNEP Collaborating Centre on Energy and Environment, Risø National Laboratory, Roskilde, DK.

Carter, T.R., M. Parry, H. Harasawa, and S. Nishioka (1994), IPCC *Technical Guidelines for Assessing Climate Change Impacts and Adaptations*, Department of Geography, University College, London.

Darwin, R (1999), "A Farmer's View of the Ricardian Approach to Measuring Agricultural Effects of Climatic Change," *Climatic Change*, **41**, 371-411.

Darwin, R., M. Tsigas, J. Lewandrowski, and A. Raneses (1995), "World Agriculture and Climate Change: Economic Adaptations", *Agricultural Economic Report*, No. 703. U.S. Department of Agriculture, Washington, DC.

Darwin, R.F. and R.J. Tol (2001), "Estimates of the Economic Effects of Sea Level Rise", *Environmental and Resource Economics* **19**, 113-129.

Downing, T. E., Butterfield, R., Cohen, S., Huq, S., Moss, R., Rahman, A., Sokona, Y. & Stephen, L (2001), *Vulnerability to Climate Change: Impacts and Adaptation,*. United Nations Environment Programme.

Fankhauser, S (1994), "Protection vs. Retreat – The Economic Costs of Sea Level Rise", *Environment and Planning*, A **27**, 299–319.

Fankhauser, S (1997), "The Costs of Adapting to Climate Change", Working Paper, No. 13, *Global Environmental Facility*, Washington, DC.

Freeman III, A. M (1994), "The Measurement of Environmental and Resource Values - Theory and Methods", *Resources for the Future*, Washington, D.C.

GAMS Development Corporation (GAMS) (1998), *GAMS, A User's Guide,* GAMS Development Corporation, Washington, DC.

Hanemann, W.M (2000), "Adaptation and Its Measurement," *Climatic Change*, **45**, 571-581.

Hurd, B. J. Callaway, P. Kirshen, and J. Smith, (1999), "Economic Effects of Climate Change on US Water Resources" in Mendelsohn, R. and J.Neumann (eds.), *The Impacts of Climate Change on the US Economy*, Cambridge University Press, Cambridge, England.

Intergovernmental Panel on Climate Change (IPCC) (2001b), *Climate Change 2001: Impacts, Adaptation and Vulnerability*, a Report of Working Group II of the Intergovernmental Panel on Climate Change.

McCarthy, J. J., Canziani, O. F., Leary, N. A., Dokken, D. J. & White, K., S. (eds.) (2001), *Climate Change 2001: Impacts, Adaptation & Vulnerability*, Cambridge University Press, Cambridge, England.

McCarl, B. and T. Spreen (1980), "Price endogenous mathematical programming as a tool for sector analysis," *American Journal of Agricultural Economics*, **62**, 87-102.

Mendelsohn, R., N.G. Androva, W. Morrison, and M.E. Schlesinger, (2000), "Country-specific market impacts of climate change," *Climatic Change*, **45**, 553–569.

Mendelsohn, R. and A. Dinar (1999), "Climate Change, Agriculture, and Developing Countries: Does Adaptation Matter?" *The World Bank Research Observer*, **14**(2), 277-293.

Mendelsohn, R., Nordhaus, W. D. & Shaw, D (1994), "The Impact of Global Warming on Agriculture: A Ricardian Analysis," *American Economic Review*, **84**, 753-771.

Mendelsohn, R. and J. Neumann (eds.) (1999), *The Impacts of Climate Change on the US Economy*, Cambridge University Press, Cambridge, England.

Parry, M., C. Rosenzweig, A. Iglesias, F. Fischer, and M. Livermore (1999), "Climate Change and World Food Security: A New Assessment," *Global Environmental Change*, **9**, S51-S67.

Rosenzweig, C. and M. Parry (1994), "Potential Impacts of Climate Change on World Food Supply", *Nature*, **367**, 133-138.

Rosenzweig, C., M. Parry, and G. Fischer (1995), "World food supply" in *As Climate Changes: International Impacts and Implications*, K.M. Strzepek and J.B. Smith (eds.), Cambridge, UK, Cambridge University Press, 27-56.

Smith, J.B. and S. Hitz (2002), Background Paper: "Estimating Global Damages from Climate Change", Environment Directorate, Environment Policy Committee, OECD. ENV/EPOC/GSP(2002)12. Paris, France.

Smit, B., (ed.) (1993), "Adaptation to Climatic Variability and Change", Report of the Task Force on Climate Adaptation, The Canadian Climate Program, Downsview, Ontario.

Smit, B., D. McNabb and J. Smithers (1996), "Agricultural Adaptation to Climatic Variation," *Climatic Change*, **33**, 7-29.

Smit, B., Burton, I., Klein, R. J. T. & Wandel, J (2000), "An Anatomy of Adaptation to Climate Change and Variability," *Climatic Change*, **45**, 223-251.

Smith, J.B. & Lenhart, S.S. (1996), "Climate Change adaptation Policy Options," *Climatic Research*, **6**(2), 193-201.

Toman, Michael (1997), "Ecosystem Valuation: An Overview of Issues and Uncertainties," in Simpson, R. David and Norman L. Christensen, (eds.), *Ecosystem Function and Human Activities: Reconciling Economics and Ecology*, Chapman and Hall, New York, NY.

Winters, P. R. Murgai, E. Sadoulet, Alain de Janvry and G. Frisvold (1998), "Economic and Welfare Impacts of Climate Change on Developing Countries," *Environmental and Resource Economics*, **12**, 1-24.

Yohe, G.W., J.E. Neumann, P.B. Marshall, and H. Ameden (1996), "The Economic Cost of Greenhouse Induced Sea Level Rise for Developed Property in the United States," *Climatic Change*, **32**, 387-410.

Chapter 5

ABRUPT NON-LINEAR CLIMATE CHANGE AND CLIMATE POLICY

by Stephen H. Schneider and Janica Lane,

Stanford University, United States

Any discussion of the benefits of greenhouse gas mitigation measures should take into consideration the full range of possible climate change outcomes, including impacts that remain highly uncertain, like surprises and other climate irreversibilities. Coupling between complex systems can cause the interconnected system to exhibit new collective behaviours known as emergent properties. Likewise, coupling of sub-models of a complex system can exhibit behaviours that are not clearly demonstrable by non-coupled sub-models. Through examples from ocean-circulation and atmosphere-biosphere interactions, this paper demonstrates that external forcings such as increases in greenhouse gas (GHG) concentrations can push complex systems from one equilibrium state to another, with non-linear abrupt change as a possible consequence. It then discusses the concept of "dangerous anthropogenic interference" (DAI) with the climate system, as detailed by Mastrandrea and Schneider (2004), showing that climate policy controls can significantly reduce the probability of DAI occurring. The paper closes with a section on policy options and our suggestion that stringent abatement policies be considered to prevent climate surprises and other irreversibilities that would likely qualify as "dangerous".

Keywords: surprise, abrupt climate change, thermohaline circulation, irreversibility, integrated assessment models, dangerous anthropogenic interference, climate sensitivity, Monte Carlo analyses.

ISBN 92-64-10831-9
THE BENEFITS OF CLIMATE CHANGE POLICIES

CHAPTER 5. ABRUPT NON-LINEAR CLIMATE CHANGE AND CLIMATE POLICY

by Stephen H. Schneider[56] and Janica Lane

1. Introduction

In the Intergovernmental Panel on Climate Change's 1996 report (IPCC, 1996), it was suggested that climate change could trigger "surprises": rapid, non-linear responses of the climate system to anthropogenic forcing, thought to occur when environmental thresholds are crossed and new (and not always beneficial) equilibriums are reached. Schneider, Turner, and Morehouse Garriga (1998) took this a step further, defining "imaginable surprises" – events that could be catastrophic but are not truly unanticipated – possibly including a collapse of the North Atlantic thermohaline circulation (THC) system, which could cause significant cooling in the North Atlantic region, with both warming and cooling regional anomalies up- and downstream of the North Atlantic;[57] and deglaciation of polar ice sheets like Greenland or the West Antarctic, which would cause (over many centuries) many meters of additional sea level rise on top of that caused by thermal expansion from the direct warming of the oceans. There is also the possibility of *true* surprises, events not yet currently envisioned (Schneider, Turner, and Morehouse Garriga, 1998), yet it is still possible to outline "imaginable conditions for surprise"—like rapid forcing of the climate system, since the faster it is forced to change, the higher the likelihood of triggering abrupt non-linear responses. Potential climate change, and more broadly, global environmental change, is likely to be replete with both types of surprises because of the enormous complexities of the processes and interrelationships involved (such as coupled ocean, atmosphere, and terrestrial systems) and our insufficient understanding of them individually and collectively.

Unfortunately, most climate change assessments rarely consider low-probability, but high-consequence extreme events. Instead, they primarily consider scenarios that

---

56 Professor, Department of Biological Sciences, Stanford University, Stanford, CA 94305-5020 U.S.A. E-mail: shs@stanford.edu.

57 For further discussion, see Stocker and Schmittner (1997) and Rahmstorf (1999).

supposedly "bracket the uncertainty" rather than explicitly integrate unlikely events from the "tails of the distribution." Thus, decision-makers reading the "standard" literature will rarely appreciate the *full* range of possible climate change outcomes, and thus might be more willing to opt solely for adaptation as a means of confronting prospective climate changes rather than attempt to avoid them through abatement than they would be if they were aware that some potentially unpleasant surprises could be lurking. (Pleasant ones might occur as well, but many individuals and policymakers, via insurance premiums, tend to insure against negative outcomes preferentially.) In fact, it is not even clear that all such surprises carry low probabilities; projections are very uncertain at this point given the state of knowledge is still evolving. The policy community needs to understand both the potential for surprises and how difficult it is for integrated assessment models (IAMs), and other models as well, to credibly evaluate the probabilities of currently imaginable "surprises," let alone those not currently envisioned, and consider policy options accordingly.

2. "Imaginable surprises": Examples of abrupt non-linear responses of the climate system

Despite the inherent complexity of most global systems, scientists frequently attempt to model them in isolation, often along distinct disciplinary lines, producing internally stable and predictable behaviour. However, real-world coupling between and among elements within those systems can cause sets of interacting systems to exhibit new collective behaviours – called "emergent properties" – that may not be clearly demonstrable by models that do not include such coupling.

Responses of the coupled systems to external forcing can be quite complicated. For example, one emergent property increasingly evident in climate and biological systems is that of irreversibility or hysteresis (partial irreversibility) – changes that persist in the new post-disturbance state even when the original level of forcing is restored. This irreversibility can be a consequence of multiple stable equilibria in the coupled system – that is, the same forcing might produce different responses depending on the pathway followed by the system. Therefore, anomalies can push the coupled system from one equilibrium state to another, each of which has a very different sensitivity to disturbances (i.e., each equilibrium may be self-sustaining within certain limits). The foregoing discussion is primarily about model-induced behaviours, but hysteresis has also been observed in nature (e.g., Rahmstorf, 1996).

Below, we outline several examples of systems that exhibit complex, non-linear behaviour due to interactions between sub-systems of the climate system, including, in one example, the socio-economic system. These include multiple stable equilibrium states of the THC in the North Atlantic Ocean and of atmosphere-biosphere interactions in Western Africa. With both of these systems, crossing thresholds can lead to unpredictable and/or irreversible changes. Such complex processes and their outcomes have implications for effective policymaking. Incorporating them into modelling of climate change policy, for example, can significantly alter policy recommendations and lead to the discovery of emergent properties of the coupled social-natural system (see Higgins *et al.*, 2002, from which much of this section is adapted).

2.1 Thermohaline circulation

The Thermohaline Current in the Atlantic brings warm, tropical water northward, raising sea surface temperatures (SSTs) about 4°C relative to SSTs at comparable latitudes in the Pacific. The SSTs in the North Atlantic warm and moisten the atmosphere, making Greenland and Western Europe roughly 5-8°C warmer than they would be otherwise and increasing precipitation throughout the region (Stocker and Marchal, 2000; Broecker, 1997).

Temperature and salinity patterns in the Atlantic create the density differences that drive THC. As warm surface waters move to higher northern latitudes, heat exchange with the atmosphere causes the water to cool and sink at two primary locations: one south of the Greenland-Iceland-Scotland (GIS) Ridge in the Labrador Sea and the other north of the GIS ridge in the Greenland and Norwegian Seas (Rahmstorf, 1999). Water sinking at the two sites combines to form North Atlantic Deep Water (NADW), which then flows to the southern hemisphere via the deep Western Boundary Current (WBC). From there, NADW mixes with the circumpolar Antarctic current and is distributed to the Pacific and Indian Oceans, where it upwells, warms, and returns to the South Atlantic. As a result, there is a net northward flow of warm, salty water at the surface of the North Atlantic.

Paleoclimate reconstructions and model simulations suggest there are multiple equilibria for the THC in the North Atlantic, including a complete collapse of circulation. Switching between the equilibria can occur as a result of temperature or freshwater forcing. Thus, the pattern of the THC that exists today could be modified by an infusion of fresh water at higher latitudes or through high-latitude warming and a concomitant reduction in the equator-to-pole temperature gradient. These changes may occur if substantial climate change increases precipitation, causes glaciers to melt, or warms high latitudes more than low latitudes, as is often projected (IPCC 1996, 2001a).

Rahmstorf (1996) presents a schematic stability diagram of THC, based on his modification of the conceptual model of salinity feedback developed by Henry Stommel (1961, 1980), that demonstrates three possible THC equilibria under different levels of freshwater forcing, and the theoretical mechanisms for switching between them. These include two classes of deep water formation, one with sinking in the Labrador Sea and north of the GIS ridge, and one with sinking north of the GIS ridge alone; and one class of complete overturning shutdown. Rahmstorf's work indicates that switching between stable equilibria can occur very rapidly under certain conditions. The paleo-climatic record supports this, suggesting rapid and repeated switching between equilibria over a period of years to decades (Bond *et al.*, 1997). In addition, complex general circulation models (GCMs) suggest that future climate change could cause a similar slowdown or even collapse in THC overturning (Wood *et al.*, 1999; Manabe and Stouffer, 1993).

Schneider and Thompson (2000) present a simplified model for THC, the Simple Climate Demonstrator (SCD), which incorporates a straightforward density-driven set of Atlantic ocean boxes that mimic the results of complex models, but is computationally efficient enough that it can facilitate sensitivity analysis of key parameters and generate a domain of scenarios that show abrupt collapse of THC (Figure 1). Model results (e.g., Stocker and Marchal, 2000; Schneider and Thompson, 2000 – our Figure 1) suggest that both the amount of greenhouse gases entering the atmosphere and the rate of build-up of those gases will affect THC overturning.

Figure 1. Equilibrium results of the Simple Climate Demonstrator (SCD) model under different forcing scenarios

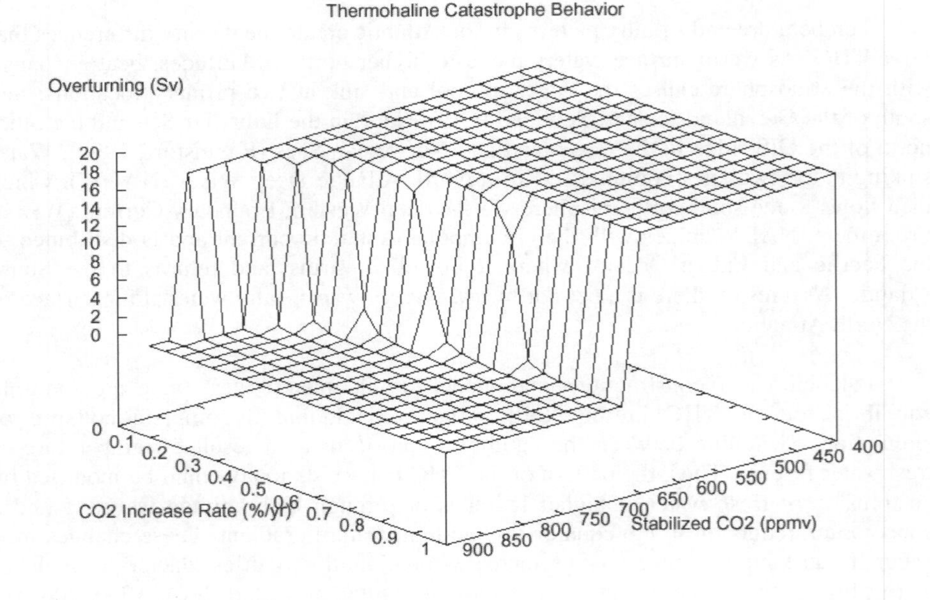

Source: Schneider and Thompson (2000).

Notes: THC overturning in Sverdrups (1 Sv = 1 million m^3/s) is shown on the vertical axis as a function of the rate of carbon dioxide (CO_2) increase in the atmosphere and the stabilization concentration. Higher stabilization levels and more rapid rates of carbon dioxide increase make a THC collapse (abrupt change from "normal" – 20 Sv – to zero Sv) more likely.

If warming reduces the ability of surface water to sink at high latitudes, the inflow of warm water from the south will be disrupted and will likely slow. This type of slowdown would cause local cooling, which would re-energize the local sinking and serve as a stabilizing *negative* feedback on the slowdown. On the other hand, the initial slowdown of the strength of the Gulf Stream would reduce the flow of salty subtropical water to the higher latitudes of the North Atlantic. This would act as a destabilizing *positive* feedback on the process by further decreasing the salinity of the North Atlantic surface water and reducing its density, continuing the inhibition of local sinking. The rate at which the warming forcing is applied to the coupled system could determine which of these opposing feedbacks dominates, and subsequently, whether a THC collapse occurs.

Some coupled models of the atmosphere and oceans (e.g., Yin *et al.*, 2004) do not produce a THC collapse from global warming, owing to some still-not-identified feedback processes in the models used. That is why it is very difficult to assign any confident probabilities to the occurrence of a THC collapse, but it is not possible to rule it out at a high level of confidence either.

Recent research efforts have attempted to incorporate THC collapse into IAMs of climate change policy. William Nordhaus' (1994a) Dynamic Integrated Climate and Economy (DICE) model is one example (though the model's damage function cannot effectively deal with nonlinear change). It is a simple optimal growth model that, when given a set of explicit value judgments and assumptions, generates an optimal future forecast for a number of economic and environmental variables. It does this through maximizing discounted utility (satisfaction from consumption), by balancing the costs to the economy of GHG emissions abatement (a loss in a portion of GDP caused by higher carbon energy prices) against the costs of damages from the build-up of atmospheric GHG concentrations. This build-up affects the climate, which in turn causes "climate damage," a reduction in GDP determined by the rise in globally averaged surface temperature due to GHG emissions. In some sectors and regions, the resulting climate damages could be negative – i.e., benefits – but DICE aggregates across all sectors and regions (see, for example, the discussions in Chapters 1 and 19 of IPCC, 2001b) and therefore assumes that this aggregate measure of damage is always a positive cost.

Mastrandrea and Schneider (2001) have developed a modified version of Nordhaus' DICE model called E-DICE, which contains an enhanced damage function that reflects the higher likely damages that would result if/when abrupt climate changes occur.[58] When climate changes are smooth and relatively predictable, the foresight afforded increases the capacity of society to adapt. Damages will be lower under this scenario than for very rapid or unanticipated changes such as "surprises" like a THC collapse. When dealing with the abrupt non-linear behaviour of the SCD model (and other "surprise" scenarios), the E-DICE model produces a result that is qualitatively

---

58 The DICE model couples a simple globally- and seasonally-average two-box climate model (introduced by Schneider and Thompson, 1981) with an economic model of similar complexity. It makes no attempt to incorporate non-linear behaviors found in more complex GCMs or observed in nature; rather, it is capable only of smooth temperature changes given a smooth CO2 increase scenario. Mastrandrea and Schneider (2001) modify the DICE model by adding a sub-model that accounts for abrupt non-linear climate changes. They add an exponent (ε) to the DICE climate damage function, which allows for a less linear, more hockey stick-shaped function. The E-DICE model determines values for ε by exchanging information with an enhanced climate model – the SCD – that simulates THC.

different from DICE, which lacks internal abrupt non-linear dynamics. As shown in Figure 1, a THC collapse is obtained for rapid and large CO_2 increases in the SCD model. An "optimal" solution of conventional DICE can produce an emissions profile that triggers such a collapse. However, this abrupt non-linear event can be prevented when the damage function in DICE is modified (as in E-DICE) to account for enhanced damages created by this THC collapse and THC behaviour is incorporated into the coupled climate-economy model.

In an optimization run of E-DICE, the coupled system contains feedback mechanisms that allow the profile of carbon taxes to increase sufficiently in response to the enhanced damages so as to lower emissions sufficiently to prevent the THC collapse. The enhanced carbon tax actually "works" to lower emissions and thus avoid future damages. Keller *et al.* (2004) support these results, finding that significantly reducing carbon dioxide emissions to prevent or delay potential damages from an uncertain and irreversible future climate change, such as a THC collapse, may be achievable and cost-effective.

The amount of near-term mitigation the DICE and E-DICE models "recommend" to reduce future damages is critically dependent on discounting (Figure 2). Discounting plays a crucial role in the economics of climate change when optimisation is the objective, yet it is a highly uncertain parameter. It is a method of aggregating costs and benefits over a long time horizon by summing across future time periods net costs (or benefits) that have been multiplied by a discount rate, typically greater than zero. If the discount rate equals zero, then each time period is valued equally (case of infinite patience). If the discount rate is infinite, then only the current period is valued (case of extreme myopia). The discount rate chosen in assessment models is critical, since abatement costs typically will be incurred in the relatively near term, but the brunt of climate damages will be realized primarily in the long term. Thus, if the future is sufficiently discounted, present abatement costs, by construction, will outweigh discounted future climate damages. The reason is, of course, that discount rates will eventually reduce future damage costs to negligible present values. (See Schneider and Kuntz Duriseti, 2002, for more information and citations to primary literature.)

In the case of THC, for low pure rate or time preference (PRTP)[59] values (less than 1.8%, in one formulation), the present value of future damages creates a carbon tax large enough to keep emissions below the trigger level for the abrupt non-linear collapse

[59] The pure rate of time preference (PRTP) is a factor proportional to the discount rate. Time preference expresses an individual's or group's preference on the timing of costs and benefits of an action (or lack thereof). In general, there is a premium placed on present versus future benefits; people typically choose to reap benefits sooner and incur costs later. The PRTP is a measure of the strength of this preference and is proportional to the discount rate. The discount rate is about double the PRTP. The higher the PRTP, the more the present is valued over the future (and, in the case of climate change, the less likely we are to spend money to reduce CO2 and other greenhouse gas emissions now, given that the benefits won't be primarily felt until the distant future) and vice versa.

of the THC a century later. A higher PRTP – and therefore discount rate – sufficiently reduces the present value of even catastrophic long-term damages so that abrupt non-linear THC collapse becomes an emergent property of the coupled socio-natural system. The discount rate is therefore the parameter that most influences the 22nd century behaviour of the modelled climate.

Figure 2. Cliff diagram of equilibrium THC overturning varying PRTP and climate sensitivity

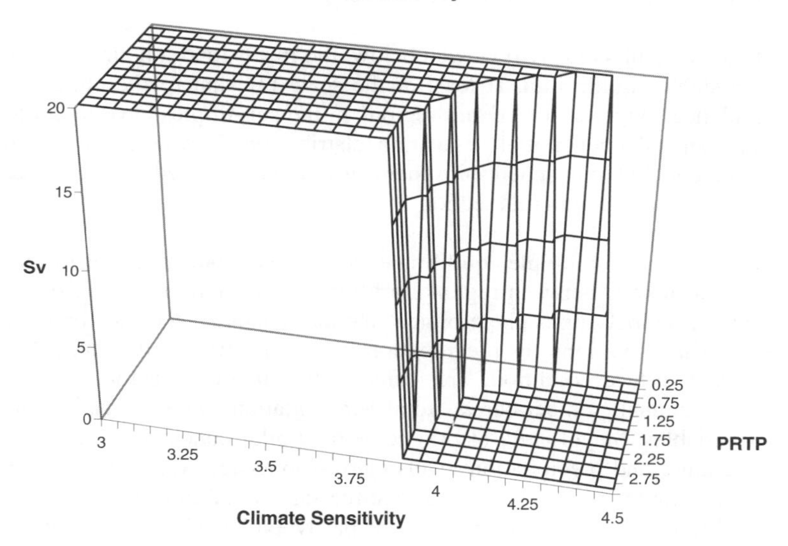

Source: Mastrandrea and Schneider (2001).

Notes: Figure shows that as PRTP increases, the climate sensitivity (how much global average temperature rises for doubling of CO_2 from present levels) threshold at which collapse of the THC occurs decreases. This is because higher discount factors imply lower present value for far future climate damages, which leads to smaller control rates on emissions. Lower control rates means more cumulative emissions and thus a greater risk of climate change sufficient to trigger abrupt non-linear responses like THC collapse in the future.

Although these highly aggregated models are not intended to provide high-confidence quantitative projections of coupled socio-natural system behaviours, we believe that the bulk of integrated assessment models used to date for climate policy analysis, few of which include any such abrupt non-linear processes, will not be able to alert the policymaking community to the importance of and risks associated with abrupt non-linear behaviours. At the very least, the ranges of estimates of future climate damages should be expanded beyond that suggested in conventional analytic tools to account for possible non-linear behaviours (e.g., Moss and Schneider, 2000).

2.2 *Vegetation cover and climate dynamics*

While several regions of the world appear to exhibit multiple stable equilibria, with the equilibrium realized depending on the initial conditions of the coupled system, other regions appear to have a single stable equilibrium, at least under current conditions. This is relevant for policymakers, since if a region has exhibited multiple equilibria in the past, it could do so again in the future if forced to change by more recent disturbances like overgrazing or greenhouse gas build-ups.

Based on the results briefly reviewed below, the forest-tundra boundary appears to have a single stable equilibrium, at least at the scale relevant to the climate system. However, evidence suggests that certain regions in the sub-tropics have multiple stable equilibria that depend upon initial vegetation distribution. The potential for multiple equilibria in the coupled atmosphere-biosphere system has received increasing attention in recent years, as evidenced in this section.

Several areas where multiple equilibria exist in the coupled atmosphere-biosphere system suggest a linkage between regional aridity and vegetation cover. For example, using a coupled global atmosphere-biome model, Claussen (1998) produces two separate equilibrium solutions for precipitation in North Africa and Central East Asia when initial land-surface conditions are different but all other factors are the same. Using average annual temperatures, total precipitation, and elevation as their independent variables, Siegel *et al.* (1995) come to similar conclusions. They find that varying temperature and precipitation levels causes noticeable variations in ecosystem areas and carbon storage in vegetation, but that the same conditions in different regions often support different ecosystems. This may be partly explained by the fact that in their modelling, Siegel *et al.* observed that mixed forest, semi-desert, and tundra ecosystems can dominate a region over a surprisingly wide range of temperature and rainfall levels.

A related study by Kleidon *et al.* (2000) compares simulations with vegetation initialized as either forest or desert. The comparison between these "green" and "desert" worlds again illustrates that some regions are sensitive to the initial vegetation while other regions retain just one set of vegetation and precipitation conditions. In particular, Kleidon *et al.* report that regions of Africa, South Asia, and Australia produce different stable atmosphere-biosphere equilibria, depending on whether the initialized vegetation is forest or desert. This means that if the system is disturbed, it may not return to its original equilibrium, and thus a large enough disturbance could cause irreversible changes. In contrast, simulations like these produce a single equilibrium for both the "green" and "desert" worlds in other regions, meaning that after a period of disturbance, they could be restored to their original conditions.

The Amazon is another candidate for multiple equilibria in the coupled climate-vegetation system. Kleidon and Heimann (1999) study interactions among vegetation type, rooting depth, and climate in the Amazon basin. During the dry season, the water transpired by plants contributes substantially to atmospheric moisture, altering the partitioning of net radiation between sensible and latent heat fluxes and increasing relative humidity. In their simulation, Kleidon and Heimann find that vegetation type determines rooting depth, which partly determines the availability of soil moisture for evapo-transpiration. Comparison between simulations that differed in rooting depth revealed that the dry season is warmer and lasts longer when vegetation with a shallower rooting depth is present than when vegetation with deeper roots is initialized.

Historical evidence suggests that two equilibria in the coupled vegetation and climate system may exist for the Sahel region of West Africa (10°N-17.5°N, 15°W-15°E) (Wang and Eltahir, 2000b), where an extended period of drought has persisted since the 1960s (Wang and Eltahir, 2000a). Experiments (Wang and Eltahir, 2000a) suggest that this drought represents a change from a self-sustaining wet climate equilibrium to another self-sustaining dry equilibrium. Initially, a SST anomaly altered precipitation in the Sahel. As a consequence, the grassland vegetation shifted to that of a drier equilibrium state. Therefore, the combination of natural climate variability (i.e., SST anomaly) and the resulting change in land cover were both necessary to alter the availability of moisture for the atmosphere in the longer term, and to determine the equilibrium state (Wang and Eltahir, 2000b).

Wang and Eltahir (2000b) detect that vegetation in their model is partly responsible for the low-frequency variability in the atmosphere-biosphere system characteristic of the Sahel and for the transition between equilibrium states. Rooting depth within the perennial grassland determines which of the equilibria the modelled system occupies at a given time. In the model, moist (i.e., favorable) growing seasons facilitate greater root growth of perennial grasses while dry (unfavorable) growing seasons lead to shallow root growth. Shallow roots lead to less evapotranspiration and less atmospheric moisture, causing a positive feedback (Wang and Eltahir, 2000b).

Other modelling studies suggest that monsoon circulation in West Africa is sensitive to deforestation, another example of coupling. However, the sensitivity of the monsoon circulation to changes in land cover depends critically on the location of the change in vegetation. Desertification along the Saharan border has little impact on the modelled monsoon circulation, while deforestation along the southern coast of West Africa results in a complete collapse of the modelled monsoon circulation, with a corresponding reduction in regional rainfall (Zheng and Eltahir, 1998). This illustrates that relatively small areas of land cover might determine the equilibrium state of the atmosphere-biosphere system of an entire region. Zickfeld (her thesis, 2004) has hypothesized that multiple equilibria may also exist in the Indian monsoon region as a result of greenhouse gas forcing.

Similar hypotheses for multiple equilibria triggered by vegetation feedbacks have been put forward for the boreal forest/tundra boundary, but as shown by Levis *et al.* (1999), these boundaries appear to exhibit a single equilibrium. In one simulation, Levis *et al.* initialize their model with the current boreal forest/tundra boundary, and in a second simulation, they initialize the model with boreal forest extended to the Arctic coast (assuming current climate conditions). In both simulations, the atmosphere-biosphere system converges to a single state, suggesting that for current conditions, there is a single stable equilibrium in the region – at least for the processes incorporated in this model and at the scale of the continent. The simulations performed by Claussen (1998) and Kleidon *et al.* (2000) are not specifically designed to test the forest-tundra boundary, but their results are consistent with a single stable equilibrium at that boundary.

The Sahara also appears to exhibit a single equilibrium. Six thousand years before present (around 4,000 B.C.), the Sahara was heavily vegetated, but over the following 1,000-2,000 years, an abrupt change in vegetation and climate occurred (Claussen *et al.*, 1999). In model simulations, Ganopolski *et al.* (1998) find that an atmosphere-ocean-vegetation coupling is better able to represent the climate of the Sahara, with the addition of vegetation increasing precipitation substantially, providing evidence of a strong positive feedback between climate and vegetation distribution. It is thought that as orbital forcing caused a slow and steady decline in summer radiation, the Sahara abruptly underwent desertification as a consequence of interactions between the orbital changes and the atmospheric and biospheric sub-systems. (Claussen *et al.*, 1999 have supported this idea through their modelling efforts.) These results suggest the Sahara of the mid-Holocene may have been prone to abrupt and irreversible changes but is currently in a single, quite stable equilibrium condition.

It must be kept in mind that results from models such as these depend on how the model aggregates processes that occur at smaller scales than are explicit in the simulation; local variations in soils, fire regimes, slope, elevation, and other characteristics may all be neglected. The extent to which it is necessary to explicitly account for smaller-scale processes, or to which they might influence conclusions about stability, remains a major debate point in all simulations that, for practical necessity, must parameterize the effects of processes occurring on small time and space scales. This suggests that using a hierarchy of models of varying complexity (and observations to test them) is the approach most likely to determine the implications of the degree of aggregation in various models and indicate whether a particular region is switching between multiple equilibria as opposed to suffering the effects of an incomplete recovery from disturbance. Most of the modelling studies briefly summarized above are suggestive of a potentially critical role that might be played by interactions between land cover and climate, but these are pioneering efforts, and a great deal more work will be needed to obtain more highly confident conclusions.

In addition, it is important to recognize that this review of multiple equilibria in the coupled climate-vegetation system is focused at the broadest scales of ecosystem structure and function as they relate to climate (e.g., albedo, transpiration, and roughness). At other biological scales (e.g., genetic, species, and population), different processes and characteristics may have multiple equilibria. For example, species or population extinction and loss of genetic diversity may occur without transitions in the climate system. Such changes clearly constitute different equilibria (e.g., with and without a particular species) that may be profoundly important biologically, but these different equilibria are not relevant at the scale of the climate system.

The key point of all these detailed examples is that even the most comprehensive coupled-system models are likely to produce unanticipated results when forced to change very rapidly by external disturbances like changes in land use, CO_2

concentrations, and aerosol levels. Some consequences could be harmful, others beneficial. Whether to trust to luck that humanity will only experience the beneficial ones or to hedge now under the supposition that harmful ones could occur as well is the risk-management problem decision-makers facing the climate issue will have to consider. In order to develop a climate policy that will lower the risk of climate catastrophes, policymakers should take into consideration rates of change in radiative forcing and possible consequences of rapid forcing, including very uncertain but highly consequential events like a THC collapse or multiple vegetation-precipitation equilibria.

3. Abrupt Events, Dangerous Anthropogenic Climate Change, and the Benefits of Climate Policies

In addition to making policymakers aware of the possibility of surprise climatic events, such as those characterized by multiple equilibria and discussed above, it is necessary for climate experts to alert them to the need to consider adopting and fully implementing policies that may prevent "dangerous" climate change.

3.1 What is "dangerous" climate change?

First, we must explore the definition of "dangerous" climate change. Article 2 of the United Nations Framework Convention on Climate Change (UNFCCC), created in 1992 and since signed by 193 nations, calls for "stabilization of greenhouse gas concentrations in the atmosphere at a level that would prevent dangerous anthropogenic interference [DAI] with the climate system" (UNFCCC, 1992). While it seems that some of the impacts of climate change discussed thus far suggest that dangerous levels of climate change may occur, the UNFCCC never actually defined what it meant by "dangerous".

In relation to climate change in general, what we *do* know is that "dangerous" is a concept that cannot simply be inferred from a set of observations or calculated by a model. In fact, it is a common view of most natural and social scientists that it is not the direct role of the scientific community to define what "dangerous" means. Rather, it is ultimately a political question because it depends on *value judgments* about the relative salience of various impacts and how to face climate change-related risks and form norms for defining what is "acceptable" (Schneider and Azar, 2001; Mastrandrea and Schneider, 2004).

This is not to say that scientists have no role whatsoever in the definition of "dangerous". Although scientists are not responsible for interpreting it themselves, they must help policymakers evaluate what "dangerous" climate change entails by laying out the elements of **risk**, which is classically defined as ***probability x consequence***. They should also help decision-makers by identifying thresholds and possible surprise events, as well as estimates of how long it might take to resolve many of the remaining uncertainties that plague climate assessments. They should make suggestions on how to avoid "surprises" and other "dangerous" climate changes, or at least limit their effects, through policies designed to bring about abatement and adaptation.

3.2 Dangerous anthropogenic interference and abrupt non-linear events

How does DAI relate to abrupt non-linear climate change, irreversibility, and surprise? Mastrandrea and Schneider (2004) suggest that DAI be defined in terms of the consequences (impacts) of climate change. The potential range of climate change impacts has been represented graphically in Figure 3. Each column in the figure represents a "reason for concern" about climate change in this century based on dozens of IPCC lead authors' examination of climate impacts literature, and thus represents a current "best estimate" of "dangerous" climate changes. The threshold temperature above which each column turns red (darker) increases from left to right. This figure, also known as the "burning embers diagram", shows that the most potentially dangerous climate change impacts (the red (darker) colours on the figure) typically occur after only a few degrees Celsius of warming.

When considering where "abrupt" climate change falls in Figure 3, it must be remembered that the definition of that term is open to interpretation and oftentimes depends on the perspective being used. A climate change event could take place rather abruptly in an absolute sense but not in a relative sense, and vice versa, or the phenomenon itself could be considered to have occurred abruptly, even though the time frame over which it was activated was very long (as is supposed in the case of THC). In the figure above, most abrupt events would likely fall into columns I, II, and V. In any case, it is likely that as global average surface temperatures increase, the likelihood of dangerous and abrupt events will increase, as indicated by the red (dark) at the top of the bars.

In defining their metric for DAI, Mastrandrea and Schneider estimate a cumulative density function (CDF) based on the burning embers diagram by assigning data points at each transition-to-red threshold and assuming that the probability of "dangerous" change increases cumulatively at each threshold temperature by a quintile, as shown by the thick black line in our Figure 3. This can be used as a starting point for analyzing "dangerous" climate change.

3.3 What is the probability of DAI and how can one assess the benefits of climate policies from it?

Mastrandrea and Schneider (2004) use 2.85°C as their median threshold for "dangerous" climate change, based on IPCC Working Group II's forecast that after "a few degrees," many serious climate change impacts could be anticipated. However, 2.85°C may still be conservative, since the IPCC also noted that some "unique and valuable" systems could be lost at warmings any higher than 1-1.5 °C.

Figure 3. Reasons for concern about climate change impacts

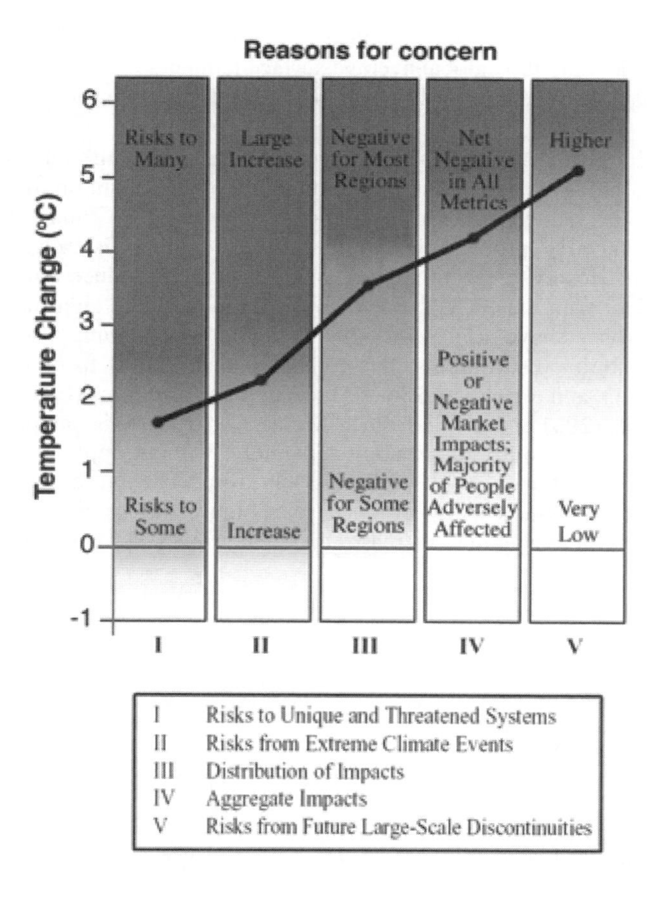

Source: Mastrandrea and Schneider (2004).

Notes: An adaptation of the IPCC "Reasons for Concern" figure (originally from IPCC 2001b, Chapter 19), with Mastrandrea and Schneider's (2004) cumulative density function (CDF) for "dangerous anthropogenic interference" (DAI) – the thick black line. They use the transition-to-red thresholds for each "reason for concern" to construct a CDF for DAI, assuming the probability of DAI is close to zero at 0°C and increases by a quintile as each threshold is reached.

Mastrandrea and Schneider apply this median 2.85°C threshold to three key parameters – climate sensitivity, climate damages, and the discount rate – all of which carry high degrees of uncertainty and are crucial factors in determining the policy implications of global climate change. To perform these calculations, they use Nordhaus' (1994a) DICE model, discussed above (in the section on THC), because it is a relatively simple and transparent IAM, despite its limitations.[60] Using an IAM allows for exploration of the impacts of a wide range of mitigation levels on the potential for exceeding a policy-relevant threshold such as DAI. Mastrandrea and Schneider focus on two types of model output: i) global average surface temperature change in 2100, which is used to evaluate the potential for DAI; and ii) "optimal" carbon taxes.

They begin with climate sensitivity, typically defined as the amount that global average temperature is expected to rise for a doubling of CO_2 from pre-industrial levels. The IPCC estimates that climate sensitivity ranges between 1.5 °C and 4.5 °C, but it has not assigned subjective probabilities to the values within this range, making risk analysis difficult. However, recent studies, many of which produce climate sensitivity distributions wider than the IPCC's 1.5 °C to 4.5 °C range, with significant probability of climate sensitivity above 4.5 °C, are now available. Mastrandrea and Schneider use three such probability distributions: the combined distribution from Andronova and Schlesinger (2001), and the expert prior (F Exp) and uniform prior (F Uni) distributions from Forest *et al.* (2001). They perform a Monte Carlo analysis sampling from each climate sensitivity probability distribution separately, without applying any mitigation policy, so that all variation in results will be solely from variation in climate sensitivity. The probability distributions they produce (our Figure 4a) show the percentage of outcomes resulting in temperature increases above their 2.85 °C "dangerous" threshold.

60 As mentioned previously (see Section 2.1), the DICE model has a somewhat weak damage function that cannot account for or prevent abrupt changes. In response to criticism on his damage function, Nordhaus performed a decision analytic survey (see Nordhaus, 1994b) to obtain experts' estimates of economic damages from climate change scenarios of varying severity. Using the survey responses published by Nordhaus (1994b), Roughgarden and Schneider (1999) created probability distributions for climate damages, which provided a range of damage functions that were both stronger and weaker than the original DICE function. To partially compensate for the DICE model's damage function, Mastrandrea and Schneider (2004) sample from Roughgarden and Schneider's (1999) probability distributions to produce a range of quadratic-form damage functions that they then use to run their Monte Carlo analyses, discussed next. Although many other damage functions could be cited, the Nordhaus survey is well-established in the literature, and allows us to demonstrate transparently and quantitatively the probabilistic framework we believe is needed to analyze the "dangerous anthropogenic interference" (DAI) issue. For extensive discussion of other weaknesses of the DICE model (i.e., the discount rate, sensitivity to changing structural assumptions) and references to other critiques, please see the supporting online material (SOM) accompanying Mastrandrea and Schneider (2004).

Figure 4. Climate sensitivity-only and joint (climate sensitivity and climate damages) Monte Carlo analyses

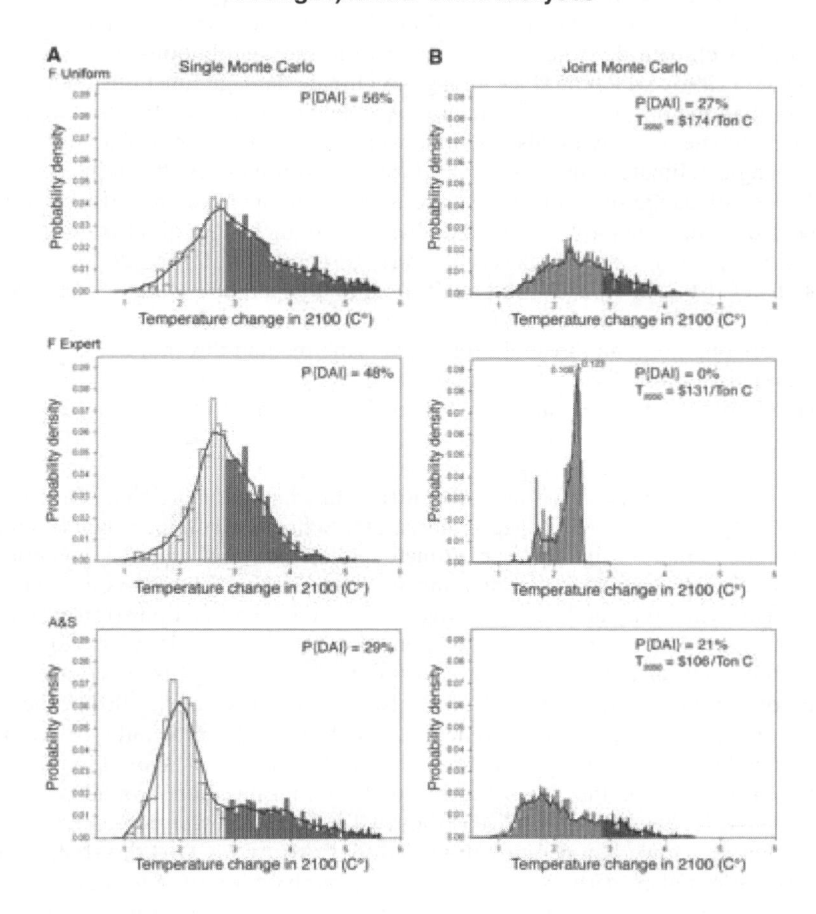

Source: Mastrandrea and Schneider (2004).

Notes: Panel a) displays probability distributions for each climate sensitivity distribution for the climate sensitivity-only Monte Carlo analyses with zero damages. Panel b) displays probability distributions for the joint (climate sensitivity and climate damage) Monte Carlo analyses. All distributions indicate a 3-bin running mean and the percentage of outcomes above our median threshold of 2.85°C for "dangerous" climate change (P{"DAI"}), and the joint distributions display carbon taxes calculated in 2050 (T_{2050}) by the DICE model using the median climate sensitivity from each climate sensitivity distribution and the median climate damage function for the joint Monte Carlo cases. Comparing the joint cases with climate policy controls, b), to the climate sensitivity-only cases with negligible climate policy controls, a), high carbon taxes reduce the potential (significantly in two out of three cases) for DAI. (However, this case uses a PRTP of 0%, implying a discount rate of about 1%. With a 3% PRTP—a discount rate of about 6%—this carbon tax is an order of magnitude less, and the reduction in DAI is on the order of 10%. See the supplementary on-line materials of Mastrandrea and Schneider, 2004 for a full discussion.)

Mastrandrea and Schneider's next simulation is a joint Monte Carlo analysis looking at temperature increase in 2100 with climate policy, varying both climate sensitivity *and* the climate damage function, their second parameter (shown in our Figure 4b). For climate damages, they sample from the distributions of Roughgarden and Schneider (1999), which produce a range of climate damage functions both stronger and weaker than the original DICE function. As shown, aside from the Andronova and Schlesinger climate sensitivity distribution, which gives a lower probability of DAI under the single (climate sensitivity-only) Monte Carlo analysis, the joint runs show lower chances of dangerous climate change as a result of the more stringent climate policy controls generated by the model due to the inclusion of climate damages. Time-varying median carbon taxes are over USD 50/Ton C by 2010, and over USD 100/Ton C by 2050 in each joint analysis. Low temperature increases and reduced probability of "DAI" are achieved if carbon taxes are high, but because this analysis only considers one possible threshold for "DAI" (the median threshold of 2.85 °C) and assumes a relatively low discount rate (about 1%), these results do not fully describe the relationship between climate policy controls and the potential for "dangerous" climate change.

Because the analysis above only considers Mastrandrea and Schneider's median threshold (DAI[50%]) of 2.85°C, Mastrandrea and Schneider continue their attempt to characterize the relationship between climate policy controls and the potential for "dangerous" climate change by calculating a series of single Monte Carlo analyses varying climate sensitivity and using a *range* of fixed damage functions (rather than just the median case). For each damage function, they perform a Monte Carlo analysis sampling from each of the three climate sensitivity distributions discussed above (one from Andronova and Schlesinger, 2001; and two from Forest *et al.*, 2001). They then average the results for each damage function, which gives the probability of DAI at a given 2050 carbon tax under the assumptions described above, as shown in our Figure 5, below. Each band in the figure corresponds to optimisation around a different percentile range for the "dangerous" threshold CDF, with a lower percentile from the CDF representing a lower temperature threshold for DAI. At any DAI threshold, climate policy "works:" higher carbon taxes lower the probability of considerable future temperature increase, and reduce the probability of DAI. For example, if climate sensitivity turns out to be on the high end and DAI occurs at a relatively low temperature like 1.476°C (DAI[10%]), then there is nearly a 100% chance that DAI will occur in the absence of carbon taxes and about an 80% chance it will occur if carbon taxes are USD 400/ton, the top end of Mastrandrea and Schneider's range. If we inspect the median (DAI[50%]) threshold for DAI (the thicker black line in Figure 5), we see that a carbon tax by 2050 of USD 150-USD 200/Ton C will reduce the probability of "DAI" to nearly zero, from 45% without climate policy controls (for a 0% PRTP).

Figure 5. Carbon taxes in 2050 and the probability of DAI

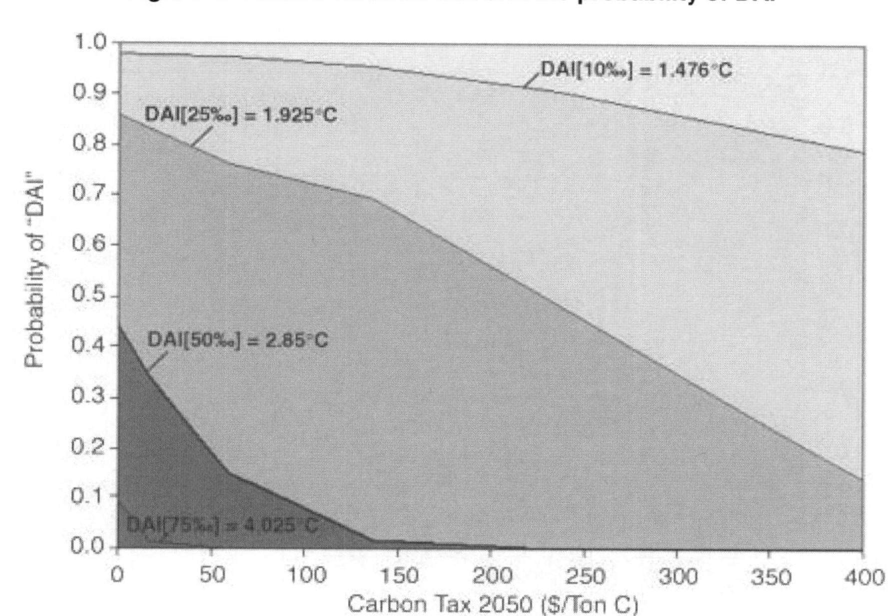

Source: Mastrandrea and Schneider (2004).

Notes: Each band represents a different percentile range for the DAI threshold CDF—a lower percentile from the CDF representing a lower temperature threshold for DAI. At any threshold, climate policy controls significantly reduce the probability of DAI. At the median DAI threshold of 2.85°C (the thicker black line above), a 2050 carbon tax of >USD 150/Ton C is necessary to virtually eliminate the probability of DAI.

Lastly, Mastrandrea and Schneider run Monte Carlo analyses varying climate sensitivity at different values for the PRTP, which illustrates the relationship between the discount rate and the probability of DAI at different temperature threshold values, as shown in our Figure 6, below. As expected, increasing the discount rate shifts the probability distribution of future temperature increase upwards; a lower level of climate policy controls becomes "optimal" and thus increases the probability of DAI. At our median threshold of 2.85°C for DAI (the thicker black line in Figure 6), the probability of DAI rises from near zero with a 0% PRTP to 30% with a 3% PRTP. A PRTP of 3% is the value originally specified in Nordhaus' DICE model. At PRTP values greater than 1%, the "optimal" outcome becomes increasingly insensitive to variation in future climate damages driven by variation in climate sensitivity.

Figure 6. PRTP and the probability of DAI

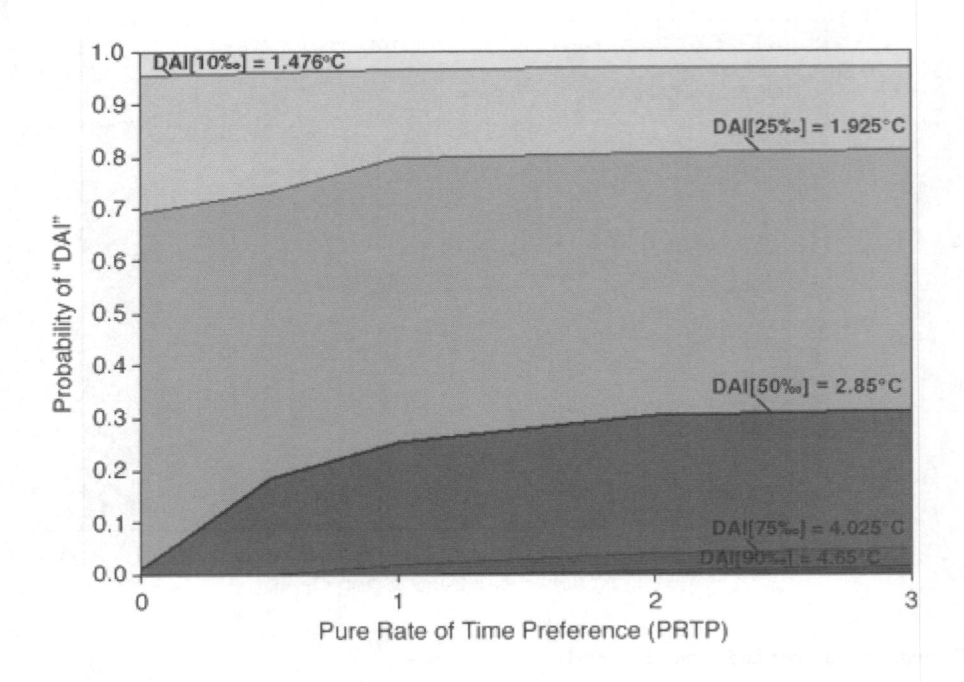

Source: Mastrandrea and Schneider (2004).

Notes: Increasing the PRTP (and hence the discount rate) reduces the present value of future climate damages and increases the probability of "DAI." At our median threshold of 2.85°C for "DAI" (thicker black line above), the probability of "DAI" rises from near zero with a 0% PRTP to 30% with a 3% PRTP, as originally specified in the DICE model.

While Mastrandrea and Schneider's results using the DICE model do not provide us with confident quantitative answers, they still demonstrate three very important issues: (1) that DAI can vary significantly, depending on its definition; (2) that parameter uncertainty will be critical for all future climate projections, and most importantly for this volume on the benefits of climate policies; (3) that climate policy controls (i.e., "optimal" carbon taxes) can significantly reduce the probability of dangerous anthropogenic interference. This last finding has considerable implications for introducing climate information to policymakers. We agree with Mastrandrea and Schneider that presenting climate modeling results and arguing for the benefits of climate policy should be framed for decision-makers in terms of the potential for climate policy to reduce the likelihood of exceeding a DAI threshold. While Mastrandrea and Schneider's quantitative results should not be taken literally, the framework and methods for assessing DAI that they use should be taken seriously, as it is an effective method for conceptualizing climate change policy decisions.

4. The Policy Challenge

4.1 The "no regrets" approach to climate change policy

To date, many policymakers have remained in the "do nothing"/"wait and see"/"perform more research" camps when it comes to climate change policy, though some have actively supported "no regrets" policies that capitalize on existing "market failures" (like inefficient energy systems, which, if replaced, would actually pay for themselves within a short period). In a few countries, officials have already implemented policies and measures to directly internalize the "externality" of carbon emissions. In Norway, for example, a tax averaging USD 21/ton across all economic sectors and fuels is levied on carbon emissions. However, the overall lack of policies and measures, particularly from the bulk of the world's biggest emitters, is of concern when considering that delay is likely to breed further delay and thus frustrate the implementation of policies (Schneider and Azar, 2001, p. 119).

4.2 Hedging

This hesitancy to implement policy seems curious given that many risk management decisions in fields unrelated to climate change in both the public and private sectors are based on strategic hedging against low-probability, high-cost events (IPCC, 2001b, p. 96). On topics other than climate change, like war, for example, policymakers and others have been known to prefer invoking the "precautionary principle" over a "wait-and-see" approach, but this stance has rarely carried over to policies to deal with risks of climate change.

4.3 Courses of policy action - abatement

Thus far, most decision-makers who have taken policy actions on climate change have focused on abatement, with the Kyoto Protocol being the most obvious example, although it has yet to go into effect. Under the Protocol, the developed (Annex I) signatory countries agreed to reduce their overall emissions by 5.2% below 1990 levels between 2008 and 2012. Developing countries were not assigned emissions targets or timetables. This is only a starting point for international-level climate policy; in order to slow down the rate of climatic changes, policymakers will need to contemplate much stronger emissions reductions than envisioned in the first commitment period of the Kyoto protocol, and eventually, all major emitters, developed and developing countries alike, will need to participate. If developing nations insist on full "catch-up" rights in per capita emissions, this atmospheric burden, when multiplied by the population of all the developing countries, which is about four times larger than that of the developed world, could lead to CO_2 tripling or more beyond the 21st century, causing warming that many would consider to be "dangerous."

It will undoubtedly be difficult economically, politically, and ethically to fashion fair, affordable, and politically-acceptable technology transfer and "leapfrogging"[61] schemes and abatement activities in all sectors and regions. One prerequisite is a spirit of international co-operation and recognition of the common destiny of the planet. It is the hope of the authors that the efforts over the past decade to fashion a collaborative international negotiation process based on cost-effectiveness and fairness can be extended into new and more climatically "safe" agreements in the decades ahead that significantly reduce the probability of DAI. In any case, decisions on the level of abatement necessary to (attempt to) prevent dangerous climate change and other possible policy actions will be aided by access to better probabilistic information on the risks of "dangerous climatic interference", as outlined in Mastrandrea and Schneider (2004) and discussed above.

4.4 Is it too expensive to mitigate CO_2 emissions?

Many policymakers seem unconvinced to implement abatement policy on the basis of risk avoidance considerations, claiming that economic costs could be severe (see, e.g., Linden, 1996). It may, therefore, be worth pointing out that substantial reductions in carbon emissions and several-fold increases in economic welfare *are* compatible goals. Using a simple model, Azar and Schneider (2002) estimated the present value (discounted to 1990, expressed in 1990 USD, and assuming a discount rate of 5%/year) of the global costs to stabilize atmospheric CO_2 at 350 ppm, 450ppm, and 550ppm over the next 100 years at 18 trillion USD, 5 trillion USD, and 2 trillion USD, respectively. The World Bank estimates that worldwide GDP in 2002 was about 32 trillion USD[62], which makes spending 18 trillion USD, or 56% of 2002 GDP, to stabilize CO_2 seem unthinkable. However, what is often forgotten is that a CO_2 stabilization cost of 10 to 20 trillion USD represents the *present value* of spending that would be done *over the entire period of the next 100 years*. Most recent economic models calculating CO_2 abatement costs assume that growth in population and the productivity of labour will drive an annual growth rate of about 2% for the worldwide economy, which amounts to a GDP-doubling time of about thirty-five years, meaning global GDP will likely reach about 240 trillion USD per annum by 2100. In that light, a present value of 20 trillion USD over the *entire century* seems relatively low-cost; in fact, if conventional economic models are remotely accurate in their 2% per year growth rate projection, then even if we were to spend those trillions of dollars on CO_2 stabilization over the next 100 years, global income levels around 2100 (some 500% higher per capita than today) would be delayed less than a decade (Schneider, 1993),

---

61 The strategy of encouraging the developing world not to mimic the Victorian industrial revolution on its road to development by increasing coal-burning and the use of internal combustion engines, but rather to jump over these outdated technologies and pursue more efficient, high-technology solutions has been called "technology leapfrogging".

62 GDP data can be found on the World Bank's website at: http://www.worldbank.org/data/databytopic/GDP.pdf.

and probably only a couple years (Azar and Schneider, 2002), behind the no-abatement-spending scenario, as Figure 7 illustrates.

Figure 7. Global income trajectories under business as usual (BAU) and in the case of stabilizing the atmosphere at 350ppm, 450ppm, and 550ppm

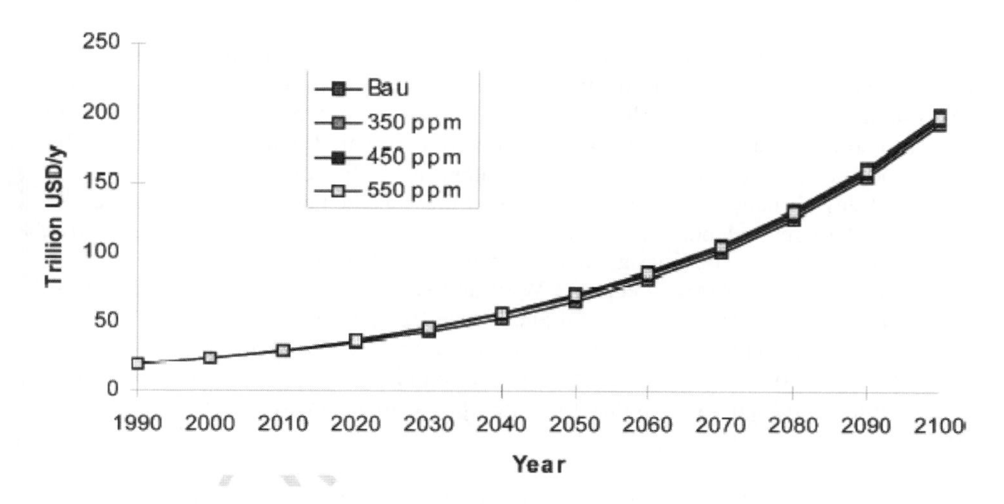

Source: Azar and Schneider (2002).

The more complex question, therefore, is not *whether* to abate—that seems essential to avoid first-decimal-point odds of "dangerous" climatic change (Mastrandrea and Schneider, 2004)—but how to fashion cost-effective incentives and schemes to share the costs fairly among nations and groups within nations, not all of which contribute equally to the dumping of gaseous wastes into the atmosphere or share equally in their adaptive capacities and vulnerabilities to climatic changes.

4.5 Courses of policy action - adaptation

Unlike abatement, adaptation is a *response to* rather than a *slowing of* global warming. The IPCC has identified two basic types of adaptation, autonomous and planned. An autonomous adaptation is a non-policy-driven reactive response to a climatic stimulus that occurs after the initial impacts of climate change are felt (IPCC, 2001b, p. 88). Its counterpart, planned adaptation, comes in two forms, as described by Schneider and Thompson (1985): passive and anticipatory. Passive adaptation, which could involve an action like buying additional water rights to offset the impacts of a drying climate after the climate has already begun to dry and the effects have been felt, is also essentially reactive in nature. Schneider, Easterling, and Mearns (2000) use the example of farming to illustrate that passive adaptation cannot be assumed to occur instantly. They question whether those in agriculture will invest heavily (e.g., change crops, update irrigation) in order to adapt their practices before demonstrable climate change materializes. While some have argued that farmers *do* adapt to changing market technology and climatic conditions, others have contended that this optimism neglects such real world problems as people's resistance to trying unfamiliar practices, problems with new technologies, unexpected pest outbreaks,[63] and the high degree of natural variability of weather (Schneider, Easterling, and Mearns, 2000, p. 206-7).

Any passive adaptation that does occur will almost certainly not be smooth or instantaneous. Rather, adaptations to slowly-evolving trends embedded in a noisy background of inherent variability are likely to be delayed by decades, as farmers attempt to sort out true climate change from random climatic fluctuations. In fact, if by dint of poor luck, there was a sequence of weather anomalies that were the opposite of slowly building climatic trends, misperception of these as the new climatic regime could actually lead to mal-adaptations.[64] Even in the face of policies to facilitate passive adjustment (regulations on sharing losses, changes in land use, changes in location, retreat from rising sea levels—see, e.g., West and Dowlatabadi, 1998), mal-adaptation can occur, which is clearly counterproductive and can be more damaging than not adapting at all, especially to developing countries and marginalized groups, who have so few financial and other resources that even one round of adaptation measures – much less multiple rounds, if the original measures prove to be mal-adaptations – will be taxing.

Anticipatory adaptation, however, appears to have considerable policy potential. Anticipatory, or proactive, adaptation could include such technical actions as the purchasing of more efficient irrigation equipment, the building of higher bridges and dams, and the engineering of seeds to make them cope better with altered climates before the climate changes actually manifest themselves. It could also include political actions, such as setting up networks to disseminate climate information and suggest potential adaptive actions, and the creation of insurance mechanisms or even transfer payments to disadvantaged groups.

With well-defined central policy coordination on a wide range of anticipatory climate change adaptation actions, mal-adaptation is more likely to be avoided, but it's not as simple as it seems. Anticipatory policy, like abatement strategy, is part of a complex cycle: Human behaviour within physical, biological, and social systems causes disturbances that propagate through natural systems and create responses that, in turn, feed back on human behaviour in the form of policies for adaptation or mitigation to the human-induced disturbances (Root and Schneider, 2001, p. 41). Most studies of anticipatory adaptation also assume that countries and groups will be able to afford it, which is unlikely to be universally true.

---

63 For more on pest outbreaks, see, for example, Ehrlich, Ehrlich, and Daily (1995).

64 See West and Dowlatabadi (1998); West, Dowlatabadi, and Small (2001); and Schneider, Easterling, and Mearns (2000).

4.6 Can adaptation be "traded off" against abatement?

It is often assumed, particularly in a cost-benefit analysis (CBA) framework, that mitigation and adaptation can be viewed as competitive strategies to deal with climate change, but we must first consider the potential implications of the oft-stated trade-off. Suppose it were cheaper for an industrialized, high-emitting nation in the political North to adapt than to mitigate. If that nation chose only to adapt, it would likely be detrimental to a poorer, less adaptable country in the South. Simply comparing mitigation and adaptation costs and aggregating the values across all nations is a "one dollar, one vote" aggregation approach, and it clearly has serious equity implications. The low-cost option for one country is most likely not synonymous with the low-cost option for its neighbours or the world at large.

5. Conclusion

Our personal value position, given the vast uncertainties in both climate science and impacts estimations, is to enact and act on policies that slow down the rate at which we disturb the climate system (i.e., abatement policies). This can both buy us time to understand better what may happen – a process that will take many more decades – and lead to the development of lower-cost decarbonisation options. That way, the costs of mitigation can be reduced well below those that would otherwise be incurred if there were no policies in place to provide incentives to reduce emissions and invent cleaner alternatives. In the face of potential surprises and irreversibilities, we must not become trapped in conventional economic wisdom that suggests we should emit now and abate later; rather, we must take action now to (at the very least) develop more cost-effective mitigation methods in the future to reduce "dangerous anthropogenic interference" (Azar and Schneider, 2002) and make adaptation possible. Slowing down the pressure on the climate system is our "insurance policy" against non-linearities, and irreversibilities like species extinction, the melting of large glaciers, or the breakdown of the Thermohaline circulation. Such non-linearities will undoubtedly be the topic of frequent debate in the coming decades, as more and more decision-makers come to understand that what we do in the next few generations may have indelible impacts on the next hundred generations to come.

References

Andronova, N.G. and M.E. Schlesinger (2001), "Objective Estimation of the Probability Density Function for Climate Sensitivity", *Journal of Geophysical Research* **106**, 22605–22612.

Azar, C. and H. Rodhe (1997), "Targets for Stabilization of Atmospheric CO_2", *Science* **276**, 1818-1819.

Azar, C. and S.H. Schneider (2002), “Are the Economic Costs of Stabilising the Atmosphere Prohibitive?” *Ecological Economics,* **42**, 73–80.

Bond, G., W. Showers, M. Cheseby, R. Lotti, P. Almasi, P. deMenocal, P. Priore, H. Cullen, I. Hajdas, G. Bonani (1997), “A pervasive millennial-scale cycle in North Atlantic Holocene and glacial climates”, *Science* **278**, 1257-1266.

Broecker, W. S. (1997), “Thermohaline circulation, the Achilles heel of our climate system: Will man-made CO_2 upset the current balance?” *Science* **278**, 1582-1588.

Claussen, M. (1998), “On multiple solutions of the atmosphere-vegetation system in present-day climate”, *Global Change Biology,* **4**, 549-559.

Claussen, M., C. Kubatzki, V. Brovkin, A. Ganopolski, P. Hoelzmann, H.J. Pachur (1999), “Simulation of an abrupt change in Saharan vegetation in the mid-Holocene”, *Geophysical Research Letters,* **26**, 2037-2040.

Ehrlich, P.R., A.H. Ehrlich, and G.C. Daily (1995), *“The Stork and the Plow”,* New York, Putnam, 364 pp.

Forest, C.E., P.H. Stone, A.P. Sokolov, M.R. Allen, and M.D. Webster (2001), “Quantifying Uncertainties in Climate System Properties Using Recent Climate Observations”, *Science,* **295**, 113-117.

Ganopolski, A., C. Kubatzki, M. Claussen, V. Brovkin, V. Petoukhov (1998), “The influence of vegetation-atmosphere-ocean interaction on climate during the mid-Holocene”, *Science,* **280**, 1916-1919.

Higgins, P.A.T., M. Mastrandrea, and S.H. Schneider (2002), “Dynamics of Climate and Ecosystem Coupling: Abrupt Changes and Multiple Equilibria”, *Philosophical Transactions of the Royal Society of London,* **357**(1421), 647-655.

Intergovernmental Panel on Climate Change (1996), *Climate Change 1995 – the science of climate change*. The second assessment report of the IPCC, contribution of working group I. J.T. Houghton, L.G. Meira Filho, B.A. Callander, N. Harris, A. Kattenberg, and K. Maskell (eds.), Cambridge, England, Cambridge University Press, 572 pp.

Intergovernmental Panel on Climatic Change (IPCC) (2001a), *Climate Change 2001: The Scientific Basis*, Third Assessment Report of Working Group I. J.T. Houghton, Y. Ding, D. J. Griggs, M. Noguer, P.J. van der Linden, X. Dai, K. Maskell, C.A. Johnson (eds.), Cambridge, England, Cambridge University Press, 944 pp.

Intergovernmental Panel on Climatic Change (IPCC) (2001b), *Climate Change 2001: Impacts, Adaptation and Vulnerability*, Third Assessment Report of Working Group II. J.J. McCarthy, O.F. Canziani, N.A. Leary, D.J. Dokken, K.S. White (eds.), Cambridge, England, Cambridge University Press, 1032 pp.

Keller, K., B.M. Bolker and D.F. Bradford (2004), "Uncertain climate thresholds and optimal economic growth", *Journal of Environmental Economics and Management* **48**, 723-741.

Kleidon, A., K. Fraedrich, and M. Heimann (2000), "A green planet versus a desert world: Estimating the maximum effect of vegetation on the land surface climate", *Climatic Change,* **44**, 471-493.

Kleidon, A. and M. Heimann (1999), "Deep-rooted vegetation, Amazonian deforestation, and climate: results from a modelling study", *Global Ecology and Biogeography,* **8**, 397-405.

Levis, S., J.A. Foley, V. Brovkin, and D. Pollard (1999), "On the stability of the high-latitude climate-vegetation system in a coupled atmosphere-biosphere model", *Global Ecology and Biogeography,* **8**, 489-500.

Linden, H. R. (1996), "The Evolution of an Energy Contrarian", *Annual Review of Energy and the Environment,* **21**, 31-67.

Manabe, S. and R.J. Stouffer (1993), "Century-scale effects of increased atmospheric CO_2 on the ocean-atmosphere system", *Nature,* **364**, 215-218.

Mastrandrea, M. and S.H. Schneider (2001), "Integrated Assessment of Abrupt Climatic Changes", *Climate Policy,* **1**, 433-449.

Mastrandrea, M. and S. H. Schneider (2004), "Probabilistic Integrated Assessment of 'Dangerous' Climate Change", *Science,* **304**, 571-575.

Moss, R.H. and S.H. Schneider (2000), "Uncertainties in the IPCC TAR: Recommendation to lead authors for more consistent assessment and reporting", in R. Pachauri, T. Taniguchi, and K. Tanaka (eds.), *Third Assessment Report: Cross Cutting Issues Guidance Papers.* Geneva, Switzerland: World Meteorological Organisation, 33-51. Available at: http://stephenschneider.stanford.edu/Publications/PDF_Papers/UncertaintiesGuidanceFinal2.pdf.

Nordhaus, W. D. (1994a), *Managing the Global Commons: The Economics of Climate Change,* Cambridge, MA, MIT Press, 232 pp.

Nordhaus, W.D. (1994b), "Expert Opinion on Climatic Change", *American Scientist,* **82**, 45-51.

Rahmstorf, S. (1996), "On the freshwater forcing and transport of the Atlantic thermohaline circulation", *Climate Dynamics,* **12**, 799-811.

Rahmstorf, S. (1999), "Decadal variability of the thermohaline circulation", in A. Navarra, (ed.), *Beyond El Nino: Decadal and Interdecadal Climate Variability*, New York, Springer, 309-332.

Root, T. L. and S. H. Schneider (2001), "Climate Change, Overview and Implications for Wildlife", in S.H. Schneider and T.L. Root (eds.), *Wildlife Responses to Climate Change, North American Case Studies,* Washington, D.C., Island Press, 1-56.

Roughgarden, T. and S.H. Schneider (1999), "Climate Change Policy: Quantifying Uncertainties for Damages and Optimal Carbon Taxes", *Energy Policy* **27**, 415-429.

Schneider, S. H. (1993), "Pondering Greenhouse Policy", *Science* **259**, 1381.

Schneider, S. H. and C. Azar (2001), "Are Uncertainties in Climate and Energy Systems a Justification for Stronger Near-term Mitigation Policies?", in *Proceedings of the Pew Center Worksh*op *on the Timing of Climate Change Policies*, 85-136. The Pew Center Workshop on the Timing of Climate Change Policies. Washington, D.C. 11-12 October 2001.

Schneider, S.H. and K. Kuntz-Duriseti (2002), "Uncertainty and Climate Change Policy" in Schneider, S.H., A. Rosencranz, and J.O. Niles, (eds.), *Climate Change Policy: A Survey.* Washington D.C., Island Press, 53-88.

Schneider, S.H. and S.L. Thompson, (1981), "Atmospheric CO_2 and climate: importance of the transient response", *Journal of Geophysical Research,* **86**, 3135-3147.

Schneider, S. H. and S. L. Thompson, (1985), "Future Changes in the Atmosphere", in R. Repetto (ed.), *The Global Possible,* New Haven, CT, Yale University Press, 363-430.

Schneider, S.H. and S.L. Thompson (2000), "A simple climate model used in economic studies of global change", in S.J. DeCanio, R.B. Howarth, A.H. Sanstad, S.H. Schneider and S.L. Thompson, *New Directions in the Economics and Integrated Assessment of Global Climate Change*. Washington, D.C., The Pew Center on Global Climate Change, 59-80.

Schneider, S. H., W. E. Easterling, and L. O. Mearns (2000), "Adaptation: Sensitivity to Natural Variability, Agent Assumptions and Dynamic Climate Changes", *Climatic Change,* **45**, 203-221.

Schneider, S.H., B.L. Turner II, and H. Morehouse Garriga (1998), 'Imaginable Surprise in Global Change Science', *Journal of Risk Research,* **1(2)**, 165-185.

Siegel, E., H. Dowlatabadi and M.J. Small (1995), "A Probabilistic Model of Ecosystem Prevalence", *Journal of Biogeography,* **22(4-5)**, 875-879.

Stocker, T. F. and O. Marchal (2000), "Abrupt climate change in the computer: Is it real?" *Proceedings of the National Academy of Sciences of the United States of America*, **97**, 1362-1365.

Stocker, T. F. and A. Schmittner (1997), "Influence of CO_2 Emission Rates on the Stability of Thermohaline Circulation", *Nature,* **388**, 862-865.

Stommel, H., (1961), "Thermohaline convection with two stable regimes of flow", *Tellus,* **13**, 224-230.

Stommel, H., (1980), "Asymmetry of interoceanic freshwater and heat fluxes", *Proceedings of the National Academy of Sciences of the United States of America,* **77**, 2377-2381.

UNFCCC (1992), "United Nations Framework Convention on Climate Change", Bonn, Germany, UNFCCC.

Wang, G. L. and E.A.B. Eltahir (2000a), "Ecosystem dynamics and the Sahel drought", *Geophysical Research Letters,* **27**, 795-798.

Wang, G. L. and E.A.B. Eltahir (2000b), "Role of vegetation dynamics in enhancing the low-frequency variability of the Sahel rainfall", *Water Resources Research,* **36**, 1013-1021.

West, J. J. and H. Dowlatabadi (1998), "Assessing Economic Impacts of Sea Level Rise", in T.E. Downing, A.A. Olsthoorn, and R.S.J. Tol (eds.), *Climate Change and Risk*, New York, Routledge, 205-220.

West, J.J, H. Dowlatabadi and M.J. Small (2001), "Storms, investor decisions, and the economic impacts of sea level rise", *Climatic Change,* **48**, 317-342.

Wood, R. A., A.B. Keen, J.F. B. Mitchell and J.M. Gregory (1999), "Changing spatial structure of the thermohaline circulation in response to atmospheric CO2 forcing in a climate model", *Nature,* **399**, 572-575.

Yin, J., M.E. Schlesinger, N.G. Andronova, S. Malyshev and B. Li (2004), "Is a Shutdown of the Thermohaline Circulation Irreversible?" *Nature*, submitted.

Zheng, X. Y. and E.A.B. Eltahir (1998), "The role of vegetation in the dynamics of West African monsoons", *Journal of Climate,* **11**, 2078-2096.

Zickfeld, K. (2004), "Modeling Large-Scale Singular Climate Events for Integrated Assessment", PhD Thesis, Potsdam University.

Chapter 6

THE SOCIAL COST OF CARBON: KEY ISSUES ARISING FROM A UK REVIEW

by Michele Pittini and Mujtaba Rahman,

Defra, UK Department for Environment, United Kingdom

Estimates of the marginal damage costs of greenhouse gas emissions (or social cost of carbon) could provide a consistent benchmark for incorporating the benefits of climate change mitigation into the assessment of a wide range of policies. This paper moves from recent reviews of the relevant literature to explore both the main drivers of variability as well as the limitations of the current social cost of carbon estimates. The sensitivity of these estimates to economic assumptions such as discount rates, equity weighting and alternative approaches to valuation is well established. It is noted here that transparent choices can be made with respect to each of these factors, which also have an ethical dimension. It is also stressed that the partial coverage of non-market impacts and the lack of coverage (with almost no exception) of low-probability catastrophic events and socially contingent effects are also issues that need to be addressed to improve the robustness of the current estimates and their relevance to policy decision-makers. The implications of uncertainty in decision-making contexts are discussed in the latter sections of the paper, which also provides a glimpse into the on-going UK review of the social cost of carbon and sets out priorities for further research.

ISBN 92-64-10831-9
THE BENEFITS OF CLIMATE CHANGE POLICIES

CHAPTER 6. THE SOCIAL COST OF CARBON: KEY ISSUES ARISING FROM A UK REVIEW[65]

by Michele Pittini[66] and Mujtaba Rahman[67]

1. Introduction

Traditionally the policy debate on climate change has been dominated by "safe minimum standards approaches" (Pearce *et al.,* 1996) largely informed by expert opinions and by assessments of the costs of mitigation. The great uncertainty on the impacts of unchecked climate change and the possibility that these impacts might be extremely severe or catastrophic has played a major role in driving this approach. Nonetheless the issue of how to explicitly place the benefits of greenhouse gas abatement in the context of costs to inform policy decision-making is becoming increasingly important in the climate change policy arena.

There are major uncertainties surrounding climate change impacts: the inevitable value judgements implied by the aggregation of the costs of climate change across different regions and different generations and the challenges associated with handling impacts which have no obvious market value. Nevertheless, economic valuation has an important role to play and could provide a consistent benchmark for the social cost of greenhouse gas emissions for use in the assessment of a wide range of policies across government.

---

65 The views expressed in the paper are those of the authors and do not necessarily reflect the views of Defra or of the UK Government. The UK review of the social cost of carbon is ongoing and this paper does not present any of its final outcomes.

66 Economic Adviser, Defra, UK Department for Environment, Food and Rural Affairs, Environment Protection Economics Division. Contact information: Michele.Pittini@defra.gsi.gov.uk ; +44 (0)20 70828592.

67 Economist, Defra, UK Department for Environment, Food and Rural Affairs, Environment Protection Economics Division. Contact information: Mujtaba.Rahman@defra.gsi.gov.uk ; +44 (0)20 70828591.

The UK Government Economic Service (GES) paper *Estimating the Social Cost of Carbon Emissions* (Clarkson & Deyes, 2002) presented a review of the literature produced up to 2001 and suggested some illustrative monetary figures for the global marginal costs of carbon emissions or social cost of carbon (SCC), which is the key measure of benefits of mitigation within a cost benefit analysis approach. It also clearly recommended periodic reviews of these illustrative figures as new evidence became available.

Recent developments in integrated assessment models (IAMs), general developments in the UK Government economic appraisal guidance and the increased policy relevance of issues that are not covered by currently available estimates for the SCC (e.g., low-probability catastrophic events and socially contingent impacts) led the UK Department for Environment, Food and Rural Affairs (Defra) to initiate a review of these figures.[68] This paper presents the background to the debate underpinning the UK review of the SCC. Section 2 provides a summary of the currently available estimates and explores the drivers in their variability. Section 3 looks at the main limitations of the current estimates, particularly the fact that they typically do not cover low-probability catastrophic impacts and "socially contingent impacts" from climate change. Section 4 raises some important issues on the use of SCC estimates in policy decision-making given the associated uncertainty. Finally, section 5 attempts to draw some conclusions, and provides a brief update on the research projects that have been commissioned by the UK Government to inform and progress the review of the SCC.

2. Available estimates and main drivers of variability

2.1 Findings of recent surveys

The body of literature that produced estimates of the social cost of carbon is significant and has been the subject of several review studies. In particular:

- In their review for the IPCC Second Assessment Report, Pearce *et al.* (1996) estimate a range of USD 5 — USD 125 per tonne of carbon (in 1990 prices, or USD 6 — USD 160/tC in 2000 prices) based on a review of existing studies and relating to carbon emissions in the period 1991-2000. For the period 2001-2010, the representative range was estimated to increase to

68 A key step in this review process was an international seminar that was held in London on 7th July 2003 and was attended by some leading modellers and environmental economists; this paper is based on many of the contributions to that seminar. Papers can be found at: http://www.defra.gov.uk/environment/climatechange/carbonseminar/index.htm

USD 7 — USD 154/tC (in 1990 prices, or USD 9-USD 197/tC in 2000 prices).[69]

- On the basis of a review of 8 major studies, the GES paper (Clarkson and Deyes, 2002) suggested a figure of around USD 100/tC (within a range of USD 50 to USD 200/tC)[70] as an illustrative estimate for the global damage cost of carbon emissions. It also suggested that these figures should be raised in real terms by approximately USD 1.5/tC per year, as the costs of climate change are likely to increase over time.

- Pearce (2003a) lists 24 estimates from 12 studies in his review, some of which have been published or peer-reviewed following the publication of the GES paper. Pearce's survey of the SCC literature leads him to conclude that a more appropriate range would be USD 6 to USD 39/tC.

- In a recent working paper that probably constitutes the most complete survey of the literature to date, Tol (2003b) counts 88 estimates from 22 published studies (see Figure 1). The mode of these estimates is USD 5/tC, the mean USD 104/tC and the 95th percentile USD 446/tC, the right skewed distribution reflecting the presence of a few estimates that place the SCC at a few hundred dollars (and in one case more than a thousand dollars) under pessimistic scenarios. After weighting the estimates[71], Tol concludes that "[...] for all practical purposes, climate change impacts may be very uncertain but it is unlikely that the marginal costs of carbon dioxide emissions exceed USD 50/tC and are likely to be substantially smaller than that."

2.2 *Main sources of variation in the current estimates*

It is apparent that the range in the SCC estimates found in the literature is rather wide. This is not surprising, if only because uncertainties (both stochastic uncertainty/variability and lack of knowledge) are key features in climate change research. But in addition to the scientific uncertainties aggregate monetary estimates of

---

69 Existing studies generally produce social cost estimates that increase through time, as on average the marginal damage of each tonne of carbon emitted tends to increase with the level of atmospheric concentration of greenhouse gases.

70 SCC estimates in the original GES paper were given as £70/tC within a range of £35 to £140/tC. A rough conversion rate of 1.45 USUSD per UK£ has been applied in this paper.

71 Tol applies weights to the to the available SCC estimates to reflect different levels of quality in the underlying studies and to account for model-dependency, i.e. for the fact that several groups of multiple results in the database are actually based on the same modelling exercise.

the damages of climate change introduce a series of economic uncertainties (see Box 1 for a typology). In fact, if we look at the sources of variation in the current SCC estimates we can trace them to two main factors:

- Methodological differences and different levels of sophistication in the representation of climatic and socio-economic systems in the models that are used to estimate the SCC.

- Economic uncertainties and implicit/explicit value judgments (on which there are varying views) associated with valuation and aggregation of impacts.

In the following sections we will consider each of these factors in turn.

Figure 1. A meta-analysis of 88 estimates of the marginal social costs of carbon dioxide

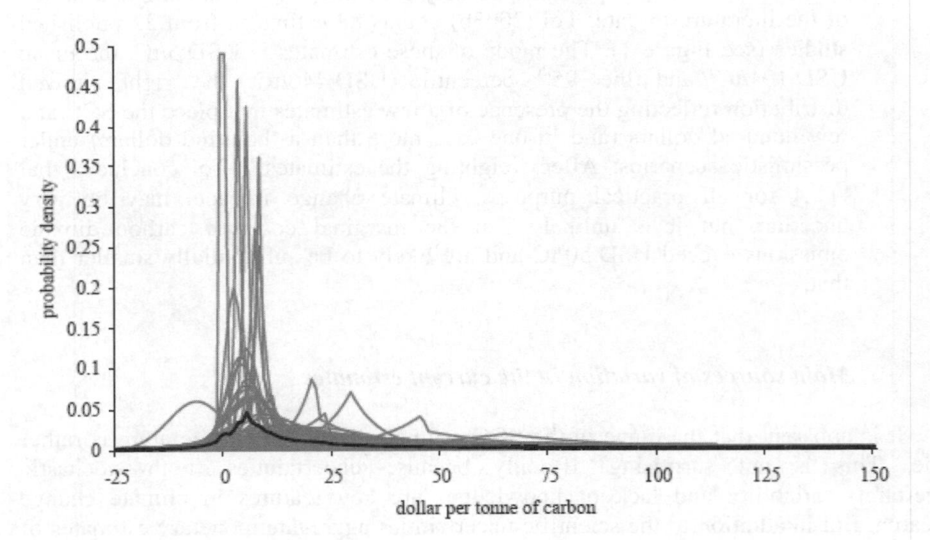

Source: Tol (2003c).

Note:
Tol (2003c) collects 88 estimates of marginal social cost of carbon dioxide figures, from 22 studies. As Tol notes, one would expect the reported estimates to vary considerably, with high to low end marginal social cost estimates ranging from USD 1666/tC through to USD 7/tC. The probability density function in gray highlights the full range of the 88 estimates. The combined probability density function appears in black.

Box 1. Uncertainties implied by global estimates of climate change damages

Scientific uncertainties:

- measurement of present, and prediction of future emissions;
- translation of emissions levels to changes in the atmospheric concentration of carbon;
- estimating the climate impact associated with an increase in atmospheric concentration;
- identification of the physical impacts resulting from climatic change;
- risk of catastrophic and socially contingent events.

Uncertainties/value judgments associated with economic valuation and aggregation:

- estimating monetary values for market and non-market impacts (i.e., those impacts for which a market price does not exist);
- predicting how the relative and absolute value of impacts will change into the future;
- determining the way in which damage estimates should be aggregated across countries and regions with different levels of national income;
- determining the rate at which the value of future impacts should be discounted to today's prices;
- basis for valuation of catastrophic and socially contingent events.

Source: Adapted from Clarkson and Deyes (2002).

2.3 Model dependency

The majority of the available estimates of the SCC are based on integrated assessment models (IAMs). Developed in the context of climate change, IAMs are those models that combine the scientific and economic aspects of climate change within a single, iterative analytical framework (see Box 2). IAMs are themselves characterized by uncertainty (see Kann and Weyant (2000) or van Asselt & Rotmans (1999) for a review) and have raised specific concerns in relation to their aggregate perspective and to the simplified representation that they often give of either the climate system, the socio-economic system or both. Nonetheless they represent a useful tool for the assessment of climate change policy and represent the most self-consistent basis for estimating the damage costs of climate change.

Box 2. Integrated Assessment Models (IAMs)

Integrated assessment models (IAMs) that estimate damage costs of climate change typically include an energy/economy/emissions module, a climate module and an impact module. The latter looks at the impacts of climate change on different sectors like agriculture, ecosystems, human health, sea level rise, etc. For a simplified mathematical description of IAMs one might wish to look at Fankhauser (1995) or Pearce (2003a), while Mendelsohn (2003) provides a useful narrative description about the key steps in IAMs modeling. Basically IAMs are characterised by the fact that climate impacts feed back to the socio-economic module thereby "closing the circle" between emissions, climate modeling, climate change impacts and the economy.

There are two main approaches to derive estimates of the social cost of carbon in IAMs (Clarkson and Deyes 2002). Some studies have looked at the optimal carbon tax within an inter-temporal optimization framework. In other words, they have determined the shadow price of emissions as the carbon tax required to keep emissions at their optimal level, corresponding to the intersection of the marginal abatement cost (MAC) curve and the marginal damage cost (MDC) curve. However, most IAMs have looked at the average incremental costs of a small perturbation in emissions from a business as usual baseline. IAMs belonging to the second group typically estimate the social cost of carbon through the following steps:

1. Produce a benchmark estimate of the overall costs of climate change in terms of loss of world GDP with respect to a hypothetical "no climate change scenario" for a concentration of twice the amount of atmospheric CO_2 with respect to pre-industrial times. Different models typically predict this doubling of CO_2 concentrations to occur at different points in time (e.g., 50, 100 years) under business as usual (BAU) emissions and to produce different increases in global mean temperature or GMT (e.g., 2 C, 2.5 C).

2. Use climate change damage functions to project this benchmark damage to different levels of increase in GMT and construct a BAU damage cost flow.

3. Run a perturbation scenario under which an additional quantum of CO_2 is emitted into the atmosphere and calculate the associated increase in GMT (and its time profile) with respect to BAU emissions.

4. Estimate the net present value of the flow of increased annual damage associated with this quantum increase in emissions.

5. Normalise to 1 tonne of carbon.

The IAMs that have been used to produce estimates of the damage costs of climate change differ with respect to their level of territorial disaggregation, complexity in their climate and/or economic components, and different level of inclusions of non-market impacts. Models of the first generation (like Nordhaus's seminal IAM) tended to focus on market impacts and to produce lower SCC estimates.

The recommendations of the GES paper (Clarkson and Deyes, 2002) were primarily based on the findings of an EC-funded ExternE report (Eyre *et al.*, 1998) which in turn reflected modeling results from FUND 1.6 by Richard Tol and the Open Framework by Tom Downing. Clarkson and Deyes expressed a preference for the ExternE report largely because Richard Tol's FUND 1.6 represented the state of the art in the published peer review literature (though a successive version of this model -

FUND 2.0 - had already been accepted for peer review and pre published). There have recently been further advances in IAMs, including:

- a revised (and more optimistic) role of adaptation to climate change;
- the inclusion of positive as well as negative impacts from climate change (see Tol and Heinzow (2003) for a discussion of these issues in relation to successive versions of the FUND model).

It should be noted that these developments do not include advances in areas that would increase costs. As a result, the new models and in particular the latest versions of FUND by Richard Tol produce central SCC estimates that are significantly lower than before (see Pearce 2003a).

Adaptation should certainly be a key feature of IAMs. If it is not properly incorporated into models there is a risk of relying on what has often been described as the "dumb farmer hypothesis", which would inevitably overestimate the damage costs from climate change (Pearce 2003a; Mendelsohn 2003).[72] However, adaptation is an essentially local and sectoral response, involving behavioural and institutional adjustments as well as technological changes. As a consequence, its incorporation in IAMs is a very complex task. The way in which autonomous adaptation is included in current models is arguably simplistic, based for instance, on speculative assumptions of farmers' reactions to climate change, simple economic models of coastal protection, historical trends in the energy sector or simple relationships between levels of economic development and vulnerability to impacts like vector-borne diseases. Members of the IAM community are of course the first in drawing attention to the limitations of the models as far as adaptation is concerned. For instance, Tol (2003b) emphasises that "various approaches are used to model adaptation (…), but they all either underestimate or overestimate its effectiveness and costs". Adaptation dynamics are therefore still an area for development in IAMs, which would also need to address more explicitly the issue of the costs of transition between different socio-climatic states.

As mentioned above, in addition to a better inclusion of autonomous and planned adaptation and to the use of different climate scenarios, the consideration of potential benefits alongside damage costs also contributes to the explanation of lower estimates of the social cost of carbon in the latest generation of integrated models. At least in some sectors/regions and up to a certain increase in global mean temperature (GMT), there are likely to be some positive impacts (e.g., net benefits in the agricultural sector in Russia or North America due to CO_2-fertilisation, potential amenity impacts in northern Europe due to warmer climate, etc.). Indeed Mendelsohn (2003) depicts an impact function based on a "hill-shaped" hypothesis whereby several sectors in

---

72 In fact, several IAMs developed in the 1990s like Tol's FUND 1.6 implicitly included some adaptation hypotheses in their impact modules (in the case of farmers for instance, by assuming different levels of "farm adaptation" like change of crops or farming practices). What they did not include are explicit adaptation dynamics.

developed countries would initially benefit from GMT increases as they climb towards their climate optimum. However a recent review of the literature on global climate change impacts by Hitz and Smith (this volume) found that all impact studies across different sectors estimate that negative impacts will prevail and become more frequent as the temperature increases beyond 3 to 4 °C in GMT. Estimates of climate change damage based on a doubling of CO_2 concentrations would not capture this threshold effect, but based on IPCC scenarios an increase in GMT of 3 to 4 °C cannot be ruled out within the time horizon that IAMs typically address (e.g. the next 100 years).[73]

2.4 Economic assumptions and value judgements

While methodological features and the current limitations of IAMs can affect SCC estimates, different choices on economic parameters like the discount rate[74] or its pure rate of time preference (PRTP) component and the presence of equity weighting can significantly affect SCC estimates[75]. Perhaps the most interesting conclusions of Tol's meta-analysis (Tol, 2003b) is that the two "ethical" parameters (PRTP and equity weighting) play a major role in explaining both the magnitude and the variability of SCC estimates. Tol's conclusion that the SCC is unlikely to exceed USD 50 tC appears to be influenced by his assessment that "although equity weighting is theoretically sound (…), it does pose an idealized worldview on the estimates". It also appears to be informed by the observation that a PRTP of 3% is "close to what most western governments use for long-term investments" and that PRTPs lower than 1% "may be morally preferable, but are clearly out of line with common practice." In fact, while the debate on these important economic assumptions remains controversial recent UK Government guidance on economic appraisal (HM Treasury, 2003) has endorsed both equity weighting and low (albeit positive) discount rates.

Equity weighting is consistent with a utilitarian view of collective welfare as a sum of individual utilities and it is theoretically justified on the basis of the "decreasing marginal utility of income"(ε). In other words (and to oversimplify) equity weighting rests on the concept that one dollar is worth more to a poor person than to a rich person and that this should be reflected in making inter-personal comparisons of utility to

73 Based on the SRES emissions scenarios, the IPCC projects an increase of GMT of 1.4 to 5.8 °C in the period 1990 to 2100 (IPCC, 2001)

74 "Discount rate" is defined here as social discount rate or social rate of time preference (SRTP). In a standard formulation this can be typically expressed as SRTP=PRTP + θg where PRTP is the pure rate of time preference (or utility discount rate), θ is the negative of the income elasticity of marginal utility of income (θ = -ε) and g is the rate of growth of consumption per capita.

75 Although these parameters might be affected by some element of genuine uncertainty (e.g., on the empirical measurement of PRTP and the elasticity of the marginal utility of income underlying equity weights), they fundamentally imply different value judgements on aggregation of welfare across time and across different regions. See also Schneider and Lane (this volume) and Tol *et al.* (2004).

inform social choices. When applied to global estimates of climate change, equity weighting provides a way of addressing the equity concerns raised by the aggregation of monetary estimates across countries and regions with very different levels of income per capita (Fankhauser *et al.,* 1997). Interestingly the latest version of the UK Treasury guidance to economic appraisal and evaluation, also known as the Green Book (HM Treasury, 2003) introduced equity weighting as a tool for assessing the distributional implications of Government projects and policies.

The relationship between weighted and un-weighted SCC estimates or "equity multiplier" (Pearce 2003a) depends on several modelling features. First of all, the equity multiplier is highly dependant on the underlying assumptions about the income elasticity of the marginal utility of income (ε) (Clarkson and Deyes, 2002; Pearce, 2003a) [76] Secondly, the equity multiplier reflects the distribution of global damages from climate change between developed countries and developing countries and the pattern of this distribution over time. This in turns makes the multiplier sensitive to the discount rate. Finally, the relative size of weighted and un-weighted estimates can be sensitive to the chosen approach to health valuation, including the choice between value of a statistical life (VSL) and value of a life year lost (VLYL) as well as the specific values assumed under the two approaches (Tol and Heinzow, 2003). Based on a standard assumption about the elasticity of the marginal utility of income ($\varepsilon = -1$) Clarkson and Deyes (2002) observed that, as a rule of thumb, equity weighted SCC estimates roughly double the un-weighted estimates. On the basis of the studies that he surveyed, Pearce (2003a) remarked that "all multipliers are contained within the bracket 0.9 to 3.6".

SCC estimates are also notoriously very sensitive to the choice of the discount rate So far most IAMs have assumed either a constant rate of discount or a constant PRTP (Wood, 2002; Pearce, 2003a). [77] However, recent work on discounting suggests that for

---

76 There is little consensus in the literature about empirical estimates of ε, which typically rely on analyses of savings behaviour or on other revealed preference studies. The IPCC (1996) stated that standard rate for ε ranged between – 1 and – 2. A survey by Cowell and Gardiner (1999) indicated a range between – 0.5 and - 4. The Green Book (HM Treasury, 2003) recommends using a value of – 1 when applying equity weighting to policy appraisal. On the basis of a thought experiment that looks at the implied distributional preferences Pearce (2003a) notes that a value of $\varepsilon = -1$ "does seem feasible" and concludes that "values of ε in the range (-) 0.5 to (-) 1.2 seem reasonable."

77 Different IAMs differ in the way they treat discounting (Clarkson and Deyes 2002; Wood 2002). Most models require an assumption about the social discount rate, which is treated as constant. Some other models (like FUND) require assumptions only on PRTP and the negative of the income elasticity of marginal utility of income (□) They then use endogenously determined values for the rate of growth of consumption per capita g to estimate the SRTP. PRTP is still typically constant in this second class of models. An exception is represented by RICE99, which assumes a 3% PRTP declining to 2.3% after 2100. A few studies have then tested the sensitivity of SCC estimates from DICE to decreasing discount rates and have been referenced in this section, including Newell and Pizer (2001) and Wood (2002). Mastrandrea and

long-term issues such as global warming it might be more appropriate to use a discount rate that declines over time (Newell and Pizer, 2001; Weitzman, 1998, 1999; OXERA, 2002). These conclusions reflect several factors including uncertainty on future PRTPs, uncertainty in future rates of growth of consumption per capita and considerations of intergenerational equity (see Box 3). The use of a lower/decreasing long-term discount rate would clearly tend to increase estimates of the social cost of carbon by assigning a higher weight to damage costs accruing at distant points in the future, thus extending the relevant time horizon for analysis before the cost of impacts converges for practical purposes to zero (Newell and Pizer, 2001; Pearce, 2003a; Wood, 2002). Lower long-term discount rates would also increase the impact on SCC estimates of low-probability, catastrophic events (discussed below in section 3.3), which are more likely to occur in the late 21st century and 22nd century. Once again, it is interesting to note that the Green Book (HM Treasury, 2003) endorses the use of a decreasing discount rate schedule when looking at policies with long time horizons.[78]

Another economic issue arising from the monetary aggregation of climate change damages relates more specifically to the way in which non-market impacts are valued. The valuation of non-market impacts of climate change has largely reflected a willingness to pay (WTP) approach, with the notable exception of valuation of health impacts that are often based on a willingness to accept compensation (WTA) approach. WTA approaches could, however, in principle be applied to the valuation of a wider range of climate change impacts (Demeritt and Rothman, 1999; Pearce, 2003b). There is a large body of evidence that suggests that the choice between WTP and WTA approaches in eliciting hypothetical values can lead to rather different results — WTA estimates being 4-20 times higher than corresponding WTP estimates (Horowitz and McConnell, 2002).

While the source of this large disparity has been often attributed to the elicitation procedures used in economic valuation studies, several explanations that rely on sound economic theory have also been put forward. While the debate on the WTP-WTA disparity is still ongoing (Pearce, 2002), it is worth noting is that SCC estimates would tend to increase if WTA approaches were chosen for valuing the non-market impacts of climate change.

Schneider (2001) have also experimented with hyperbolic discounting using the standard DICE model and a modified version of DICE that allows for abrupt change (see here at section 3.3).

78 The Green Book recommends the following social discount rate (SRTP) schedule: 0-30 years 3.5%, 31-75 years 3%, 76-125 years 2.5%, 126-200 years 2%, 201-300 years 1.5%, 300 + years 1%.

Box 3. Changes in economic assumptions that could increase SCC estimates

Decreasing discount rates

Different reasons have been given in the literature for adopting a time varying (decreasing) discount rate[79], including:

- *Observed individual choice*

  Empirical observations as to how people actually behave when discounting the future would seem to indicate that individuals tend not to discount the future at constant rates, but rather at decreasing rates, so that values in the near future are discounted at a higher rate than values in the distant future.

- *Uncertainty about future economic magnitudes*

  In the first instance, there is uncertainty regarding the social weight to be attached to future costs and benefits (and so the discount rate itself). In the second instance, there is uncertainty concerning the future state of the economy and subsequent levels of consumption.

- *Future fairness*

  Normative assumptions are adopted which show that a reasonable and fair balance of interests can be reached between current and future generations on the basis of a time-declining discount rate. This argument is based on intergenerational equity and concerns the fact that constant discounting represents the current generations lack of willingness to consider the welfare of future generations.

Difference between WTA and WTP estimates of non-market impacts

In order to explain the variation between the two techniques alluded to, several explanations have been put forward. The most important of these include:

- *Income and substitution effects*

  According to pure economic theory, one reason for the observed difference in WTP and WTA arises as a result of differences in real income (WTP is necessarily constrained by the individuals income). However, Hanemann (1991, 1999) has demonstrated how a substitution effect may also be interacting with an income effect to increase the disparity between WTP and WTA in an environmental context. The lower the substitutability of the environmental good in question (i.e. the more it is unique in nature), the higher will be the additional disparity between WTP and WTA.

- *Behavioural issues relating to the design of the questionnaire*

  Other explanations look at behavioural patterns in respondents to stated preference surveys as a reason for the differences in results arising from the WTP and WTA studies. Respondents have an incentive to attempt to maximise what they would receive as compensation rather than state the minimum they would actually require to keep their original level of welfare unchanged — so underestimating maximum WTP and overestimating minimum WTA compensation. Furthermore, if respondents are unsure as to the value they would place on an environmental improvement or loss, and the information is costly to acquire due to a number of transaction costs, then WTP which is stated will tend to be below the true WTP value, whilst stated WTA compensation will exceed true WTA.

- *Endowment effect*

  The endowment effect argument suggests that individuals are far more averse to losing something that they already have as opposed to forfeit gaining something they do not yet possess. Therefore, in the context of a given quantity of a certain good/service losses are weighted far more heavily than gains. This provides another rationalisation as to why WTA tends to lead to higher estimates than the WTP.

Sources:: Oxera, 2002; Pearce, 2002.

---

79 Note that the different arguments summarised here do not necessarily imply the same functional form for the variation of the discount rate over time.

Interestingly, WTA would be consistent with an allocation of property rights under which individuals have a right not to suffer climate change-related damages (e.g., a right not to suffer increased frequency of flooding, or a right not suffer the loss of valuable ecosystems). At the extreme, one could think of a "right to an anthropogenically unaltered climate", although whether Article 2 of the UNFCCC recognises such a right is open to debate.[80] Alternatively (and perhaps more meaningfully) one could think of a right not to suffer specific climate change impacts However the status quo allocation of these rights is likely to differ across countries and across sectors and for many non-market impacts property rights are typically not well established.

Finally, while both WTP and WTA for non-market goods are a function of personal income (as other things being equal we can expect wealthier individuals to attach more value to environmental quality[81]), WTA is not bound by ability to pay. This could have attractive equity implications, although equity weighting (which could in principle also apply to WTA estimates) already corrects to some extent for differences in income.

2.5 *Distributional issues and the social cost of carbon*

It is worth making explicit the relationship between aggregate estimates of climate change damages and distributional impacts of climate change policies. The literature on climate change impacts shows that the latter will vary significantly across sectors and regions over time and that those living in the poorest (and most vulnerable) societies as well as future generations are likely to suffer the most. By contrast SCC estimates typically reflect aggregate costs that are estimated on the basis of a limited number of macro-regions and then aggregated at a global level and over time. SCC estimates therefore reflect potential compensation between costs and benefits faced by different people.

Equity weighting is the most obvious means through which distributional concerns can be incorporated in SCC estimates. While equity weighting *per se* does not resolve the issue of inequality in the global distribution of income and its implications for the climate change policy debate (Demeritt and Rothman, 1999), aggregation procedures that incorporate equity weighting would arguably reflect a sympathetic view of poorer individuals compared to a benchmark case where equity weighting was not adopted. In a similar manner a low/decreasing discount rate would assign a greater weight to the welfare of future generations. However, identification of those who stand to gain or lose under alternative climate scenarios is important information in the context of international climate change negotiations. There is therefore an issue of how to address

---

80 This comment was made by David Pearce in his keynote speech at the international seminar that the UK Department for Environment, Food and Rural Affairs (Defra) hosted in London on 7th July 2003.

81 Technically, the marginal rate of substitution between income and environmental quality is likely to be decreasing with increasing income.

broader distributional issues alongside aggregate monetary values and cost-benefit bottom lines in assessing climate change policy at a strategic level. While this issue is unlikely to be satisfactorily addressed through IAMs because of their scale, other approaches to impact assessment (e.g., multi-criteria assessment framework informed by regional, sectoral and bottom-up modelling approaches) could be employed alongside IAMs to usefully inform the debate about the distributional implications of climate change policy.

3. Limitations of the current estimates

3.1 Two main caveats

Allowing for the inherent uncertainties and for the current limitations in IAMs, the main message from the recent developments in the modelling literature seems to be that the socio-economic systems might react better than the climate change research community initially thought to slow and moderate changes in climate. This is reflected in relatively lower estimates for the SCC. However, these conclusions are subject to some major caveats:

- First of all, there is scope for improvement in treating those impacts (both market and non-market) that SCC estimates already try to incorporate to some extent.

- More importantly, with almost no exception[82], current SCC estimates do not cover two main categories of impacts that are increasingly becoming of key importance to the policy debate on climate change, i.e. "low-probability catastrophic events" and "socially-contingent effects".

3.2 Impacts that are already covered by IAMs

Allowing for a simplistic representation of adaptation and vulnerability dynamics, the impacts of slow, moderate climate change on sectors like agriculture, flooding and health are generally well represented in IAMs. However other categories of impact (particularly non-market impacts) are still covered rather unevenly. In particular the impacts of climate change on biodiversity and ecosystem functions have begun to be systematically covered only in the most recent versions of IAMs. Recent impacts literature shows that the impacts of climate change on ecosystems could be severe even for small changes in GMT (see for instance Leemans and Eickhout, 2004). There is no straightforward established methodology for valuing the loss of ecosystem functions but the total economic value (TEV) of the associated damages could be substantial. This

---

[82] The only exception is the experimental approach used by Nordhaus and Boyer (2000), which is discussed here in section 3.3.

area is "work in progress" primarily for climate change impacts science and for economic valuation, but IAMs should ideally reflect the state of the art in both fields.

Another issue of concern is that current IAMs typically cover a subset of the types of (non catastrophic) extreme weather events that are likely to become more frequent as global mean temperatures and climate variability increase. Eyre *et al.* (1998) listed seven categories of extreme weather events whose frequency and/or severity is likely to be affected as a result of climate change including frost and cold spells[83], heat waves, drought, riverine floods, mid-latitude windstorms, tropical cyclones and "other hazards" (e.g. lightning, hail and tornadoes). They then pointed out that most of those IAMs that provide a detailed description of the sectoral impacts of climate change only include a subset of the first six categories, while the residual weather events have not received much attention in the impacts literature. Difficulties arise due to the lack of consensus in climate change scenarios on the likely effect of climate change and increased climate variability on the frequency, severity, sequence and location patterns of climate hazards. Impacts and vulnerability to the latter are also better understood at a local scale (see Eyre *et al.,* 1998, for a discussion). Nonetheless, the apparent time-lag with which the likely changes in climate extremes that were listed in the IPCC Third Assessment Report (IPCC, 2001) are being incorporated into IAMs is an issue that would seem to deserve the attention of modellers.

3.3 Climate catastrophes

With almost no exception the current SCC estimates typically exclude any consideration of the probability of "climate catastrophes" or "abrupt climate change" for example, the melting of the West Antarctic ice sheet, collapse of thermohaline circulation (THC) in the North Atlantic, methane hydrate destabilisation, etc. Partially because the risk of abrupt climate change has only recently received widespread recognition and partially because generating scenarios of abrupt climate change is inherently difficult, research on the impacts of a changing climate has tended to focus on scenarios of slow, gradual change (Alley *et al.,* 2003). As a result, agreed probability distributions for linking the probability of these impacts to atmospheric concentrations of greenhouse gases are currently not available in the scientific literature.

Against this background, several authors have used subjective probability distributions and ad-hoc assumptions on potential costs (or, at best, emerging insight from climate models and rough cost estimates) in order to incorporate low-probability catastrophic events into IAMs and thereby explore the potential consequences on policy-relevant modelling results. Nordhaus and Boyer (2000) is the only study which

83 It should be noted that by contrast with all the other categories of extreme events, the frequency of frost and cold spells is projected to decrease in the future, with both positive impacts (e.g., decrease cold-related morbidity and mortality, reduced heating energy demand and reduced risk of damage to a number of crops) and some negative impacts (e.g., increased risk to some other crops and extended range of activity of some pest and disease vectors) (IPCC, 2001)

attempts to incorporate climate catastrophes in the estimates of the damage cost of climate change and the social cost of carbon, more accurately, in the context of an optimal carbon tax. The authors' approach was to factor the costs of major environmental risks into Nordhaus' RICE model on the basis of a previous survey of a small number of experts. The latter were faced with two different global warming scenarios (+ 3 C° in GMT 2090 and + 6 C° in GMT in 2175) and were asked to state their subjective probabilities that the world might suffer a permanent loss of 25% of global income. To reflect growing concerns about catastrophic impacts in the second half of the 1990s, Nordhaus and Boyer more than doubled the probability of a climatic catastrophe for a 2.5 C increase in GMT and doubled the equivalent probability for a 6 C increase. They also increased the damage cost estimate by 20%. Finally, they estimated WTP to avoid catastrophic impacts based on the assumptions that countries are risk-averse. On this basis, Nordhaus and Boyer estimated that in expected utility terms catastrophic impacts could account for 1% of global GDP for a GMT increase of 2.5 C° and could be as high as 7% of global GDP for a GMT increase of 6 C°.

Keller *et al.* (2000) used Nordhaus's DICE model to investigate the impact of the risk of a collapse of thermohaline circulation (THC) in the North Atlantic, which is one of the major low-probability catastrophic events. Having constrained concentrations of greenhouse gases at the level beyond which THC is supposed to collapse, they found that a much higher level of abatement is required compared to the optimal path when the risk of THC collapse is ignored. They also find that even relatively low (and in their view plausible) damage costs associated with THC collapse (less then 1% of global GDP) would justify this higher level of abatement in cost-benefit terms than when THC collapse is ignored.

In a more recent paper, Keller *et al.* (2003) adopted a similar constrained optimization approach to analyse the impacts on optimal emissions paths of a constraint to avoid a different climate catastrophe (i.e., the West Atlantic ice sheet disintegration) and the widespread loss of unique ecosystems (through coral bleaching). They found that preventing either of these events would require limiting GHG concentration below a much lower threshold than the optimal threshold when these risks are ignored (as in the standard RICE). They also found that while adopting policies to reduce the risk of a West Antarctic ice sheet disintegration allow for a smoother abatement path, trying to prevent widespread coral bleaching would call for much more drastic (and costly) cuts in emissions while possibly being an unavoidable outcome. However in this latest work by Keller et al no attempt was made to explicitly estimate the costs associated with these events.

Mastrandrea and Schneider (2001) coupled Nordhaus's DICE model to a climate-ocean model that included the risk of a THC collapse. Using ad-hoc assumptions about the loss of global GDP that this event might cause (i.e., a 1% loss of GDP under an optimistic scenario and a 5% loss of GDP under a more pessimistic scenario) they then moved on to assess the potential impacts on an optimal carbon tax. While the authors themselves warned against attaching much importance to specific values, they show how the inclusion of this climate catastrophe can have a significant impact on optimal carbon taxes but only when a zero PRTP (or a low, hyperbolic

PRTP) is used. This is because the risk of reversal in THC tends to become significant only towards the end of the 21st century or the early 22nd century.

Baranzini *et al.* (2003) add ad-hoc assumptions on the risk of climate catastrophes to the IAM developed by Cline (1992) and assess the potential implications within a "real option" approach to policy assessment. Their results show that while the combined effect of uncertainty and learning is likely to delay the adoption of abatement policies, the introduction of low-probability catastrophic events obtains the opposite effect by increasing the discounted stream of benefits from emissions abatement.

Finally, although they do not explicitly report impacts on damage cost estimates or optimal carbon taxes, Azar and Lindgren (2003) demonstrate how the inclusion of an illustrative (and equally subjective) low-probability catastrophic damage function in a stochastic version of Nordhaus's DICE model would significantly affect the optimal emissions path in a situation of sequential decision-making under uncertainty.

3.4 *Socially-contingent effects*

In addition to catastrophic events, current estimates of the SCC do not account for the so-called "socially contingent effects" of climate change. The underlying hypothesis is that environmental stress induced (or aggravated) by climate change – in interaction with a complex of social, economic and political factors — could contribute to generate impacts along a chain that goes from distress migration to severe societal, economic and political crisis, leading ultimately to conflict (Barnett, 2003; Brauch, 2002). This includes the possibility that either the increased frequency of extreme weather events or "high" climate change projections could lead to destabilisations of small to medium economies in sensitive regions of the world (e.g., South East Asia, Central America, small Pacific islands, etc.).

Barnett (2003) emphasised that while the theoretical and empirical debate has reached some degree of consensus on the presence of some links between environmental change and conflict, it has not yet reached the state where confident predictions can be made. This is precisely because climate change-induced environmental stress interacts with several socially-contingent factors in determining the risk of conflict. For instance, even if environmental stress might indeed produce episodes of mass migration, the possibility of this turning into conflict (whether political or violent) will crucially depend on the political and institutional responses in the host countries. Overall, Barnett concludes that "a research programme looking to empirically investigate climate-conflict linkages in greater detail would be most effectively targeted at the sub-state level in countries where governance systems are in transition, levels of inequality are high, social-ecological systems are highly sensitive to climate change, and which have a history of large-scale migration."

While in the long-term the research agenda proposed by Barnett could help improve the understanding of socially-contingent effects, their context-dependent nature is likely to continue make their inclusion in IAMs, and hence their coverage in SCC estimates, challenging. Nonetheless socially-contingent effects are important to policy decision makers (also in terms of the associated geopolitical risks). It is therefore crucial to identify a suitable way of incorporating concern for these effects into the assessment of climate change policies, while keeping in mind that these impacts are not reflected in current SCC estimates and will probably not be robustly covered in medium-term estimates either.

4. Using SCC estimates in policy decision-making

4.1 Key issues for policy decision-makers

The interest of UK Government departments and agencies in the SCC has been driven by an increasing demand among policy decision-makers for monetary benchmarks that can be applied in a transparent and consistent way to policy assessment and to inform the setting of economic instruments. This has been part of a more general trend to a wider endorsement of economic valuation of non-market impacts in policy appraisal, as reflected for instance in the latest UK Government guidance to economic appraisal (HM Treasury, 2003). Illustrative SCC estimates have been applied in a number of economic analyses and regulatory impact assessments (RIAs) across several policy areas including energy, road transport, aviation and waste management. Nonetheless, the significant uncertainties and the partial coverage of SCC estimates stand in the way of a more systematic use of SCC estimates to inform policy decisions.

The discussion in section 2 has highlighted the sensitivity of SCC estimates to economic assumptions and value judgements in terms of approaches to valuation and aggregation over time and across different regions While the theoretical and empirical debate on issues like discounting, equity weighting and valuation techniques is still open, choices can be made on these critical economic assumptions. For instance, UK appraisal guidance recommends a decreasing discount rate schedule for long-term decisions (starting from 3.5% for the first thirty years) and appears to endorse equity weighting. In any case we believe it is extremely important that policy decision-makers are fully informed about these underlying drivers of SCC estimates, which should not be lost among modelling details.

The discussion in section 3 has further highlighted that in terms of climate change impacts and dynamics two further issues need to be addressed to increase the confidence of policy decision makers in SCC estimates:

- Even with reference to the impacts that are currently being covered, single SCC estimates or even fairly wide ranges might not fully represent the major underlying uncertainties and risks which generally tend to become more significant the more we move away from a market context.

- SCC estimates tend to only represent a subtotal of the true costs of emissions. In particular, the way in which these estimates and their underlying models address important, policy-relevant issues like low-probability catastrophic events and socially-contingent effects deserves greater attention.

Catastrophic events probably represent the single major gap in coverage of current SCC estimates, also in consideration of the fact that abrupt, unpredictable changes make adaptation of human as well as ecological systems much more difficult. Until low-probability catastrophic events are better covered in SCC estimates the latter will only represent a subtotal of climate change impacts. In turn this has implications for the way in which they can be used to inform policy decisions.

The lack of coverage of climate catastrophes also has implications in terms of communication to policy decision-makers of IAMs modelling results and of any associated monetary estimates. Azar and Schneider (2001) argued that when low-probability catastrophic events are excluded from a model then it should be stated clearly that the analysis has omitted one of the key concerns in relation to climate change. We believe that this warning should be extended to SCC estimates that are based on such models.

4.2 From narrow ranges to broad ranges

Tol and Downing (2002) recognise that current SCC estimates "do not necessarily cover the wider range of social and environmental values that might be considered appropriate at a global or regional level".

Based on the insights from Tol and Downing, some members of the ExternE team suggested in a recent report to the European Investment Bank (AEA Technology, 2003) that instead of focusing on single values a more appropriate way for including SCC estimates in project appraisal would be to test the sensitivity of NPVs to a low value of 5 Euro/tC (USD 5.5/tC or £3/tC) and a high value of 125 Euro/tC (USD 138/tC or £80/tC). However Tol and Downing also suggest that higher values might be adopted at a programme or strategic level, where the analysis takes a longer-term perspective and should therefore consider more diverse impacts and values.

Finally, Tol and Downing suggest that a wider range of valuations (often looking at non-marginal effects) might be employed when looking at environmental sustainability issues in a way that incorporates maximum probable losses and irreversible impacts. Tol and Heinzow (2003) explicitly suggested in recent work for the GreenSense project that different sustainability paradigms (weak, medium and strong) might be more relevant in different decision-making contexts (respectively, project appraisal, programme appraisal and the setting of long-term stabilisation targets) and that higher values than those produced by integrated models might be acceptable when looking at strategic choices.

But even without invoking strong sustainability there are good reasons to think that dealing with fundamental strategic choices in climate change policy might be different than dealing with relatively minor changes in emissions. SCC estimates are typically measured in IAMs with reference to relatively small changes in emissions with respect to a business-as-usual emission pattern. There is an issue, however, of whether these estimates are suitable to assess non-marginal changes that are potentially implied by different long-term stabilisation targets. On the one hand there is a classic cost-benefit analysis issue that for non-marginal changes in prices (including shadow prices) stop representing a good approximation of changes in utility and therefore raise the need for fully-specified utility functions. Perhaps more importantly in this case, the physical damage function might be characterised by discontinuities above certain GMT thresholds and non- marginal changes in emissions might significantly increase the probability of triggering those impacts that are currently not covered in SCC estimates (i.e., low-probability catastrophic events and socially-contingent impacts).

4.3 *The trade-off between representation of uncertainty and consistency*

Overall, the combined impacts of uncertainty, dependency on value judgments and lack of coverage of important impact categories would suggest the use of broad ranges for the SCC in economic assessment. It might possibly also suggest that more strategic decision-making contexts should make use of broader ranges of value (see Figure 2).

Figure 2. Drivers of climate change damages and their policy application

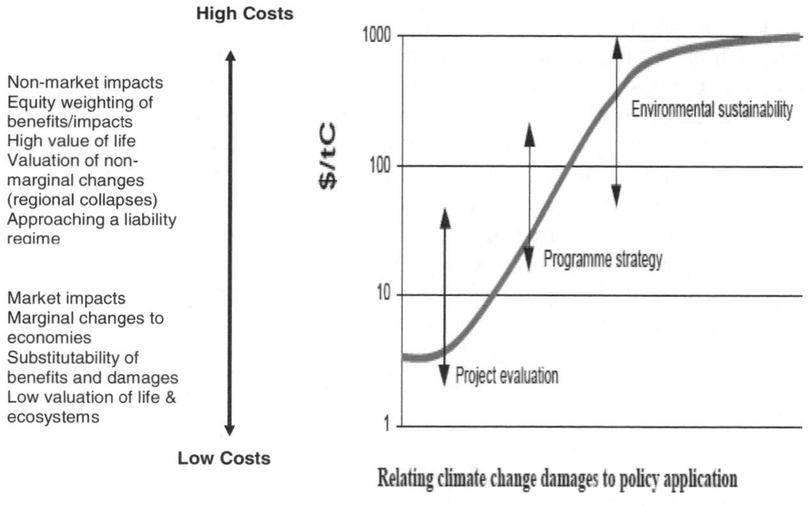

Relating climate change damages to policy application

Source: Adapted from Downing and Watkiss (2003).

Note:

As Downing and Watkiss (2003) note, studies that derive either low or high marginal social costs make use of a number of in-built assumptions. Those which derive low marginal social costs are likely to focus only on market impacts. In contrast, studies that derive higher marginal social costs are likely to consider the impacts of non-market damages, as well as address concerns for equity among world-wide impacts. As the authors highlight, certain policy decision contexts can typically be associated with a range of marginal social costs of climate change.

One of the key attractions of SCC estimates is that they can be used as a consistent benchmark for decision making at all levels, with associated benefits in terms of allocation of resources. In this sense, the use of broad SCC ranges discussed above might raise two issues of consistency:

- "Horizontal consistency" where broad ranges introduce a discretionary element in ranking of policies and measures;
- "Vertical consistency" where implicit adoption of higher values for strategic decisions introduces an element of discrepancy between different decision making levels.

There is therefore a trade-off between adopting approaches that adequately respond to the main concerns for policy decision makers in relation to the SCC and the benefits from adopting a consistent approach across policies for valuing carbon savings. Pushing this trade-off to its limit, the efficiency gain of adopting a consistent decision rule could be considerable in spite of the risk of consistently using an excessively optimistic or pessimistic value.

4.4 SCC estimates in setting economic instruments

One context in which views on the SCC ultimately need to converge to a single number is the setting of "Pigouvian" economic instruments aimed at internalising the carbon externality. Clarkson and Deyes (2002) described the "marginal cost" and the "cost-benefit" approach (CBA) to the estimation of the social cost of carbon. Instead of directly looking at the marginal cost of each tonne of carbon emitted under the BAU, the latter approach adopts an intertemporal optimisation framework and seeks to determine the level of a carbon tax that would deliver the socially optimal level of emissions through time. The required optimal tax per tC is the one that bridges the gap between private marginal damage and social marginal damage in correspondence of the intersection of the marginal damage cost curve (MDC) with the marginal abatement cost curve (MAC). The value of the optimal carbon tax will generally differ from the marginal cost of emissions under the BAU. Arguably, this is the value that we should look at when designing economic instruments. However the available CBA studies are based on very simplistic representations of both the MAC and MDC curves.

4.5 Towards a formal treatment of risk and uncertainty

Independently of the different views that different commentators might have on the use of broad ranges or of different ranges of values at different levels of decision making, a better representation and a formal assessment of the uncertainty surrounding SCC estimates should certainly figure highly on the IAMs development agenda.

More specifically, Ulph and Ingham (2003) observe that most IAMs model are based best-guesses or estimated values of uncertain parameters, and then compute the SCC estimates as if these parameters were known with certainty, at most carrying out some kind of simple sensitivity analysis. The same authors argue that IAMs could instead adopt a more sophisticated form of sensitivity analysis on key parameters, based on random sampling from fully specified probability distributions for the latter. They also find that as climate change is a "right – skewed" issue, (characterised by the fact that negative surprises are more likely than positive ones (Tol 2003c), a greater incorporation of uncertainty is *ceteris paribus* likely to increase SCC estimates.

For instance, in a recent paper looking at the uncertainty surrounding climate change, Tol (2003a) applies a Monte Carlo analysis on key variables in his FUND model to show that there is a very small - yet not zero - probability that due to water scarcity economic growth might reverse in some regions of the world (namely Central and Eastern Europe and the former Soviet Union) until they regress to subsistence levels. Under such a scenario the negative growth in GDP per capita would translate in a negative discount rate and, as these economies collapse, the discount factor and hence the SCC would tend to be infinite. It should be noted however that Tol cautions against drawing any policy implications from these result, the focus of his paper being on whether uncertainty about climate change is too large to apply CBA as a policy decision-making rule in a situation in which the variance of the expected SCC is infinite.[84].

Probabilistic modelling can also help to explicitly assess the value associated with those courses of action that keep future policy options alive (including the value of learning about climate change impacts) in a decision-making context characterised by risks and irreversibility. There are several studies, including Azar and Lindgren (2003) or Baranzini *et al.* (2003) mentioned above, that have tried to assess empirically how strategic climate change policy choices might vary when uncertainty, learning and irreversibility are explicitly taken into account, particularly in order to deal with low-probability catastrophic events. In order to be operational, these "real option" approaches rely on subjective probability distributions for the future states of the world as well as on ad-hoc assumptions about the associated pay-offs and on the timing required for the uncertainty about climate change to be solved. Nonetheless they might provide a way of investigating the precautionary principle, much invoked in the climate change policy arena.

It is worth noting that while inclusion of catastrophic impacts into IAMs is likely to raise SCC estimates, Ulph and Ingham (2003) conclude that a closer look at the issue of irreversibility and learning *per se* through the lens of economic analysis reveals the direction of these effects on optimal abatement paths (and therefore on the implicit

---

84 The commentaries on Tol by Yohe (2003), Azar and Lindgren (2003) and Howarth (2003) included the same issue of *Climatic Change* provide interesting discussions of this broader issue of whether stochastic CBA (based on the maximisation of discounted expected net benefits) is an appropriate decision-making framework for climate change.

"optimal value" of carbon savings) to be ambiguous. This is because the prospect of obtaining better information in the future may actually translate into a lower level of abatement compared to a situation where uncertainty, learning and irreversibility are ignored. On the other hand, Ulph and Ingham acknowledge that the findings of the studies that they reviewed are sensitive to the underlying modelling assumptions (e.g., the choice of a functional form specification for utility functions) and also rely on simplified representations of uncertainty, learning and irreversibility.

Whilst the development of IAMs into probabilistic frameworks looks like a promising area of work and might provide new insights on the SCC, a better treatment of the uncertainty surrounding the valuation of climate change impacts would also require a more transparent representation of how summary values are derived and of the confidence that can be assigned to their constituent elements. The risk-based approach to valuation proposed by Downing and Watkiss (2003) relies on a matrix that combines different elements of uncertainty on climate change and climate change impacts with different degrees of confidence in economic valuation for different impact categories. Along the climate change dimension the matrix moves from those impacts which are generally well understood in global climate change models (e.g., predicted gradual increase in GMT and consequent loss of dry land) through to those which stem beyond current means of estimation and where little consensus exists (e.g., major shifts in the climate system and surprises). Along the valuation dimension, the matrix moves from market impacts through to non-market and socially contingent impacts, the latter being those impacts for which little confidence can be attached to monetary values. Downing and Watkiss argue that while current SCC estimates reflect a good understanding of the "top-left" quadrant of the matrix (i.e., market impacts associated with slow, gradual change), confidence in monetary estimates of the damage of climate change tends to decrease as one moves to the outer quadrants of the matrix (i.e. in a 'South-East direction' through the matrix). The also suggest that consideration of all aspects of the risk matrix would lead to fundamentally different SCC values. This framework allows for a transparent, disaggregate representation of the current coverage of SCC estimates and of the associated uncertainties. At the same time, it provides some structure for further research aimed at reducing these uncertainties and produce more robust SCC estimates.

Downing and Watkiss (2003) suggest that the most up-to-date estimates of damages relate to climate risk modeling for projections and market values (A), whilst some improvements are possible for non-market costs (B). The authors conclude that it is unlikely that significant progress can be made in estimates of the social and economic costs of surprise 'abrupt' climate change and for socially-contingent damages in the medium term.

Figure 3. A risk-based matrix for the valuation of climate change impacts

		Market	Non-Market	Socially contingent
A	**Projection**	Lower projections recently	Some improvements	No real quantitative assessment
B	**Bounded risks**	Some improvements possible	Model improvements, little empirical basis	No improvement likely in near term
C	**System change and surprise**	Speculative assessment	Not included in near term	Not likely to be valued in medium term

A: region of high confidence in both climate impacts and valuation of damages
B: possible to explore range of values with formal risk assessment techniques, but empirical basis of estimates will be weak and unstable (subject to new models)
C: very difficult to quantify even using subjective estimates due to diverse views of climate impacts (and adaptive capacity) and potential for low probability-high impact scenarios

Source: Downing and Watkiss (2003).

5. Conclusions and key issues for a review of the SCC

5.1 Key conclusions

This paper has dealt with the causes that have led to variations in the observed estimates of the social cost of carbon (SCC). In particular, two major areas of concern have been identified.

In the first instance, economic assumptions which have been extensively covered in the literature are still open to the question of value judgements, and the way in which these affect SCC estimates. Several key parameters were discussed in this context and their contribution to the variability on final estimates analysed. These included:

- The notion of equity weighting.
- The determination of the rate at which the value of future impacts should be discounted to today's prices, and in particular, the idea that a decreasing discount rate is more appropriate for global warming given the long-term nature of the issue, and the functional form that such a decrease should take.
- The inherent uncertainty involved in the valuation of impacts of climate change for which no obvious market price exists. Realisation of the importance of property right allocation bound up with the willingness to pay/accept compensation approaches was identified in this respect.

Furthermore, the need for the modelling community to improve the way in which adaptation is factored into IAMs was identified, given that current models either under or over estimate the extent to which adaptation drives the social costs of carbon.

In the second instance, limitations in the coverage of several key climate change issues was seen as limiting the usefulness of any social cost of carbon estimates. These relate to areas that are either covered to some extent in current IAMs or not covered at all. With reference to the former, there is a need to make sure that IAMs reflect the latest understanding on the impacts of climate change and their economic valuation, particularly in areas like biodiversity, ecosystems, extreme weather events and abrupt change. With regard to the latter two, the two issues explicitly highlighted were those of low-probability catastrophic events and socially-contingent impacts. Until these are better understood, and subsequent probability distributions factored into SCC estimates, any monetary estimate will only represent a subtotal of the true value of the marginal damage cost of carbon dioxide emissions. This clearly has implications for the extent to which current estimates may be used as a basis to inform policy decisions.

5.2 *Issues for further review*[85]

The longer-term improvement of our understanding of the SCC is likely to require several developments in IAMs and in the underlying impact science. Nonetheless some key areas have emerged as priorities for further modelling work aimed at improving the treatment of those impacts that SCC estimates already try to incorporate to some extent, as well as at expanding their coverage to include catastrophic and abrupt change. In particular these include:

- The need to scrutinize and possibly improve adaptation assumptions across all categories of impacts where these are perceived to be simplistic.

- The need to ensure the appropriate coverage and valuation of non-market impacts in IAMs, particularly of damages to ecosystem functions.

- The need to better address the costs associated with climate variability and extreme events and to explore ways in which they could be better represented in IAMs.

- The need to explore ways of incorporating low-probability, catastrophic events into IAMs. In the medium term this is likely to require subjective probability distributions for low-probability catastrophic events and ad-hoc assumptions about damages.

---

85 See also the Annex which outlines current research supported by the UK Defra in this area.

- The need to address concerns about socially-contingent effects by exploring ways in which the associated risks could be quantitatively or qualitatively incorporated in SCC estimates.

- The case for looking at time varying discount rates and their interactions with the treatment of low-probability catastrophic events.

- The need to quantify and represent uncertainties more fully in IAMs through probabilistic sensitivity analysis on key model parameters, whether they are parameters for which agreed probability distributions exist in the literature or parameters for which subjective assumptions are adopted within a risk-based approach.

However, even if modeling work in the medium and longer-term might successfully address some of the major concerns on the coverage and robustness of SCC estimates, these are likely to remain characterised by considerable uncertainty. Indeed, a better representation of risks and uncertainty and a broader coverage of impacts in IAMs might well result in broader ranges of SCC estimates, moving away from small ranges to wider ranges, probability distributions or even bounded values (i.e., higher and lower bounds with no specified probability distribution in between). However, this raises issues on how best SCC estimates could be incorporated in policy decision-making, particularly in terms of:

- The trade-offs between adopting wide ranges for the SCC and consistency of decisions both in terms of implied SCC values and in relation to marginal abatement costs estimates.

- The need to adopt SCC values in the setting of economic instruments in the face of uncertainty and taking account the principle of precautionary action.

Finally, in order to address those categories of impacts that are currently not being covered by IAMs and that are not likely to be covered in the foreseeable future, SCC estimates could be nested within extended, multi-criteria approaches to policy decision making. This type of framework – which would be particularly suitable to strategic climate policy decisions – could also be adapted to offer a transparent representation of the distributional implications of the costs of climate change, which inevitably tend to be lost when looking at figures for globally aggregated damages.

Annex: Ongoing UK research

At the time of writing an inter-departmental group led by the UK Department for Environment, Food and Rural Affairs (Defra) has recently commissioned two major research projects in the context of the social cost of carbon review that should start addressing several of the issues that have been highlighted above. In particular:

- A research consortium led by the Stockholm Environment Institute and including AEA Technology Environment, the Centre for Policy Modelling at Manchester Metropolitan University, the University of Hamburg, Imperial College and Metroeconomica is carrying out a study of the SCC with focus on exploring the nature of uncertainties and the extent of consensus around both central estimates and the range of estimates. Possible IAM developments are being explored as part of this analysis.

- A research consortium lead by AEA Technology Environment and including the Stockholm Environment Institute, the University of Hamburg and Metroeconomica is looking at how best to incorporate SCC estimates in relevant decision making contexts, given the uncertainty which affects monetisation of global damage.

- The risk-based matrix discussed above will constitute the framework within which estimates of the SCC are developed and understood.

Both projects are expected to report in 2004.

Acknowledgements

The first version of this paper was prepared as a background document to inform a UK review of the social cost of carbon and an international seminar that the UK Department for Environment, Food and Rural Affairs (Defra) hosted in London on 7th July 2003. The current text has been revised and developed in light of the seminar papers and discussion. We are grateful to Jan Corfee-Morlot at the OECD for her comments and for pointing us to some important references. We are also grateful to colleagues in Defra and at the Department of Trade and Industry for their valuable comments on previous drafts of the paper.

References

AEA Technology, (2002), "Quantification and Valuation of Environmental Externalities: The Case of Global Warming and Climate Change", a Report produced for the European Investment Bank, Culham (Oxon), UK, August.

Alley, R. B., J. Marotzke, W.D. Nordhaus, J.T. Overpeck, D.M. Peteet, R.A. Pielke Jr., R.T. Pierrehumbert, P.B. Rhines, T.F. Stocker, L.D. Taltey and J.H. Wallace (2003), "Abrupt Climate Change", Review, *Science*, **299** (March).

Azar, C. and K. Lindgren (2003), "Catastrophic Events and Stochastic Cost-Benefit Analysis of Climate Change — Editorial", *Climatic Change*, Issue 56, 245-255.

Azar, C. and S.H. Schneider, (2001), "Are Uncertainties in Climate and Energy Systems a Justification for Stronger Near-Term Mitigation Policies?", paper prepared for the PEW Centre on Global Climate Change workshop on the timing of climate change policies.

Baranzini, A., M. Chesney and J. Morriset (2003), "The Impact of Possible Climate Catastrophes on Global Warming Policy", *Energy Policy*, No. 31, 691-701.

Barnett, J. (2003), "Security and Climate Change", *Global Environmental Change*, No. 13, 7-17.

Brauch, H.G. (2002), "Climate Change, Environmental Stress and Conflict", AFES-PRESS report for the Federal Ministry for the Environment, Nature Conservation and Nuclear Safety, Berlin, Germany.

Clarkson, R. and K. Deyes (2002), "Estimating the Social Cost of Carbon Emissions", Government Economic Service Working Paper 140, London, UK.

Cline, William R. (1992), "The Economic Benefits of Limiting Global Warming", in *The Economics of Global Warming*, Chapter 3, Washington, DC, Institute for International Economics, 81-138.

Cowell, F.A. and K. Gardiner (1999), "Welfare Weights", Report to the UK Office of Fair Trading, STICERD, London School of Economics, Economics Research Paper 20, August 1999.

Demeritt, D. and D. Rothman., (1999), "Figuring the Costs of Climate change: an Assessment and Critique", *Environment and Planning*, **31**, 389-408.

Downing, T. and P. Watkiss (2003), "The Social Costs of Carbon in Policy Making: Applications, Uncertainty and a Possible Risk Based Approach", paper presented at the Defra International Seminar on the Social Cost of Carbon, 7th July 2003, London, UK.

Eyre, N., Downing, T., Rennings, K. and R.S.J. Tol (1998), "Global Warming Damages", report to the European Commission, Externalities of Energy, **8**, September, European Commission, Brussels, Belgium.

Fankhauser, S. (1995), Valuing Climate Change: the Economics of the Greenhouse, Earthscan, London, UK.

Fankhauser, S., R.S.J. Tol and D.W. Pearce (1997), "The Aggregation of Climate Change Damages: A Welfare Theoretic Approach", *Environmental and Resource Economics*, **10**, 249-266.

Hanemann, M. (1991), "Willingness to pay and willingness to accept: how much can they differ?" *American Economic Review,* **81,** 635-647.

Hanemann, M. (1999), "The economic theory of WTP and WTA", in I. Bateman and K. Willis (eds.), *Valuing Environmental Preferences: Theory and Practice of the Contingent Valuation Method in the US, EU and Developing Countries*, Oxford: Oxford University Press, 42-96.

Hitz, S. and J. Smith (2004), "Estimating global impacts from climate change", in *The Benefits of Climate Change Policies*, OECD, Paris.

HM Treasury, (2003), *The Green Book – Appraisal and Evaluation in Central Government*, The Stationery Office, London, UK.

Horowitz, J.K. and K.E. McConnell (2002), "A Review of WTA/WTP Studies", *Journal of Environmental Economics and Management* **44**(3), 426-447, R.B. Howarth (2003), "Catastrophic Outcomes in the Economics of Climate Change — An Editorial Comment", *Climatic Change*, Issue 56, 257-263.

Howarth, R. (2003), "Catastrophic Outcomes in the Economics of Climate Change", *Climatic Change* **56**, 257-263.

IPCC (1996), *Climate Change 1995: The second assessment report of the Intergovernmental Panel on Climate Change*, Cambridge University Press, Cambridge, UK.

IPCC (2001), *Climate Change 2001: Synthesis Report: Contribution of Working Groups I, II and III to the Third Assessment Report of the Intergovernmental Panel on Climate Change*, Cambridge University Press, Cambridge, UK.

Kann, A. and J.P. Weyant, (2000), "Approaches for Performing Uncertainty Analysis in Large-Scale Energy/Economic Policy Models", *Environmental Modelling and Assessment*, **5**(1), 29-46, Stanford, USA.

Keller, K., K. Tan, F. Morel and D. Bradford (2000), "Preserving the Ocean Circulation: Implications for Climate Policy", *Climatic Change*, **47**, Issue 1-2, 17– 43.

Keller, K., M. Hall, S.R. Kim, D.F. Bradford and M. Oppenheimer (2003), "Avoiding dangerous anthropogenic interference with the climate system", (submitted to *Climatic Change*, October 2003).

Leemans, R. and B. Eickhout (2004), "Another reason for concern: regional and global impacts on ecosystems for different levels of climate change", *Global Environmental Change: Special Edition on the Benefits of Climate Policy* **14**: 219-228.

Mastrandrea, M.D. and S.H. Schneider (2001), "Integrated Assessment of Abrupt Climate Changes", *Climate Policy*, No. 1, 433-449.

Mendelsohn, R. (2003), "The Social Cost of Carbon: An Unfolding Number", paper presented at the Defra International Seminar on the Social Cost of Carbon, 7th July 2003, London, UK.

Newell, R. and W. Pizer (2001), "Discounting the benefits of climate change mitigation: How much do uncertain rates increase valuations?", paper prepared for the PEW centre on Global Climate Change, Washington D.C.

Nordhaus, W. and J. Boyer (2000), "Warming the World: Economic Models of Global Warming". Cambridge, Mass: MIT Press, USA.

OXERA (2002), "A Social Time Preference Rate for Use in Long-Term Discounting", Report for the Office of the Deputy Prime Minister, the Department for Transport and the Department for Environment, Food and Rural Affairs, London, UK, December.

Pearce, D.W., W.R. Cline, A. Achanta, S. Fankhauser, R. Pachauri, R. Tol and P. Vellinga (1996), "The social costs of climate change: greenhouse damage and the benefits of control", in *Intergovernmental Panel on Climate Change, Climate Change 1995: Economic and Social Dimensions of Climate Change*, Contribution of Working Group III to the Second Assessment Report of the Intergovernmental Panel on Climate Change, Cambridge University Press, Cambridge, UK, 183-224.

Pearce, D.W. (2002), "The Role of 'Property Rights' in Determining Economic Values for Environmental Costs and Benefits", report to The Environment Agency (Draft 1), Bristol, UK.

Pearce, D.W. (2003a), "The Social Costs Of Carbon And Its Policy Implications", Oxford Review of Economic Policy, **19**(3), London, UK.

Pearce, D.W. (2003b), "International Seminar on The Social Cost of Carbon: Rapporteur's Summary", presented at the Defra International Seminar on the Social Cost of Carbon, 7th July 2003, London, UK.

Schneider, S.H. and J. Lane (2004), "Abrupt non-linear climate change and climate policy", in *The Benefits of Climate Change Policies*, OECD, Paris.

Tol, R.S.J. (2003a), "Is the Uncertainty about Climate Change too Large for Expected Cost-Benefit Analysis?", *Climatic Change*, Issue 56, 265-289.

Tol, R.S.J. (2003b), "The Marginal Costs of Carbon Dioxide Emissions: an Assessment of the Uncertainties", Working Paper FNU-19, Hamburg University, Germany.

Tol, R.S.J. (2003c), "The Marginal Costs of Carbon Dioxide Emissions", paper presented at the Defra International Seminar on the Social Cost of Carbon, 7th July 2003, London, UK.

Tol, R.S.J. and T. Downing, (2002), "Appendix 4: Marginal Costs of Greenhouse Gas Emissions", in *AEA Technology, Quantification and Valuation of Environmental Externalities: The Case of Global Warming and Climate Change*, a report produced for the European Investment Bank, August 2002.

Tol, R., T. Downing, O. Kuik, and J. Smith (2004), "Distributional aspects of climate change impacts", *Global Environmental Change: Special Edition on the Benefits of Climate Policy* **14**: 259-272.

Tol, R.S.J. and T. Heinzow (2003), "External and Sustainability Costs of Climate" in A. Markandya (ed.) GreenSense Final Report, University of Bath, Bath, UK.

Ulph, A. and A. Ingham, (2003), "Uncertainty, Irreversibility, Precaution and the Social Cost of Carbon", paper presented at the Defra International Seminar on the Social Cost of Carbon, 7th July 2003, London, UK.

Van Asselt, M.B.A. and J. Rotmans (1999), "Uncertainty in Integrated Assessment Modelling: A Bridge Over Troubled Water", paper for the EFIEA and SEI 1999 Matrix Workshop on Uncertainty, International Centre for Integrative Studies, Maastricht University, The Netherlands.

Weitzman, M. (1998), "Why the far distant future should be discounted at its lowest possible rate", *Journal of Environmental Economics and Management* **36**, 201-208.

Weitzman, M. (1999), "Just keep on discounting, but…", in P.Portney and J.Weyant (eds), *Discounting and Intergenerational Equity*, Washington DC: Resources for the Future.

Wood, D. (2002), "Estimating the Social Cost of Carbon: Models, decreasing Discount Rates and Sensitivity Analysis", MSc Ec. dissertation, University of Surrey.

Yohe, Gary W. (2003), "More Trouble for Cost benefit Analysis", *Climatic Change* **56**, 235-244.

Chapter 7

MODELLING CLIMATE CHANGE UNDER NO-POLICY AND POLICY EMISSIONS PATHWAYS

by Tom M.L. Wigley,

National Centre for Atmospheric Research, USA

Future emissions under the SRES scenarios are described as examples of no-climate-policy scenarios. The production of policy scenarios is guided by Article 2 of the UN Framework Convention on Climate Change, which requires stabilization of greenhouse-gas concentrations. It is suggested that the choice of stabilization targets should be governed by the need to avoid dangerous interference with the climate system, while the choice of the pathway towards a given target should be determined by some form of cost-benefit analysis. The WRE concentration profiles are given as examples of stabilization pathways, and an alternative 'overshoot' pathway is introduced. Probabilistic projections (as probability density functions – pdfs) for global-mean temperature under the SRES scenarios are given. The relative importance of different sources of uncertainty is determined by removing individual sources of uncertainty and examining the change in the output temperature pdf. Emissions and climate sensitivity uncertainties dominate, while carbon cycle, aerosol forcing and ocean mixing uncertainties are shown to be small. It is shown that large uncertainties remain even if the emissions are prescribed. Uncertainties in regional climate change are defined by comparing normalized changes (i.e., changes per 1°C global-mean warming) across multiple models and using the inter-model standard deviation as an uncertainty metric. Global-mean temperature projections for the policy case are given using the WRE profiles. Different stabilization targets are considered, and the overshoot case for 550ppm stabilization is used to quantify the effects of pathway differences. It is shown that large emissions reductions (from the no-policy to the policy case) will lead to only relatively small reductions in warming over the next 100 years.

ISBN 92-64-10831-9
THE BENEFITS OF CLIMATE CHANGE POLICIES

CHAPTER 7. MODELLING CLIMATE CHANGE UNDER NO-POLICY AND POLICY EMISSIONS PATHWAYS

by Tom M.L. Wigley[86]

1. Introduction

It is now widely accepted that human activities have contributed substantially to climate change over the past century and that such activities will be the dominant cause of change over the 21st century (and probably longer). The Intergovernmental Panel on Climate Change (IPCC), in their Third Assessment Report (TAR), states that "most of the warming observed over the last 50 years is attributable to human activities" (Houghton *et al.*, 2001, p. 10). The observed warming of global-mean temperature over the past 100 years has been about 0.6°C, and the best estimate of the human component of this is 0.4-0.5°C. For the future, the IPCC TAR gives 1.4-5.8°C as the range of warming resulting from human activities over 1990-2100 (Cubasch and Meehl, 2001, Figure 9.14).

The purpose of the present paper is to assess possible future changes in more detail, paying particular attention to uncertainties. These uncertainties accrue through uncertainties in: human population growth, economic growth and technology changes, which in turn determine the emissions of gases that may directly or indirectly affect the climate; in the changes in atmospheric composition that will occur for any given emissions scenario; and in the changes in climate that may occur in response to a given set of atmospheric composition changes.

On the emissions side, two types of pathway into the future are usually distinguished: either we may do nothing to halt the progress of human-induced ('anthropogenic') climate change and simply adapt to the changes that occur – 'no-climate-policy' pathways; or we may decide to follow a set of policies designed to reduce future climate change. Both types of pathway are considered here. There is, of course, a wide range of possibilities for the no-policy case. The 'IPCC Special Report on Emissions Scenarios' (SRES – see below) gives insight into this range of possibilities. Equally, for any given no-policy case there is a wide range of possible

86 National Centre for Atmospheric Research, Boulder, CO 80307, USA.

policy responses. Together, therefore, the no-policy and policy cases form a continuous and overlapping spectrum of possibilities.

On the climate side, global-mean temperature changes are frequently used as an indicator of the magnitude of future change, but it is through the regional details of future changes in temperature, precipitation, storminess, etc. that we will experience the impacts of changes in global-mean temperature. I will consider both global-mean and regional changes, and their uncertainties.

In the following Sections I will consider no-climate-policy scenarios in Sections 2 through 5 and policy cases in Sections 2 and 6. Section 2 considers future emissions under both no-policy and policy assumptions. Section 3 considers the implied changes in global-mean temperature for the no-policy case, while Section 4 elaborates on sources of uncertainty at the global-mean level. In both Sections, a probabilistic approach is used. In Section 5, I discuss the spatial details of future changes in temperature and precipitation and quantify regional uncertainties by investigating differences between different climate models. Section 6 considers global-mean temperature projections under the WRE CO_2 concentration-stabilization pathways. A summary and conclusions are given in Section 7.

2. Future emissions

The primary source of information about future emissions under the 'no-climate-policy' assumption is the set of SRES scenarios from the Special Report on Emissions Scenarios (Nakićenović and Swart, 2000), produced as part of the IPCC TAR. These scenarios are based on four different narrative 'storylines' (labelled A1, A2, B1 and B2) that determine the driving forces for emissions (population growth and demographic change, socioeconomic development, technological advances, etc.) Briefly, the A/B distinction corresponds to an emphasis on (A) market forces, or (B) sustainable development. The 1/2 distinction corresponds to: (1) higher rates of economic growth, and economic and technological convergence between developing and more developed nations; versus (2) lower economic growth rates and a much more heterogeneous world.

Of the 35 complete scenarios, six were selected by IPCC as illustrative cases: one 'marker' scenario from each storyline; and two others from the A1 storyline that are characterized by different energy technology developments. The marker scenarios are labelled A1B-AIM, A2-ASF, B1-IMAGE and B2-MESSAGE, and the two other illustrative scenarios are A1FI-MiniCAM and A1T-MESSAGE. Here, the appended acronyms refer to the assessment models used to calculate the emissions scenarios (details in Nakićenović and Swart, 2000). The technology options in the A1 scenarios are designated by FI for Fossil-fuel Intensive, T for Technological developments focused on non-fossil fuel sources, and B for a Balanced development across a range of technologies. Note that, although they span a wide range of emissions, the illustrative scenarios do not span the full range represented by the 35 complete scenarios. In the analyses below, I use the full set of scenarios.

The SRES scenarios do not incorporate any direct, climate-related emissions control policies. They do, however, incorporate, both directly and indirectly, emissions control policies arising as assumed responses to other environmental concerns. In line with recent research (Grübler, 1998; Smith *et al.*, 2001b), emissions controls for sulphur dioxide (SO_2) are accounted for as responses to acidic precipitation and urban air pollution problems. Non-climate policies, however, have not been accounted for in a fully consistent way. For example, emissions of tropospheric ozone precursors in the scenarios are too high in many scenarios because they imply unreasonably high future levels of ozone in urban environments (Wigley *et al.*, 2002).

For all gases, the SRES scenarios give a wide range of future emissions. Figure 1 illustrates this for CO_2. Wide emissions uncertainties imply correspondingly wide ranges for possible future concentration levels. Figure 2, corresponding to Figure 1, gives the SRES range for CO_2 concentrations (using a 'best-estimate' carbon-cycle model). The extremes do not necessarily correspond to the same scenario over the whole period. Over 2050-2100, the upper extreme is for the A1C-AIM scenario ('C' here indicates a coal-based version of the FI group) and the low extreme is for the B1T-MESSAGE scenario. The different assumptions underlying these scenarios have been described above. Uncertainties like those shown in Figures 1 and 2 lead to uncertainties in future climate change, as will be described further below.

Figure 1. Extremes of fossil CO2 emissions from the full set of 35 complete SRES scenarios

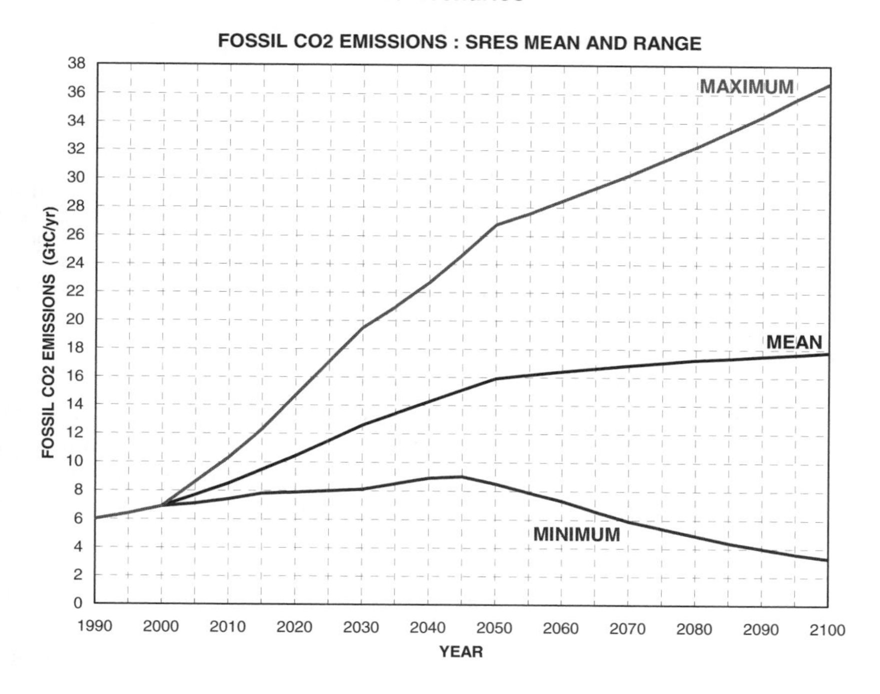

Figure 2. CO2 concentration projections corresponding to Figure 1

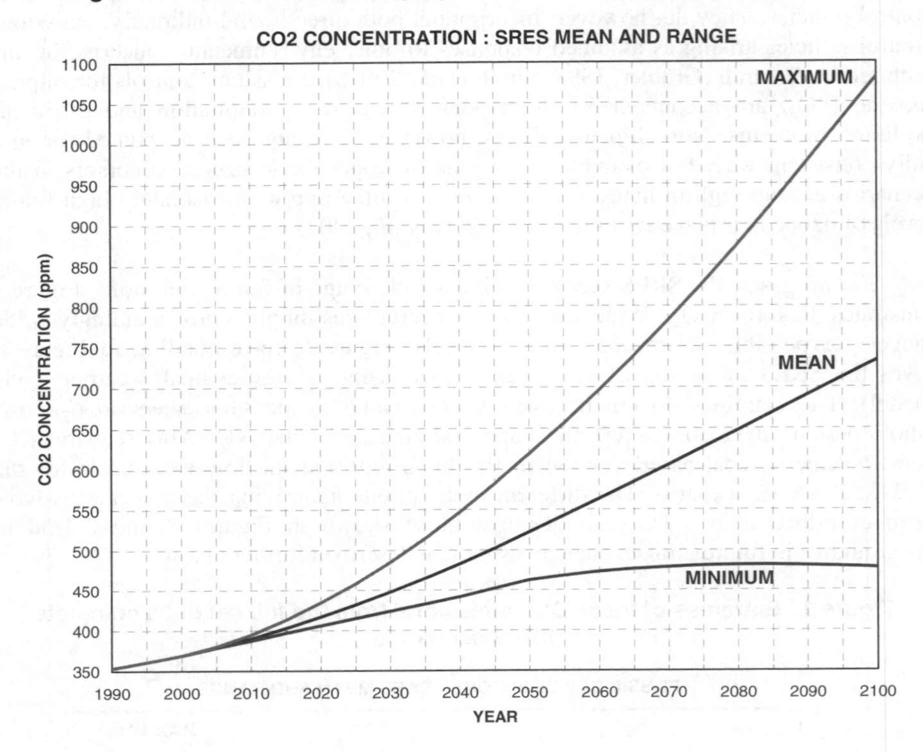

Note:

These results account for the full range of SRES emissions, but they use only central estimates for the concentration for any given emissions scenario. Because of this, they do not span the full range of concentration possibilities, since they do not take account of carbon cycle uncertainties. For example, with strong climate feedbacks, substantially higher concentration values might obtain. In climate projections given in the IPCC TAR (Chapter 9) only central CO_2 concentration projections were used, as in this Figure. In the analyses of Wigley and Raper (2001), carbon cycle uncertainties were accounted for.

For the policy case, there is no equivalent to the SRES scenarios; but there is nevertheless a wide range of possibilities. The guiding principle for policy is Article 2 of the UNFCCC, which has as its ultimate objective "…stabilization of greenhouse gas concentrations in the atmosphere at a level that would prevent dangerous interference with the climate system … achieved within a time frame … to enable economic development to proceed in a sustainable manner".

The key words here are 'stabilization' and 'dangerous interference'. The choice of a target for stabilization depends on what is considered to be dangerous interference with the climate system, a particularly difficult question that is beyond the scope of the present paper. We circumvent the problem by considering different stabilization targets. Note that the goal is to stabilize the concentrations of <u>all</u> greenhouse gases, which would effectively stabilize the climate system. Of course, the climate system is never strictly stable, being continually subject to internally-generated variability and to the influences of natural external forcing factors, such as volcanic eruptions and changes in solar irradiance. We should therefore interpret stabilization of the climate system to mean 'within the limits of natural variability'.

Most studies of possible policy pathways have focused on stabilization of CO_2 concentration alone, since CO_2 has been and will continue to be the primary cause of anthropogenic climate change – a notable exception is the work of Manne and Richels (2001). For CO_2, a widely-used set of CO_2 concentration stabilization pathways (or 'profiles') has been devised by Wigley, Richels and Edmonds (1996), updated versions of which (Wigley, 2004a) are shown in Figure 3. These profiles, commonly referred to as the 'WRE' profiles, were designed to take account of the economic costs of reducing CO_2 emissions below a no-policy baseline ('mitigation') by assuming that the departure from the no-policy case is initially very slow (negligible in the idealized WRE cases). For higher stabilization targets, the need for early departure from the baseline is less, a factor that is accounted for in the WRE profiles by assuming a later departure date.

The WRE profiles sample only a relatively small part of the space of possible concentration stabilization pathways for CO_2. They assume, for example, only a single no-policy baseline, viz. the median of the SRES scenarios in the updated profiles (denoted 'P50 BASELINE' in the Figures). Different concentration trajectories would arise for different baselines. The lower/higher the baseline, the easier/harder it would be to achieve a given stabilization target. The effect of different baselines is considered further in Wigley (2004a). The WRE profiles, except for the WRE350 case (350ppm target, not shown here) also assume that concentrations approach the target along a continually increasing trajectory. In principle it may be advantageous to follow an 'overshoot' pathway for <u>any</u> target. An example is shown in Figure 3 for the 550ppm stabilization case. The implications of an overshoot pathway will be considered in detail below.

The critical issues for CO_2 stabilization (ignoring the non-CO_2 gases) are: the implied emissions, which, when compared to the baseline, determine the mitigation costs; and the implied benefits of the reduction in climate change below the baseline. Ideally, a decision on the stabilization target and pathway should consider both costs and benefits (as a liberal interpretation of the above-quoted extract from Article 2 of the UNFCCC might imply). It is not, however, possible to do this today in any credible way. There are considerable uncertainties in quantifying the costs of mitigation, and even larger uncertainties in quantifying the benefits side of the equation. Benefits uncertainties arise because there are large uncertainties in defining how a reduction in emissions will affect the regional details of future climate change, and large uncertainties in translating these climate-change details into impacts.

For CO_2, concentration stabilization is not the same as emissions stabilization. Figure 3 shows the concentration changes that would occur if we were to stabilize fossil-fuel emissions at the present (year 2000) level. Concentrations would increase almost linearly, at about 100ppm per century, for many centuries. Stabilization must therefore require, eventually, a reduction in CO_2 emissions to well below present levels. This is true even if one accounts for non-CO_2 gases, since CO_2 is the dominant factor in causing future climate change.

The emissions requirements corresponding to the profiles in Figure 3 are shown in Figure 4. There are considerable uncertainties in these emissions estimates, due to uncertainties in the carbon cycle model (Wigley, 1993) on which they are based. These uncertainties include those due to the possible effects of climate on the carbon cycle, which are not considered here. Nevertheless, the main qualitative features of these emissions results are robust to these uncertainties.

Figure 4 shows the following. First, for stabilization targets above about 450ppm, emissions can rise substantially above present levels and still allow stabilization to be achieved. Second, after peak emissions, rapid reductions in emissions are required to achieve stabilization, implying a rapid transition from fossil to non-fossil energy sources and/or a rapid reduction in carbon intensity (CO_2 emissions per unit of energy production). Third, as already noted, emissions must eventually decline to well below present levels.

The first point is crucial – if a stabilization target of 550ppm or higher were deemed acceptable, then the fact that an immediate reduction in emissions is not required may give us time to develop the infrastructure changes and new technologies required to achieve the rapid reductions in emissions that would eventually be required. But this is still a daunting task. As Hoffert *et al.* (1998, 2002) have pointed out, the technological challenges raised by continued growth in global energy use while, at the same time, moving away from fossil energy sources to meet the demands of stabilization are enormous.

It is worth noting that, *from the standpoint of emissions requirements alone*, there are conflicts between the results of Figure 4 and the demands of the Kyoto Protocol. The Kyoto Protocol requires immediate emissions reductions, whereas the WRE profiles show that stabilization at levels of 550ppm and above could still be achieved even if there were no immediate reductions relative to the no-policy case. A challenging short-term target, following the Kyoto Protocol, may motivate an awareness of the long-term problem – but a target that appears unnecessarily stringent can lead to outright rejection, as has been the case for the USA. Equally, blind acceptance of the WRE results would be unwise since they only provide a qualitative assessment of the economic aspects.

The WRE stabilization profiles and their implied emissions raise a number of other issues, not least being the question of how to choose a stabilization target. Article 2 provides some confusing guidelines. The above quote, and other statements in the UNFCCC, imply that a cost-benefit framework should be applied. However, Article 2 is more specific with regard to the constraint of 'dangerous interference', implying a risk avoidance approach that might be inconsistent with a conventional economic optimization or cost-benefit analysis. One interpretation is that risk avoidance might be used to select a target (with the option of changing the target as our knowledge of the risks improves), while some form of cost-benefit analysis might be used in selecting the pathway towards that target. This is essentially the approach used by, for example, Manne and Richels (2001), although these authors employ cost-optimization only and do not consider benefits.

Figure 3. Updated versions of the WRE concentration stabilization profiles for CO2

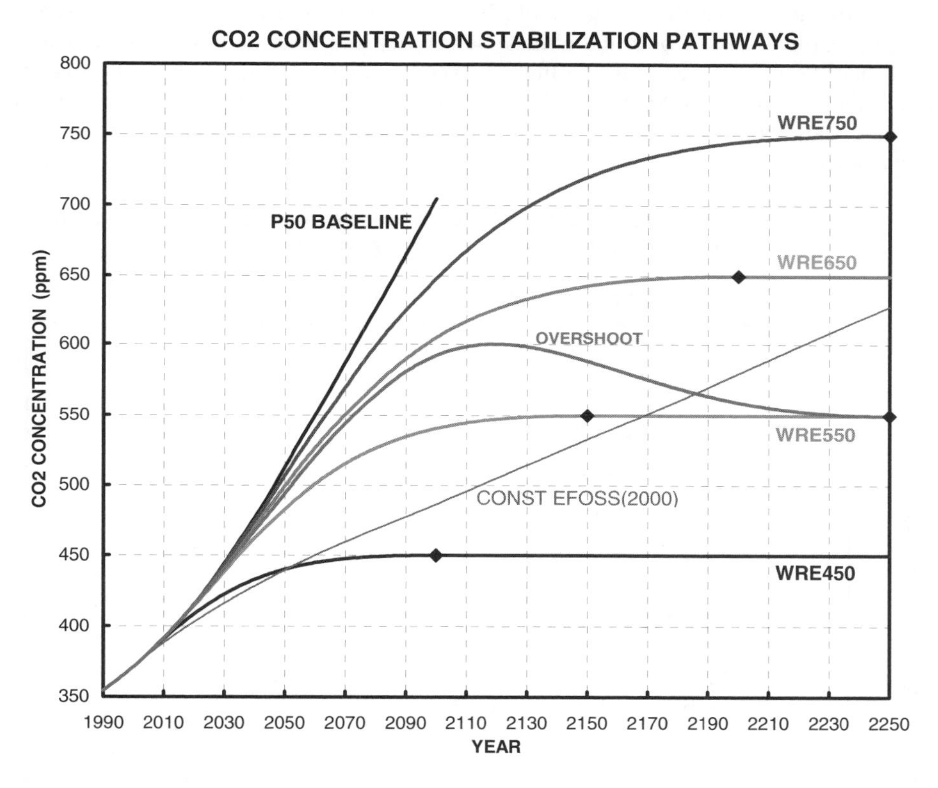

Note:

An alternative 'overshoot' profile is also shown for the 550ppm stabilization case. Diamonds show the stabilization dates. The diagonal line shows how concentrations would evolve if fossil-fuel emissions were kept constant at their 2000 level. The assumed no-policy baseline (P50) is the median of the full set of SRES scenarios. Stabilization profiles depart from the baseline in 2005 (WRE450), 2010 (WRE550), 2015 (overshoot case and WRE650) and 2020 (WRE750).

The issue of choosing a CO_2 stabilization target has been addressed probabilistically by Wigley (2004b). He notes that it is impossible to define a single dangerous interference threshold, because what is dangerous in one economic sector or geographical region may be far less consequential elsewhere, because of uncertainties in defining sector- or region-specific thresholds, and because development tends to reduce a society's vulnerability over time. Given these differences and uncertainties the best one can do is specify the dangerous interference threshold as a pdf. He then uses the literature to specify a threshold pdf. Given this, to determine the corresponding CO_2 stabilization level requires specifying the climate sensitivity and the radiative forcing from non-CO_2 sources as pdfs, and then combining these to give a pdf for the stabilization level. This leads to a wide range of possible stabilization targets: for example, a high climate sensitivity or large forcing from non-CO_2 sources would require a low CO_2 target to compensate. For the input pdfs assumed by Wigley, there is a 17% probability that the target should be less than the current CO_2 level (taken as 370ppm). Conversely, if both the climate sensitivity and future non-CO_2 forcing were small then the CO_2 target could be high. In Wigley's analysis, the probability that the target might be above 1000ppm is 9% -- but he notes that such high CO_2 levels would change ocean chemistry and biology in ways that may be even more damaging than large changes in future climate.

Figure 4. Total CO2 emissions (fossil plus land-use change) for the WRE and overshoot concentration profiles given in Figure 3

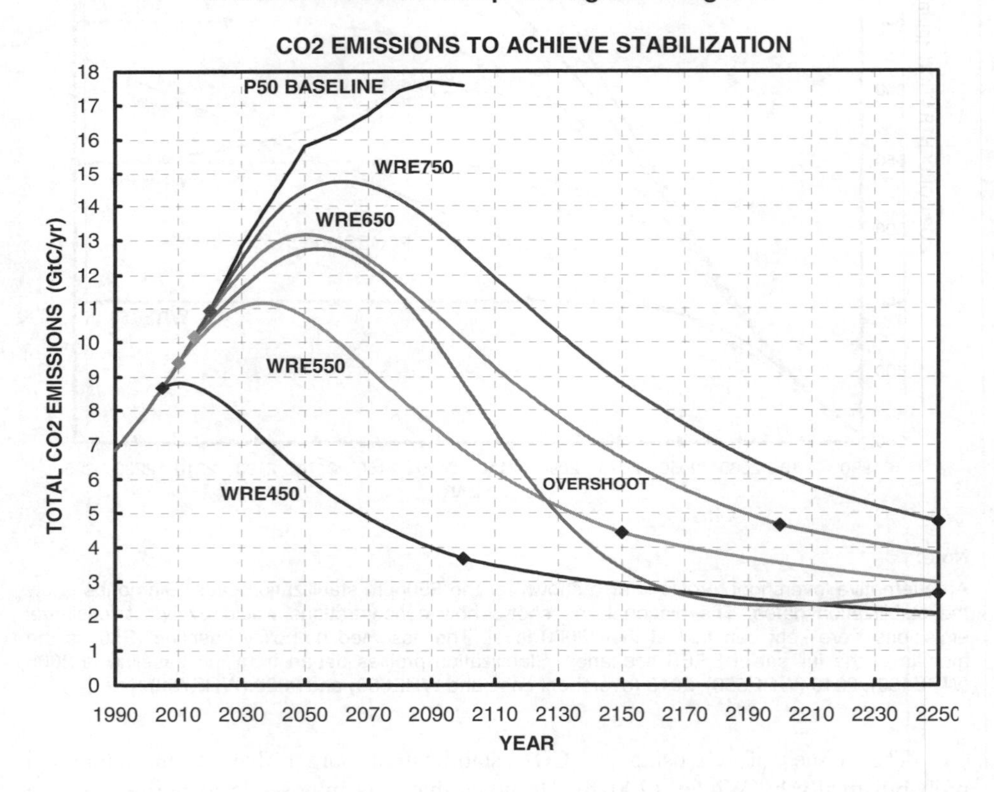

3. Future changes in global-mean temperature

23I will consider first the no-climate-policy case, beginning with global-mean temperature and then (in Section 4) the regional details.

The factors that control future climate change are the emissions of various gases, the attendant changes in atmospheric composition, and the way the climate system responds to these composition changes. All three aspects are considered here.

The relevant emissions are those of: a range of greenhouse gases (CO_2, CH_4, N_2O, and a large number of halocarbons and related species); certain other gases that affect the build-up of greenhouse gases and/or that lead to the production of tropospheric ozone, which is a powerful greenhouse gas (primarily the so-called 'reactive gases', CO, NOx and VOCs); and particles or gases that can produce aerosols (small particles) in the atmosphere (of which SO_2 is arguably the most important).

Greenhouse gases affect the balance between incoming solar radiation and outgoing long-wave radiation by absorbing outgoing radiation, while aerosols affect both the incoming and outgoing fluxes to varying degrees depending on the aerosol type. The imbalances so produced are called 'radiative forcing', and the climate system responds by trying to restore the balance either by warming the atmosphere in the case of greenhouse gases or by cooling the atmosphere in the case of aerosols – although some aerosols, notably carbonaceous (soot) aerosols act in more complex ways and may cause warming or cooling depending on aerosol type.

In terms of the climate response, emissions may be divided into those that produce long-lasting changes (i.e., timescales of decades or longer) in atmospheric composition (CO_2, CH_4, N_2O, and most halocarbons), and those that have short-term effects (timescales of order days to weeks). SO_2, leading to the production of sulphate aerosols, and the reactive gases, which produce tropospheric ozone and affect the lifetime of CH_4, fall into the second category. Long-lived gases produce spatially-uniform changes in atmospheric composition, and their effects are often lumped together with CO_2 to give an 'equivalent-CO_2' concentration. Short-lived gases affect mainly the regions near to their points of emission and so lead to spatially-heterogeneous changes in atmospheric composition and more complex patterns of climate change than the long-lived gases.

As noted above, and illustrated in Figure 1 for CO_2, future emissions of all gases are subject to large uncertainties.

Translating emissions changes to changes in atmospheric composition requires the use of 'gas-cycle' models, of which carbon-cycle models (for CO_2) are one example. (Such models may also be used in reverse or 'inverse' mode to determine the emissions required to follow a prescribed concentration pathway, as illustrated by Figures 3 and 4.) All gas-cycle models have inherent uncertainties, the most important of which, because of the dominant role CO_2 has in affecting the climate, are those in the carbon cycle. These include direct uncertainties associated with the transfer of CO_2 between the various carbon reservoirs (the land and terrestrial biosphere, the oceans, and the atmosphere), and indirect uncertainties associated with feedbacks within the carbon cycle and between the carbon cycle and the climate system. The main internal carbon-cycle feedback arises from the effect of CO_2 on plant growth. At higher CO_2 levels plants absorb more CO_2, providing a negative feedback that tends to slow the growth of atmospheric CO_2 – the effect is called 'CO_2 fertilization'. Climate feedbacks occur because the transfers between the carbon reservoirs are climate-dependent. Most important here are transfers involving the terrestrial biosphere – higher temperatures, for example, lead to increases in both plant productivity and respiration, with the latter tending to dominate. This is a net positive feedback, increasing the rate of growth of atmospheric CO_2 concentration.

In determining the climate consequences, the first step is the translation of atmospheric composition changes to radiative forcing. This is done internally from first principles in more complex models (such as General Circulation Models – GCMs). Simpler models (such as the MAGICC model used below), which use radiative forcing as an input parameter, must make use of empirical relationships to relate concentrations or emissions to radiative forcing (see, e.g., Harvey *et al.*, 1997). There are uncertainties involved in calculating greenhouse-gas radiative forcing from concentrations, but these are probably no more than +/- 10% of the central estimates. The biggest forcing uncertainties are those for aerosols. Large forcing uncertainties, however, do not necessarily translate to large uncertainties in global-mean temperature, as will be demonstrated below.

For given radiative forcing, the global-mean temperature response is determined primarily by the 'climate sensitivity'. Strictly, this is the equilibrium (i.e., eventual) temperature change per unit of radiative forcing; but it is often characterized by the equilibrium response to a doubling of the CO_2 concentration, denoted ΔT2x. ΔT2x is still subject to considerable uncertainty – Wigley and Raper (2001) estimate that its value lies in the range 1.5-4.5°C with 90% confidence, although other studies suggest an even wider range (e.g. Andronova and Schlesinger, 2001).

Global-mean temperature response is also determined by the thermal inertia of the climate system, which, in turn, is largely determined by how rapidly heat is mixed down into the ocean. These two aspects, the sensitivity and the inertia, act like a motor vehicle responding to the accelerator. The accelerator's position determines the vehicle's eventual top speed and is akin to the sensitivity, while the vehicle's mass (inertia) determines how rapidly this speed is reached. Just as different vehicles have different top speeds, so too do different climate models have different sensitivities – reflecting uncertainties in the climate sensitivity of the real-world climate system. We do not yet know whether the climate system is a Porsche or a Volkswagen.

Projected future changes in global-mean temperature are shown in Figure 5. The range shown here combines the lowest and the highest radiative forcing values from the full SRES emissions set (the CO_2 component of which is shown in Figure 2), with low (1.5°C) and high (4.5°C) climate sensitivities. The bounds shown here differ slightly from those given in the IPCC TAR (Cubasch and Meehl, 2001, Figure 9.14) even though the same climate model has been used (viz. the MAGICC model: Wigley and Raper, 1992; Raper *et al.*, 1996). This is mainly because a wider sensitivity range has been employed here (the TAR range was 1.7-4.2°C, whereas the range used here is 1.5-4.5°C). Figure 5 does not account for all uncertainties, just those in emissions and the climate sensitivity. Other uncertainties are discussed below.

Figure 5 provides a graphic illustration of both the large magnitude of potential changes in global-mean temperature (even the lower bound represents a warming rate of about double that of the past century; the upper bound is more than ten times the past rate) and the overall uncertainty range.

Figure 5. Global-mean temperature projections for the IPCC SRES emissions scenarios

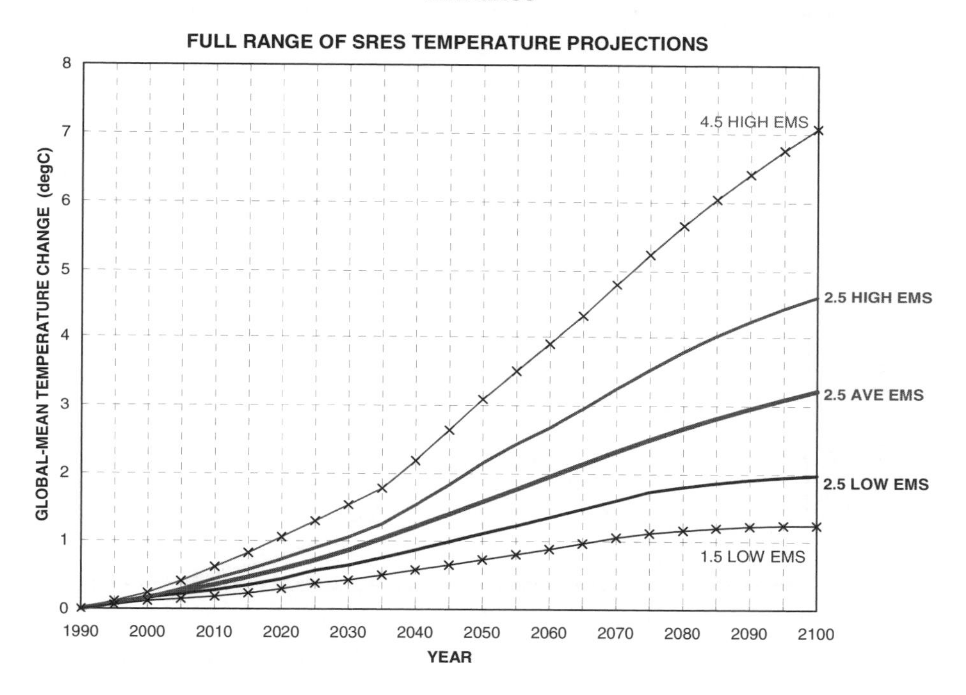

Note:

The outer curves combine sensitivity and forcing extremes (4.5 HIGH EMS and 1.5 LOW EMS). The inner curves show the range of results for different emissions for a climate sensitivity (ΔT2x) of 2.5°C. For all model parameters other than sensitivity, best-estimate values following Wigley and Raper (2001) are used.

It is better for policy assessment purposes, however, to try to present these results probabilistically (as explained by Schneider, 2001). This is done in Figure 6, following Wigley and Raper (2001). In a probabilistic framework, temperature changes are presented as probability density functions (pdfs) where the area under the curve to the left of a given temperature value (indicated along the abscissa) gives the probability that the warming will be less than that value. Figure 6 considers additional sources of uncertainty beyond the emissions and sensitivity uncertainties considered in Figure 5; viz. uncertainties in ocean mixing rates, in aerosol forcing, and in carbon-cycle feedbacks (for details, see Wigley and Raper, 2001). For emissions, the Wigley and Raper analysis assumed that each of the SRES emissions scenarios was equally likely, since the SRES team did not assign probabilities to the individual scenarios. For the other four uncertainty factors, each was input to the MAGICC climate model as a pdf, and some 110,000 simulations are carried out to span the range of input uncertainties.

Figure 6. No-climate-policy probabilistic projections for global-mean temperature change over different time intervals

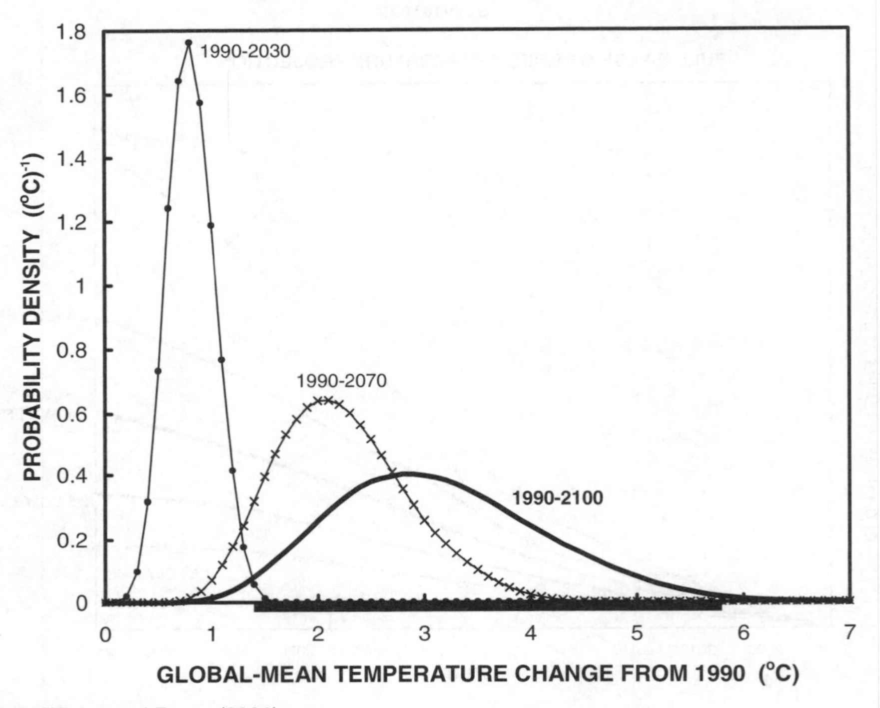

Source: Wigley and Raper (2001).

Note:

The bar beneath the abscissa (x-axis) shows the range given in the IPCC TAR. The pdfs account for uncertainties in ocean mixing, aerosol forcing, the carbon cycle, emissions (using the full set of 35 complete SRES scenarios) and the climate sensitivity.

4. Sources of uncertainty in global-mean temperature projections

Figure 6 quantifies the overall uncertainty in global-mean temperature in probabilistic terms, consistent with the state of the science that is represented by the IPCC TAR. It does not, however, give information about the individual sources of uncertainty and their relative importance. One way to do this is to successively eliminate sources of uncertainty (i.e., replace that particular input pdf by a single – e.g. median – value) and see how the pdf for global-mean temperature changes. A small change in the output pdf would indicate that, compared to other sources of uncertainty, the omitted factor was relatively unimportant.

Figure 7. Effect of removing carbon-cycle model uncertainties on the pdf for 1990-2100 global-mean temperature change

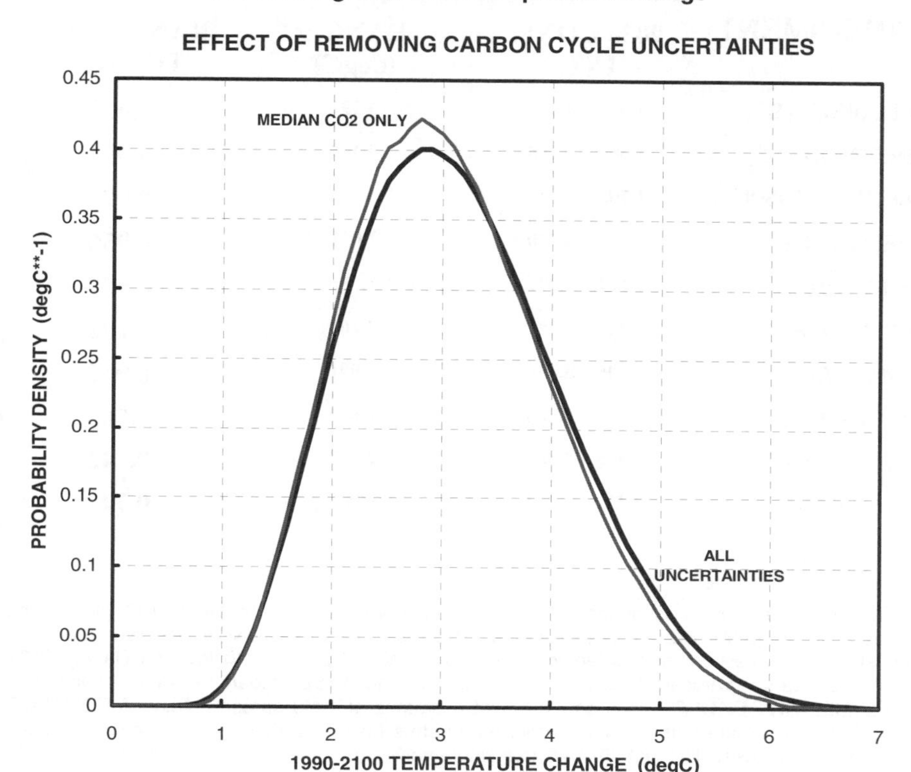

Note:

The base case (ALL UNCERTAINTIES) is the same as in Figure 6.

This has been done for each of the uncertainty factors incorporated in Figure 6. For ocean mixing and aerosol forcing uncertainties, their influences are negligible (see Table 1). For carbon cycle uncertainties, the results are shown in Figure 7 and Table 1. Contrary to common belief, the influence of carbon cycle uncertainties can be seen to be relatively small. This may seem puzzling since, in terms of projected increases in CO_2 concentration, the uncertainties can be quite large. For example, in the present analysis, the 90% confidence interval for concentration in 2100 under the A1FI emissions scenario is 900-1240ppm, with a median value of 990ppm. For global-mean temperature, however, when considered in conjunction with other uncertainties, these carbon cycle uncertainties are overwhelmed. Of course, if we were to reduce the other uncertainties substantially, or to consider particular applications where they were less important (such as in defining the emissions requirements for a given concentration stabilization profile), then carbon-cycle uncertainties would become more important.

Table 1. Sources and magnitudes of uncertainties

EXPERIMENT	90% CONF. INT.	RANGE (degC)	RANGE REL. TO 'ALL'
ALL UNCERTS	1.683 – 4.874	3.191	1.000
MEDIAN Kz	1.686 – 4.855	3.169	0.993
MEDIAN AEROSOL	1.705 – 4.871	3.166	0.992
MEDIAN CO2	1.679 – 4.730	3.051	0.956
EMIS. = B1	1.187 – 2.971	1.783	0.559
EMIS. = A1B	1.821 – 4.081	2.260	0.708
EMIS. = A1FI	2.805 – 5.796	2.991	0.937
$\Delta T2x = 1.5^{\circ}C$	1.251 – 2.864	1.613	0.505
$\Delta T2x = 2.6^{\circ}C$	1.985 – 4.353	2.369	0.742
$\Delta T2x = 4.5^{\circ}C$	2.737 – 5.743	3.006	0.942

Note:

The Table shows 90% confidence intervals and interval ranges for 1990-2100 global-mean temperature change for different uncertainty factor combinations. ALL UNCERTS gives the results when all uncertainties are accounted for, duplicating values given in Wigley and Raper (2001). MEDIAN Kz refers to vertical diffusivity in the model, and means that ocean mixing uncertainties are ignored. MEDIAN AEROSOL uses only the best estimates of 1990 aerosol forcing. MEDIAN CO2 uses only best-estimate carbon cycle model parameters (including those that quantify fertilization and climate feedbacks). The last six rows use single values for either the emissions scenario or the climate sensitivity.

The uncertainties that overwhelm the carbon cycle uncertainties are those in emissions and the climate sensitivity. The effect of emissions uncertainties is shown in Figure 8. For emissions, it is not sufficient to consider only the median emissions case as a way to eliminate these uncertainties, since the residual uncertainties depend on the selected pathway for future emissions. Figure 8 gives results for a low (B1-IMAGE), mid (A1B-AIM) and high (A1FI-MiniCAM) emissions scenario, and shows that the residual uncertainties are greater for higher emissions (see also Table 1). An interesting additional result is that removing emissions uncertainties has only a small effect on the spread of the global-mean temperature pdf – in other words, even if we knew what future emissions were going to be, the uncertainties surrounding any prediction of global-mean temperature would still be large. What changes, of course, is the central value of the distribution.

Figure 8. Effect of removing emissions uncertainties on the pdf for 1990-2100 global-mean temperature change

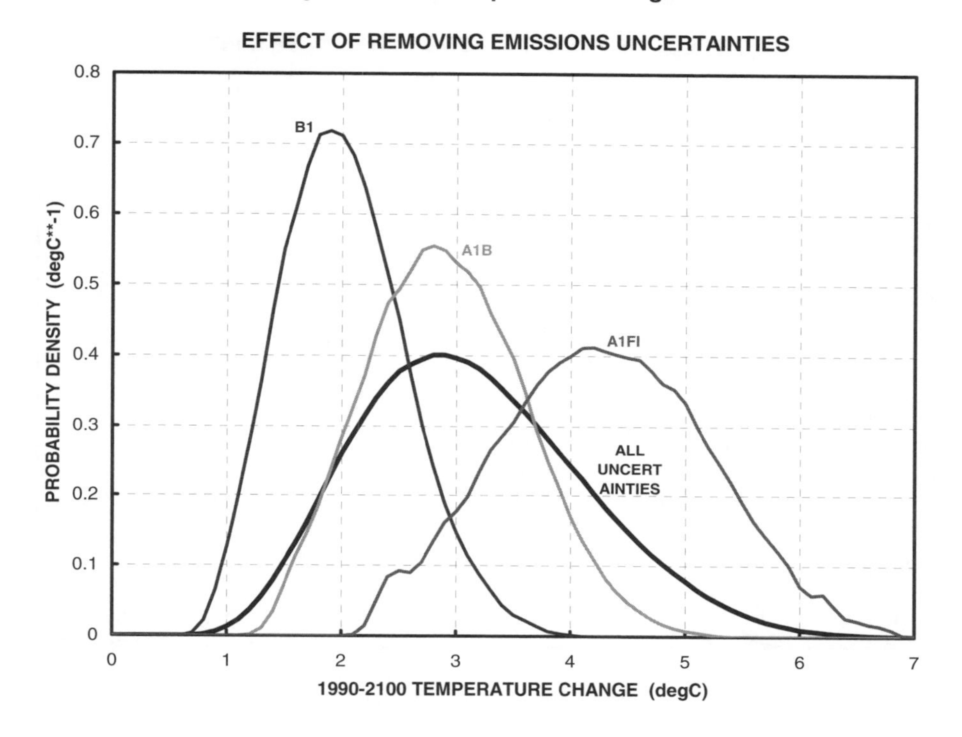

Note:

The base case (ALL UNCERTAINTIES) is the same as in Figure 6.

This highlights an often neglected facet of uncertainty analysis. Such analyses must consider, not only the uncertainty range, but also the position of the pdf. As Figure 8 clearly demonstrates, the primary effect of removing (or reducing) a source of uncertainty may well be to alter the best estimate of future change rather than reduce the spread about this best estimate.

A similar situation arises in the case of climate sensitivity uncertainties. Figure 9 (see also Table 1 in the Annex) shows uncertainties in 1990-2100 global-mean warming for three different values of the climate sensitivity. For low and high sensitivities, the position of the pdf is changed radically from the general case, but the spread of the pdf remains large – due primarily to emissions uncertainties.

Figure 9. Effect of removing emissions uncertainties on the pdf for 1990-2100 global-mean temperature change

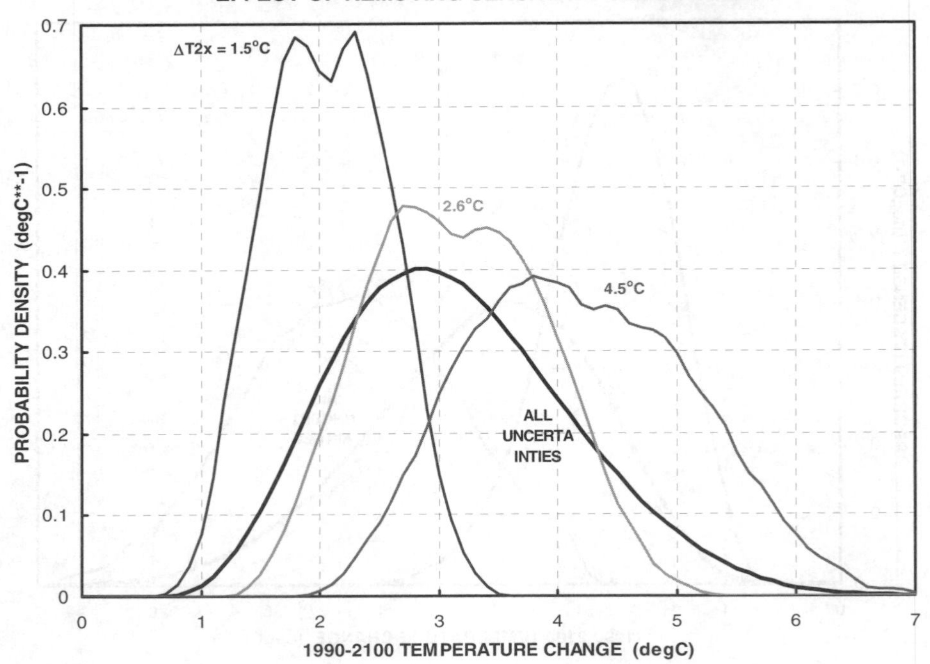

Note:

The base case (ALL UNCERTAINTIES) is the same as in Figure 6.

5. Spatial patterns of climate change

Figures 6-9 show uncertainties in global-mean temperature projections for the no-climate-policy case. It should be noted that Figure 8 gives some important insights into the effects of policy. For example, one could suppose that the B1 scenario was achieved by implementing policies against the A1B scenario as a no-policy baseline. If this were so, then there would be clear benefits, since the median of the global-mean temperature pdf is shifted substantially towards lower warming. However, the spread of the pdf is reduced only minimally, a result that will apply to any policy case.

It is, of course, impossible to make a *detailed* assessment of the benefits, in terms of avoided impacts, of any mitigation policy using just global-mean temperature. The impacts of climate change occur at the local to regional level, and depend on many variables in addition to temperature. Because of this, and because of the uncertainties that surround projections of climate change at the regional level that will be described below, quantifying benefits is a daunting task. It is made even more difficult by the fact that most impacts models have large inherent uncertainties that compound uncertainties in their inputs.

The focus of this Section will be on uncertainties in the regional climate change *signal* that might arise from a mitigation policy. In reality, any such signal will be embedded in the noise of the climate system's natural variability. For a given scenario, this adds another element of uncertainty. While natural variability is ignored here, it may be that it cannot be ignored in some impacts sectors. For example, in ecosystem impacts where changes in disturbance frequency can be of paramount importance, it may be essential to express these impacts in probabilistic terms generated by considering ensemble means spanning a range of (unpredictable!) future natural variability realizations. These variability issues make the problem of assessing uncertainties even more difficult. Indeed, even when restricting the discussion to a deterministic signal (i.e., ignoring the effects of internally-generated variability or noise), there has been little work of any value on quantifying uncertainties in regional climate.

So, how do we quantify these uncertainties in regional climate? One approach is to compare results from different models. To facilitate this, we employ the scaling method developed by the present author and first described in Santer *et al.* (1990). The fundamental assumption of this method is that future changes in climate may be decomposed into global-mean and spatial pattern components. Instead of considering the 'raw' patterns of climate change, therefore, we consider patterns of change per unit (1°C) of global-mean warming (which we refer to as 'normalized' patterns). This has many advantages, not least that it allows us to compare results from models that have very different climate sensitivities. Normalization essentially factors out the effects that are directly related to climate sensitivity differences. We can then consider the effects of sensitivity uncertainties separately at the global-mean level, as done in the preceding Sections, and concentrate on the underlying patterns of regional climate. By quantifying inter-model differences in the *normalized* patterns of change we obtain more fundamental insights into regional uncertainties.

Here, I use coupled atmosphere/ocean GCM results from the CMIP model-intercomparison project (Covey *et al.*, 2003). The CMIP project has compiled data sets for unforced ('control run') experiments and for a standard forced ('perturbation') experiment for most of the world's AOGCMs. (The perturbation experiment is one in which CO_2 is increased at a compound rate of 1% per year corresponding closely to a linear increase in radiative forcing.)

In interpreting the CO_2 increase results, it is important to define the signal correctly. A problem here is that, in terms of the spatial patterns of climate change, many AOGCMs show a significant 'drift'. In other words, if the model is run in control mode (i.e., not subject to any external forcing), and even if the model's global-mean temperature exhibits no long-term trend under such circumstances, the patterns of climate change may show substantial trends on the 100-year timescale. If this is the case, comparing future and initial climate states in a perturbation experiment (referred to as the Definition 1 method for defining the signal; Santer *et al.*, 1994) will not give the true signal, but will give a mix of signal plus drift. To account for this, a standard procedure is to compare the perturbation experiment with a parallel control run (with both starting with the same initial conditions). If the drift is common to both, and there is strong evidence from tests with AOGCMs that it is, then subtracting the control run from the perturbation experiment will give a better estimate of the underlying signal. This is referred to as the Definition 2 method, and is the method employed here.

Normalized signals for annual-mean temperature and precipitation change, averaged over 17 AOGCMs, are shown in Figure 10. (Note that annual means are used here simply for illustrative purposes. Seasonal results would be of more direct relevance for impacts, but to span the seasons would have required four times as many maps. The information given here comes from the user-friendly software package 'MAGICC/SCENGEN' (downloadable from www.cgd.ucar.edu) that allows the user to produce a wide range of similar results).

Figure 10 shows a number of well-realized features: amplified warming in high northern latitudes; warming minima in the North Atlantic and around Antarctica associated with regions where deep water production occurs; increased precipitation over the tropical oceans and in high latitudes; and reduced precipitation in the subtropics (which are already dry) and around the Mediterranean. In most areas, the model-average changes in precipitation are quite small (less than 3% change per degree C of global-mean warming), but this statistic hides large differences between the seasons and between models.

Uncertainties can be quantified by comparing these normalized signal patterns with the 'noise' of inter-model variability (i.e., by dividing the signal by the inter-model standard deviation on a grid point by grid point basis to form patterns of an inter-model signal-to-noise ratio (SNR)). These results are also shown in Figure 10 – also from MAGICC/SCENGEN. High SNR values frequently occur where the signal is large; although high SNR may also occur where the signal is small, but inter-model differences are even smaller. Low SNR generally means that model results differ widely.

In standard statistical terms, an SNR above 2 would indicate that the model-mean signal was statistically significant at the 5% level – i.e., that there was only one chance in 20 that the signal was zero (using a two-tail test). For temperature changes, SNR values exceed 2 for all regions except those around and downwind of the regions where deep water formation occurs. Low SNR values in these regions reflect smaller signals (see Figure 10) and larger differences between models. In contrast, for precipitation, SNR values are less than 2 for most of the globe. The exceptions are the high latitude regions where the signals are highest (Figure 10). Interestingly, there is some indication that the regions where we can have *most* confidence in the temperature change signals are also the regions where we have <u>least</u> confidence in precipitation signals. The temperature and precipitation SNR patterns are negatively correlated, but the pattern correlation is not high, about -0.4.

In practical terms, however, we might accept a much less stringent significance level. An SNR of +1 corresponds to the 16% significance level (one-tail test) and a probability of only one in six that the signal is zero or less. This would not be enough to satisfy normal statistical testing criteria, but, since there is still a high likelihood of a non-zero change, such SNR values may be of considerable 'significance' to the policy maker.

The phraseology used here suggests an alternative way of presenting SNR results. Instead of SNR values, if one assumes the distribution of models results to be Gaussian, then one can use the model mean and inter-model variability to calculate the probability of a change exceeding a specified threshold, such as the probability of a precipitation increase (where the threshold is zero) – see Santer *et al.* (1990). MAGICC/SCENGEN presents results in this way as well as through SNR values.

Would such a result (i.e., an SNR value around 1, or a probability of 84% that precipitation would increase) warrant action by a policy maker? Deciding on an SNR or probability threshold in the decision-making context depends on what is judged to be an acceptable risk. This in turn depends on the costs associated with two types of error relative to the benefits of action, responding to a supposed climate change signal when the true signal is zero, and failing to respond when there is a substantially non-zero true signal.

6. Temperature projections under CO_2 stabilization

In Section 2 I considered the SRES no-climate-policy emissions scenarios and showed, as policy examples, the WRE CO_2 concentration stabilization profiles and the implied emissions requirements. Sections 3, 4 and 5 considered climate changes under the no-policy scenarios. Here I consider climate changes under the WRE policy scenarios. Although Article 2 of the UNFCCC has as its goal stabilization of all anthropogenic greenhouse gases, I will consider only the effects of CO_2 stabilization here. Very little work has been done on the climate implications of multi-gas stabilization scenarios.

Non-CO_2 gases cannot be ignored, however. To isolate the effects of CO_2 stabilization I assume just a single scenario for non-CO_2 gases, namely the P50 scenario used as a baseline for the updated WRE profiles. P50 emissions, however, are specified only to 2100, while the CO_2 concentration profiles go to 2250 and beyond. In order to give climate projections to 2250, I therefore assume emissions of non-CO_2 gases to remain constant at the 2100 level given in the P50 scenario. The radiative forcing from these gases therefore increases after 2100, albeit only slightly.

To further constrain the analysis I consider only a single set of climate model parameters, a climate sensitivity of 2.5°C and 'best guess' values for all other parameters (i.e., the values used in the IPCC TAR).

Figure 10. Model-average patterns of normalized annual-mean temperature change (i.e., changes per 1oC global-mean warming)

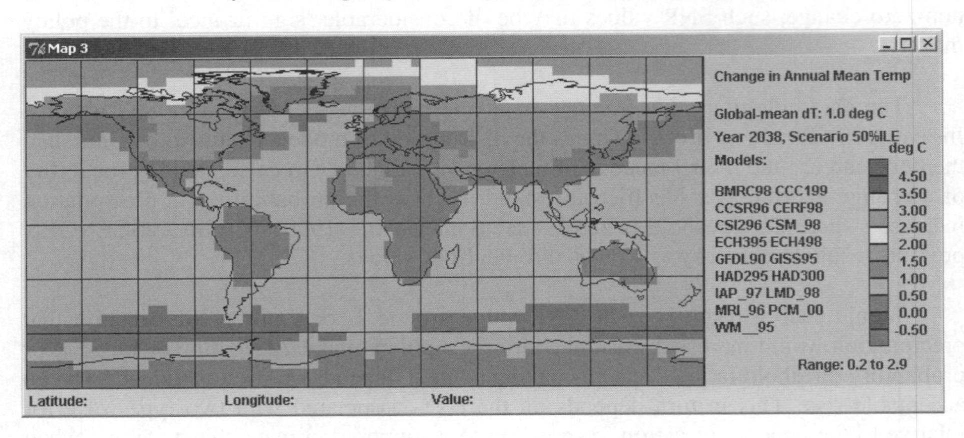

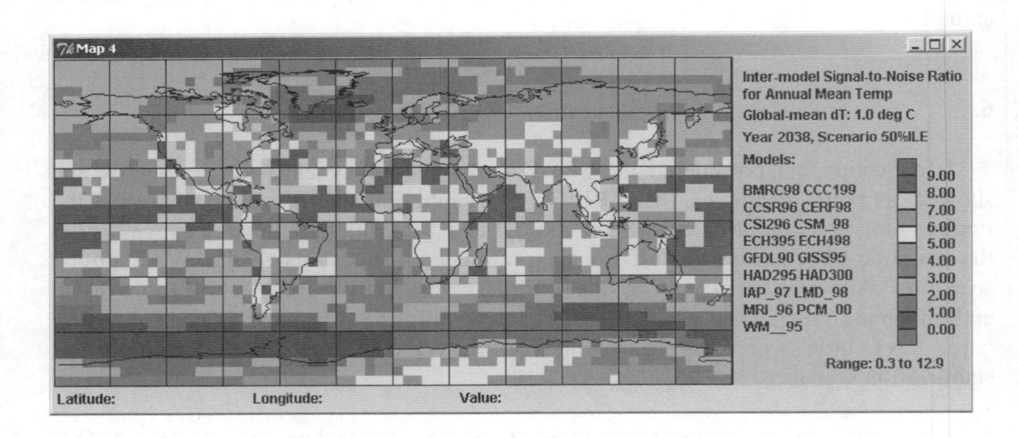

Note:

Averaged over 17 models from the CMIP data base for 1% compound CO_2 increase perturbation experiments (Definition 2 changes) – top map, and corresponding inter-model signal-to-noise ratios (model-mean normalized change divided by inter-model standard deviation) – bottom map.

Figure 10 (continued)

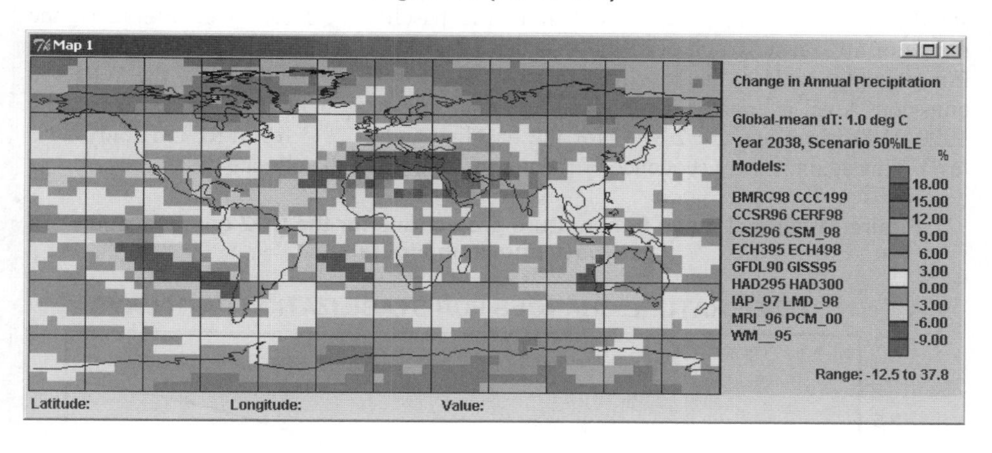

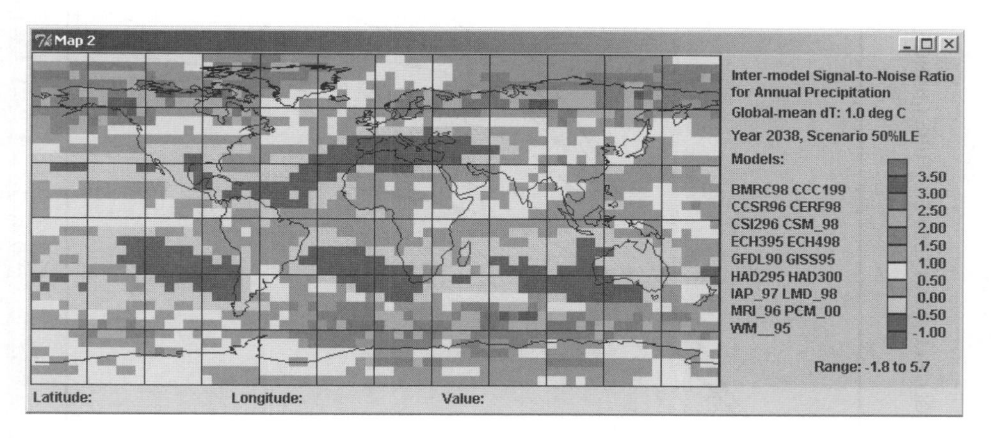

Note:

Averaged over 17 models from the CMIP data base for 1% compound CO_2 increase perturbation experiments (Definition 2 changes) – top map, and corresponding inter-model signal-to-noise ratios (model-mean normalized change divided by inter-model standard deviation) – bottom map.

Figure 11 shows changes in global-mean temperature for the stabilization cases considered in Figure 3, together with the constant 2000-level fossil CO_2 emissions case. There are a number of points to note. First, the greatest separation is between the 450ppm and 550ppm cases, while the least is between the 650ppm and750ppm cases. This results mainly from the logarithmic dependence of CO_2 radiative forcing on concentration, which means that as the CO_2 level increases, the forcing increment for a 100ppm concentration increment decreases. Second, although CO_2 concentrations stabilize in all cases (as early as 2100 in the WRE450 case), warming continues beyond 2250. This is partly due to the influence of the non-CO_2 gases, and partly because of the large thermal inertia of the climate system. Third, even though the emissions in the CO_2 stabilization cases are much less than in the baseline P50 case (see Figure 4), the reduction in warming achieved through these emissions reductions is relatively small. For example, in 2100 the baseline emissions level is 17.57GtC/yr, while the WRE550 emissions level is approximately 60% less than this at 6.85GtC/yr. The corresponding 1990-2100 warmings are 2.81°C for the baseline and 2.22°C for WRE550 (only 20% less). This again is a consequence of the thermal inertia of the climate system.

Figure 11. Global-mean temperature projections for the CO2 concentration pathways shown in Figure 3

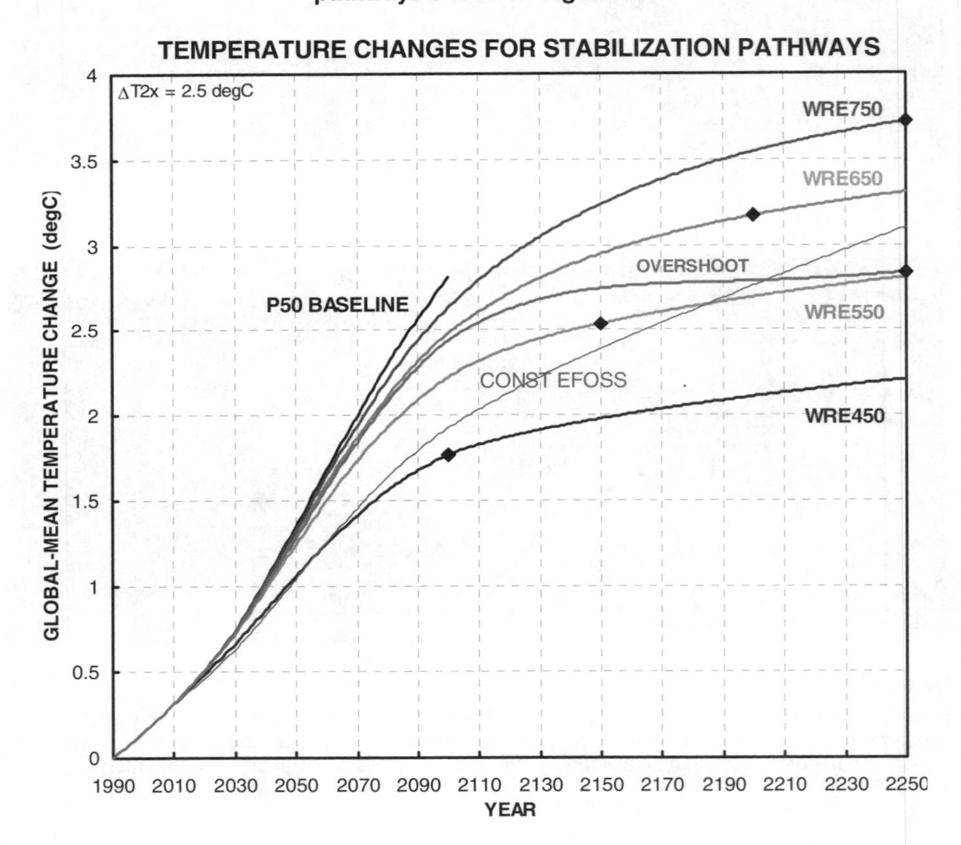

Note:

Results are for a climate sensitivity of 2.5°C and IPCC TAR central estimates of all other model parameters. Non-CO_2 emissions are assumed to follow the P50 scenario to 2100 and remain constant thereafter. Black diamonds show the dates at which CO_2 concentrations stabilize.

Finally, by 2250 warming under the overshoot 550ppm concentration pathway has returned almost to the standard 550ppm case. Overshoot allows emissions to rise 14% above the no-overshoot case, while the maximum increase in global-mean temperature change relative to 1990 is somewhat less, around 10%. It is likely that the overshoot case would lead to reduced mitigation costs, while the attendant increased warming would certainly lead to a reduction in the benefits of averted climate change. This raises the question of whether the reduced mitigation costs are enough to offset the reduction in climate-change benefits. While answering this question is beyond the scope of the present work, it is clear that overshoot CO_2 stabilization pathways warrant further investigation.

7. Conclusions

This paper considers future changes in climate and the uncertainties in these changes at the global-mean and spatial pattern levels for both no-climate-policy and policy cases. Where possible, a probabilistic approach is used.

Emissions projections, as the primary drivers of anthropogenic climate change, are considered first, for both the no-policy case (characterized by the SRES scenarios) and the policy case (following the WRE CO_2 stabilization profiles). The most important feature of the WRE profiles is that, for stabilization targets above about 450ppm, they do not require immediate reductions in emissions below the baseline no-policy case. Eventually, emissions must drop below present levels, but, depending on the chosen stabilization target, they may remain above present levels for decades to more than a century. The possibility of following an overshoot pathway to stabilization (where the peak concentration exceeds the eventual stabilization target) is introduced.

Probabilistic projections for global-mean temperature under the no-policy assumption produced by Wigley and Raper (2001) are broken down by uncertainty factor in order to quantify the relative importance of different sources of uncertainty (ocean mixing, aerosol forcing, the carbon cycle, emissions, and the climate sensitivity). A new method is employed to do this: removal of individual sources of uncertainty and comparison of the resulting pdf with that based on the full uncertainty assessment. Small changes in the pdf imply that uncertainties in the factor removed are relatively unimportant. For temperature changes over 1990-2100, emissions and climate sensitivity uncertainties are shown to be by far the most important. Removing either emissions or sensitivity uncertainties, however, has only a small effect on the spread of the global-mean temperature pdf – in other words, the effect of emissions or sensitivity uncertainties alone is similar to their combined effect.

Uncertainties in the regional patterns of climate change are examined using normalized patterns of change (i.e., patterns of change per 1°C global-mean warming). This method, proposed many years ago but rarely used, allows us to separate the effects of climate sensitivity uncertainties from those inherent in the patterns of change. Pattern uncertainties are based on the differences between 17 coupled atmosphere-ocean general circulation models (AOGCM) in the Common Management Information Protocol (CMIP) data base, and quantified using a signal-to-noise ratio (model-mean signal divided by inter-model standard deviation). Temperature uncertainties are shown to be greatest at high latitudes and least in low latitudes, while precipitation shows the opposite pattern.

For mitigation policies, Article 2 of the UNFCCC implies that both mitigation costs and the benefits of averted climate change should be considered. On the cost side, for CO_2, it is sufficient to consider only changes at the global-mean level. For benefits, regional climate change information is needed. This almost certainly makes the assessment of benefits more difficult than the assessment of costs.

References

Andronova, N.G. and M.E. Schlesinger (2001), "Objective estimation of the probability density function for climate sensitivity", *Journal of Geophysical Research* **106**, 22605-22611.

Covey, C., K.M. Achuta Rao, U. Cubasch, P.D. Jones, S.J. Lambert, M.E. Mann, T.J. Phillips and K.E. Taylor (2003), "An overview of results from the Coupled Model Intercomparison Project (CMIP)", *Global and Planetary Change* **37**, 103-133.

Cubasch, U. and G.A. Meehl (2001), "Projections for future climate change", *Climate Change 2001: The Scientific Basis*, (eds. J.T. Houghton, *et al.*), Cambridge University Press, Cambridge, U.K., 525–582.

Grübler, A. (1998), "A review of global and regional sulphur emission scenarios", *Mitigation and Adaptation Strategies for Global Change* **3**, 383-418.

Harvey, L.D.D., J. Gregory, M. Hoffert, A. Jain, M. Lal, R. Leemans, S.B.C. Raper, T.M.L. Wigley and J. de Wolde (1997), *An introduction to simple climate models used in the IPCC,* Second Assessment Report, IPCC Technical Paper 2, J.T. Houghton, L.G. Meira Filho, D.J. Griggs and M. Noguer (eds.), Intergovernmental Panel on Climate Change, Geneva, Switzerland, 50 pp.

Hitz, S. and J. Smith (2004), "Estimating global impacts from climate change", in *The Benefits of Climate Change Policies*, OECD, Paris.

Hoffert, M.I., K. Caldeira, A.K. Jain, E.F. Haites, L.D.D. Harvey, S.D. Potter, M.E. Schlesinger, S.H. Schneider, R.G. Watts, T.M.L. Wigley and D.J. Wuebbles (1998), "Energy implications of CO_2 stabilization", *Nature* **395**, 881–884.

Hoffert, M.L., K. Caldiera, G. Benford, D.R. Criswell, C. Green, H. Herzog, A.K. Jain, H.S. Kheshgi, K.S. Lackner, J.S. Lewis, H.D. Lightfoot, W. Mannheimer, J.C. Mankins, M.E. Mauel, L.J. Perkins, M.E. Schlesinger, T. Volk and

T.M.L. Wigley (2002), “Advanced technology paths to global climate stability: Energy for a greenhouse planet”, *Science* **298**, 981-987.

Houghton, J.T., *et al.*, (eds,) (2001), *Climate Change 2001: The Scientific Basis.* Cambridge University Press, xxx pp.

Manne, A.S. and R.G. Richels (2001), “An alternative approach to establishing trade-offs among greenhouse gases”, *Nature* **410**, 675-677.

Nakićenović, N. and R. Swart (eds.) (2000), *Special Report on Emissions Scenarios*, Cambridge University Press, 570 pp.

Raper, S.C.B., T.M.L. Wigley and R.A. Warrick (1996), “Global sea level rise: past and future”, in *Sea-Level Rise and Coastal Subsidence: Causes, Consequences and Strategies* J. Milliman and B.U. Haq (eds.), Kluwer Academic Publishers, Dordrecht, The Netherlands, 11–45.

Santer, B.D., T.M.L. Wigley, M.E. Schlesinger and J.F.B. Mitchell (1990), *Developing Climate Scenarios from Equilibrium GCM Results*, Max-Planck-Institut für Meteorologie Report No. 47, Hamburg, Germany, 29 pp.

Santer, B.D., W. Brueggemann, U. Cubasch, K. Hasselmann, H. Hoeck, E. Maier-Reimer and U. Mikolajewicz (1994), “Signal-to-noise analysis of time-dependent greenhouse warming experiments”, *Climate Dynamics* **9**, 267-285.

Schneider, S.H. (2001), “What is dangerous climatic change?” *Nature* **411**, 17-19.

Schneider, S.H. and J. Lane (2004), “Abrupt non-linear climate change and climate policy”, in *The Benefits of Climate Change Policies*, OECD, Paris.

Smith, J.B., H.-J. Schellnhuber and M.M.Q. Mirza (2001a), “Vulnerability to climate change and reasons for concern: A synthesis”, in *Climate Change 2001: Impacts, Adaptation, and Vulnerability,* Contribution of Working Group II to the Third Assessment Report of the Intergovernmental Panel on Climate Change, J. McCarthy, O. Canziana, D. Dokken and K. White (eds.), Cambridge University Press, Cambridge, U.K., 913–967.

Smith, S.J., H. Pitcher and T.M.L. Wigley (2001b), “Global and regional anthropogenic sulphur dioxide emissions”, *Global and Planetary Change* **29**, 99-119.

Wigley, T.M.L. (1993), “Balancing the carbon budget. Implications for projections of future carbon dioxide concentration changes”, *Tellus* **45B**, 409–425.

Wigley, T.M.L. (2004a), “Stabilization of CO_2 and other greenhouse gas concentrations”, *Climatic Change* (accepted for publication).

Wigley, T.M.L. (2004b), "Choosing a stabilization target for CO_2", *Climatic Change* (in press).

Wigley, T.M.L. and S.C.B. Raper (1992), "Implications for climate and sea level of revised IPCC emissions scenarios", *Nature* **357**, 293–300.

Wigley, T.M.L. and S.C.B. Raper (2001), "Interpretation of high projections for global-mean warming", *Science* **293**, 451-454.

Wigley, T.M.L., R. Richels and J.A. Edmonds (1996), "Economic and environmental choices in the stabilization of atmospheric CO_2 concentrations", *Nature* **379**, 240–243.

Wigley, T.M.L., S.J. Smith and M.J. Prather (2002), "Radiative forcing due to reactive gas emissions", *Journal of Climate* **15**, 2690-2696.

Chapter 8

MANAGING CLIMATE CHANGE RISKS

by Roger Jones,

CSIRO Atmospheric Research, Australia

Issues of uncertainty, scale and delay between action and response mean that 'dangerous' climate change is best managed within a risk assessment framework that evolves as new information is gathered. Risk can be broadly defined as the combination of likelihood and consequence; for climate change risk, the latter is measured as vulnerability to greenhouse-induced climate change. The most robust way to assess climate change damages in a probabilistic framework is as the likelihood of critical threshold exceedance. Because vulnerability is dominated by local factors, global vulnerability is the aggregation of many local impacts being forced beyond their coping ranges. Several case studies, generic sea level rise and temperature, coral bleaching on the Great Barrier Reef and water supply in an Australian catchment, are used to show how local risk assessments can be assessed then expressed as a function of global warming. Impacts treated thus can be aggregated to assess global risks consistent with Article 2 of the UNFCCC. A 'proof of concept' example is then used to show how the stabilisation of greenhouse gases can constrain the likelihood of exceeding critical thresholds at both the local and global scale. In terms of managing climate change risks, adaptation is most effective at reducing vulnerability likely to occur at low levels of warming. Successive efforts to mitigate greenhouse gases will reduce the likelihood of reaching levels of global warming from the top down, with the highest potential temperatures being avoided first, irrespective of contributing scientific uncertainties. This implies that the first cuts in emissions will always produce the largest economic benefits in terms of avoided impacts, the sum of these benefits depending on the sensitivity of the climatic response and the damage function of the respective impacts. The major benefit of the structure presented in this paper is that risk can be translated across local and global scales, linking both adaptation and mitigation within a framework consistent with the aims of the UNFCCC.

This Chapter is available in printed form with black and white graphics; however, it is also available as an e-book or electronic file (pdf) with colour, higher resolution graphics.

ISBN 92-64-10831-9
THE BENEFITS OF CLIMATE CHANGE POLICIES

CHAPTER 8. MANAGING CLIMATE CHANGE RISKS

by Roger Jones[87]

1. Introduction

In trying to understand and mitigate the enhanced greenhouse effect, scientists and policy makers are undertaking a global-scale risk assessment (e.g. Beer, 1997 and others). By framing criteria to identify, assess, prioritise and manage risks, the structure of Article 2 of the United Nations Framework Assessment on Climate Change (UNFCCC) is compatible environmental risk assessment frameworks, therefore many of the tools developed to assess and manage environmental risks can be used in support of the UNFCCC (Jones, 2001). The requirement to stabilise greenhouse gases at levels sufficient to prevent dangerous anthropogenic climate change sets the criteria for assessment, while maintaining food security, allowing sustainable economic development and allowing ecosystems to adapt naturally set the criteria for management. This task is complicated through issues of uncertainty, complexity, scale and delays between action and response.

Risk is the combination of the likelihood of an event and its consequences. Impact assessments have widely examined the consequences of climate change but have been less able to attach likelihoods to those outcomes. Pervasive uncertainties limit most assessment to using scenarios that present alternative futures without being able to determine which of those futures may be more likely (Carter and La Rovere, 2001). The use of likelihoods in climate change assessments is increasing. Utilising guidance from Moss and Schneider (2000), the Intergovernmental Panel on Climate Change (IPCC) applied a more structured approach to assessing uncertainty in its Third Assessment Report (2001a–c). Binary true/false statements were given confidence levels based on expert assessment of the evidence that attached words such as *very likely* or *high confidence* to ranges of probability (IPCC, 2001a, p44). However, these guidelines were

87 CSIRO Atmospheric Research, PMB1 Aspendale, Victoria 3195, Australia, roger.jones@csiro.au.

not used consistently across the different working groups and chapters (Reilly *et al.*, 2001).

The following conclusions drawn from the Third Assessment Synthesis Report were considered by the Core Writing Team to be robust (IPCC, 2001a). *Projected climate change will have beneficial and adverse environmental and socio-economic effects, but the larger the changes and rate of change in climate, the more adverse effects predominate* (IPCC, 2001a, p 67). These consequences were addressed by IPCC using mean global warming as a common metric (IPCC, 2001a; Smith *et al.*, 2001). Above a few °C relative to 1990, impacts are predominantly adverse, so net primary benefits of mitigation would become positive (IPCC, 2001a). Adaptation is necessary due to climate change that has already occurred and to prevent further climate change that cannot be mitigated. Adaptation is most suited to modest and/or gradual changes in climate (IPCC, 2001a).

This chapter describes methods for risk assessment that aim to reconcile the needs of the UNFCCC across the different scales on which climate risks will manifest themselves and will need to be managed. This framework concentrates on bottom-up methods that have been developed in a series of earlier papers (Jones 2000a & b; New and Hulme, 2000; Jones, 2001; Jones and Page, 2001). While Article 2 of the UNFCCC makes no mention of scale, there is often an implicit assumption that dangerous climate change should be assessed at the global scale. While not disagreeing with the need to address dangerous climate change as global issue, this paper deals with how risk assessments conducted at the spatial scale most appropriate to a specific activity can be addressed in such a way that they can contribute to global assessments.

The scale of assessment is a key consideration. In this paper, the terminology of bottom-up and top-down is used in a scale-specific manner. Bottom-up methods are local in scale and tend to address assessments through consequences experienced at that scale, working up towards the global atmospheric scale. Top-down methods usually conduct assessments at the global scale (emissions or climate) working down towards the local scale. Most (though not all) impacts are local in nature, so are best analysed using bottom-up methods suited to particular activities and locations. For example, climate hazards, measured as changes in the magnitude and frequency of climate variability and extremes, and their resulting vulnerabilities, need to be assessed at the local scale (Jones *et al.*, in review). The likelihood of exceeding critical thresholds under well-quantified ranges of change can then be assessed without, then with, risk management in the form of adaptation and mitigation.

The scale of risk management options also affects the scale at which an assessment takes place. Mitigation and adaptation are complementary strategies but affect risk on different spatial scales. Whereas the primary benefits of adaptation are generally local, mitigation will reduce climate change impacts at the global scale. (Although emission reduction and greenhouse gas sequestration is likely to be undertaken across a range of scales). Planning and policy decisions will also address different scales. For example, individual country parties to the UNFCCC will address their responses based on their exposure to climate risks weighed against the perceived risks of undertaking adaptation and mitigation actions (i.e. based on national self-interest). One can argue that this exercise is currently being carried out in a very ad hoc fashion, where policy is remote from the science. Because climate risks can potentially manifest on any scale from local to global, individual national or regional responses (e.g. European Union, small island states) will also require frameworks for aggregation.

Aggregation of climate damages assessed at local scales is therefore one of the steps needed to understand risk at the global scale. Aggregation of the exposure to climate risks and also of the benefits of adaptation and mitigation can be achieved by expressing local outcomes as a function of mean global warming. A generic example and two case studies are used to demonstrate methods of analysing risk at the local scale. The results are then expressed as a function of mean global warming, allowing comparison and aggregation using a common metric. The studies are: a generic example using global temperature and sea level rise, coral bleaching on the Great Barrier Reef and water supply changes in eastern Australia. Although the process of expressing different outcomes in this metric increases the uncertainty associated with each individual impact, the benefit of achieving a more integrated outlook outweighs the disadvantages. This methodology is complementary to global, or 'top-down', assessments of impacts such as those reviewed by Hitz and Smith (this volume).

A final example shows how risk can be assessed at the global scale using a similar framework (assessing the likelihood of exceeding a critical threshold linked to the notion of dangerous climate change). This example quantifies climate risks at the global scale by estimating the likelihood of stabilising the atmospheric concentration of CO_2 below a series of equilibrium temperatures. Because current knowledge of what constitutes dangerous climate change is regarded as being too uncertain (e.g. Dessai *et al.*, 2004), successive levels of warming at equilibrium are investigated (also allowing different levels of risk tolerance and applications of the precautionary principle to be tested). Bayesian inference under a range of prior assumptions can be tested by assessing the likelihood of threshold exceedance under different input conditions. Outcomes that occur under a wide range of prior assumptions are likely to be robust, i.e. they are insensitive to underlying assumptions so that they can be considered more likely to occur. Using such methods changing risks can be assessed under different policy assumptions and can be updated as new information becomes available.

A robust aspect of the framework is that critical thresholds exceeded at low levels of global warming will be those that can be given high priority for adaptation because they are highly likely to be exceeded no matter what decisions are taken on mitigation. The risk of experiencing more severe and adverse consequences (those occurring at a global scale, or widespread local damages aggregated from the bottom up at higher levels of global warming) expressed as a function of global warming, although less likely to occur, provide the impetus for mitigation.

2. Basic structure of risk

Risk is the combination of the likelihood of an event and its consequences (e.g. Beer and Ziolkowski, 1995; USPCC RARM, 1997). Likelihood can be attached to the hazard (i.e. risk equals the likelihood of a hazard and its consequences) or to the consequences (i.e. risk equals the likelihood of exceeding a given level of damage), distinguished as the natural hazards-based approach and the vulnerability-based approach (Jones and Boer, 2004). A hazard can be broadly described as an event with the potential to cause harm. Climate hazards at the local scale include the direct effects of climate and immediate impacts arising from climate events (e.g. secondary climate hazards such as flood and fire). Climate hazards at the global scale include global mean warming, global mean sea level rise and large-scale singularities, such as cessation of thermohaline circulation and collapse of large ice sheets (Schneider and Lane, this volume). Vulnerability is the degree to which a system is susceptible to harm and is measured in terms that express a measure of value. This measure may be monetary but can utilise any type of value-based criteria, such as the five numeraires of Schneider (2003) or the somewhat larger set of measures suggested by Jacoby (this volume). Global vulnerability is the aggregation of costs from climate risks and the benefits of managing those risks.

If we apply this construction of risk to the IPCC, changing climate hazards are dealt with by Working Group I (WGI; climate) and in part by Working Group II (WGII; impacts), and risk management is dealt with by Working Groups II and III. There are two ways to manage the risks of climate change in a planned manner. Adaptation will reduce the vulnerability to a given climate hazard or hazards, as assessed by WG II; the mitigation of greenhouse gases will reduce the magnitude and frequency of climate hazards, as assessed by WG III.

2.1 Uncertainty, complexity and probability

Climate change assessment is dominated by uncertainty, affecting the choice of method and the confidence that can be attached to the results. Uncertainties can be distinguished according to those that may be reduced with improved knowledge and those that remain due to fundamental system uncertainty (Hulme and Carter, 1999; Moss and Schneider, 2000). Improved knowledge will, in theory, make forecasting possible, but not if fundamental uncertainties persist. This is likely to be the case under global warming, which is subject to a number of fundamental uncertainties, including multiple feedbacks resulting from interactions between biophysical and socio-economic systems (Dessai and Hulme, 2003). Different parts of the climate system are affected by different uncertainties; those uncertainties will be reduced at different rates and only as far as knowledge barriers allow. New knowledge can actually increase quantified ranges of uncertainty as the 'unknown' becomes known.

All of these factors need to be taken into account in communicating outcomes. In order of decreasing certainty, a result can be expressed: a) as a central prediction; b) as a central prediction with error bars; c) as a known probability distribution function (PDF);

d) as a bounded range with upper and lower limits but no known probability distribution; e) as a bounded range within a larger range of unknown possibilities; f) as individual scenarios with plausibility but no further aspect of likelihood; and g) as a hypothesis with unknown levels of plausibility (see Morgan and Henrion, 1990 for examples).

Both complex behaviour and incomplete knowledge mean that central predictions and well-calibrated probability distributions for climate change are not possible. Climate change assessments are limited to expressing outcomes drawn from lower down the above list, such as ranges of uncertainty, scenarios and hypotheses. However, it is incumbent on a risk assessment to minimise unquantified uncertainty by utilising as wide a range of uncertainty as possible (Jones, 2000a, b).

Figure 1a illustrates how ranges of uncertainty propagate through an assessment. Figure 1b shows the relationship between individual scenarios and ranges of change that can be constructed from a set of scenarios. Note that the range in Figure 1 is not portrayed with a probability distribution. This is consistent with the range of global warming provided by the IPCC (2001b) which has an upper and lower limit but has no quantitative probability attached to it, explicit or implicit. Fundamental system uncertainty will not allow objectively derived probabilities to be attached to outcomes such as global warming (e.g. see Dessai and Hulme, 2003). Nevertheless the use of Bayesian probabilities (showing the impact of a range of assumptions or prior information on the results) can offer a number of insights (New and Hulme, 2000; Schneider, 2001 and 2002; Wigley, 2003).

Any strategy to manage uncertainty by sampling ranges or relating different model-based projections using other strategies (weighting etc) will contain assumptions about how those projections relate to each other. For example, when sampling estimates of regional rainfall change from different climate models, are those models independent to each other, should they be weighted by their performance in simulating current climate or can ranges be constructed from the entire sample with a probability distribution applied independently? There is no clear answer to these questions, and any particular choice will affect the outcome.

Recent commentaries (Schneider, 2001 and 2002; Schneider and Moss in Giles, 2002) argue that assessments undertaken by the IPCC should attach probabilities to their conclusions using methods drawing on expert opinion and Bayesian statistics. The principal thesis is that if guidance on likelihoods is not given, policymakers will attach their own likelihoods in an *ad hoc* manner. This is preferable to relying on current scenario-driven approaches, but the nature of the framework within which probabilities are used is critical. For example, Patt and Schrag (2003) show that if the public are given probabilistic estimates of climate change, they will interpret these according to their own interpretations of the consequences. Therefore, climate forecasts are interpreted as risk, but in a disorganised manner.

Figure 1. Schematic portrayal of climate change uncertainties

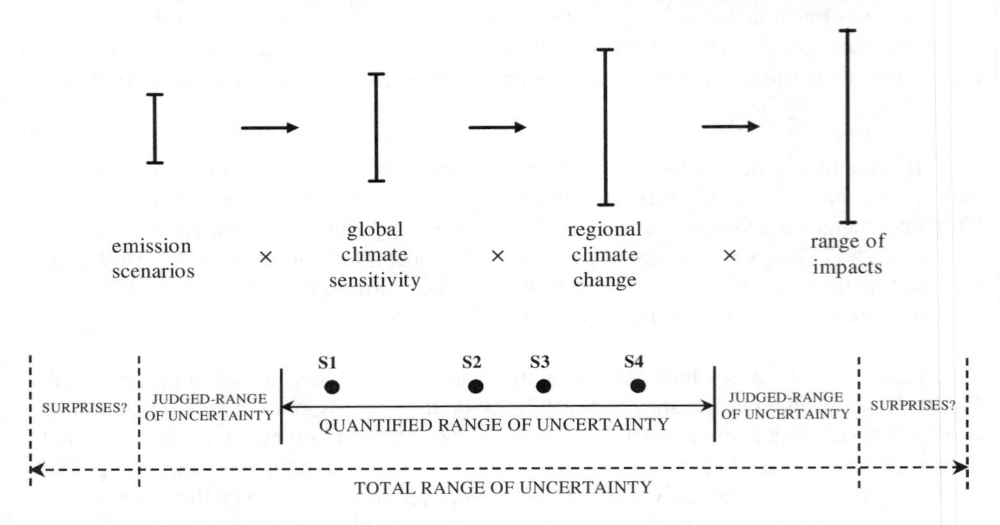

Source: from Jones (2000) and Schneider and Kuntz-Duriseti (2002).

Note: The relationship between (a) ranges of uncertainty cascading through an assessment, and (b) between individual scenarios, S1 to S4, and resultant ranges of uncertainty. .

In this chapter I build on those arguments, proposing that risk assessment is a way of using probabilities rather than relying on forecasting methods, e.g. climate forecasting as an outgrowth of weather forecasting. Risk assessment can consider a range of potential outcomes under different policy assumptions, whereas predictive frameworks are much less flexible. Both methods are preferable to relying on individual scenarios with no further application of probabilities to distinguish the likelihoods of various outcomes, though these latter approaches will continue to be useful to explore a wide range of possible futures.

The act of attaching probabilities to outcomes based on complex socio-economic relationships is controversial (Dessai and Hulme, 2003). For example, Grübler and Nakicenovic (2001) argue that because conditional probabilities cannot be attached to the underlying socio-economic processes driving greenhouse gas emission scenarios, probabilities cannot be attached to any outcomes dependent on those scenarios ("good scientific arguments preclude determining 'probabilities' or the likelihood that future events will occur"; Grübler and Nakicenovic (2001)). This implies a serial dependence where probabilities follow the time-dependent sequence of emissions, climate change, impacts and adaptation/mitigation utilised by scenario-based assessments. However, if probabilities do break down at the weakest link in a chain of consequences, the optimal adaptation strategies are those that provide benefits across a wide range of possible outcomes (e.g. Lempert and Schlesinger, 2001).

Ideally, research into the adverse impacts of climate change (i.e. vulnerability) should be conducted independently of how likely those adverse impacts are to occur (Grübler and Nakicenovic (2001)). This is possible if vulnerability is assessed independently of climate scenarios, for example, the construction of critical thresholds for use in risk assessments (Pittock and Jones, 2000; defined loosely as a tolerable limit of harm). Critical thresholds can then be assessed for their likelihood of exceedance under ranges of plausible climate change (Jones, 2000a, 2001). This is a probability over threshold approach widely used in engineering, but in this case, thresholds are used much more widely, covering biophysical and socio-economic criteria and as such, are often normative.

Although the point where criticality occurs for any type of hazard cannot be addressed accurately in many situations, using consensus it is be possible to identify a point where it is agreed that significant harm will happen. In other cases, different groups will have very different constructions of what constitutes criticality – this is typical of the planning issues affecting any contentious proposal, and much may be learnt from investigating these different constructions. Another area of criticality that has been little explored is where one activity ceases and others become possible. In many cases, this point will be impossible to predict *a priori*, but may have major consequences.

If likelihoods are to be quantified, a crucial step is whether PDFs can be attached to ranges of uncertainty or developed from a set of samples. Probabilities constructed in this way are not predictions in the conventional sense but encompass a range of possibility that contains a single unknown outcome (sometimes known as single-event uncertainties). Probabilities are constructed by utilising individual scenarios, ranges or probability distribution functions to explore uncertainties. Ranges can be constructed using a set of scenarios (see Figure 1b), using Delphi analysis (formalised methods for gauging expert opinion; e.g., Morgan and Henrion, 1990; Nordhaus, 1994; Morgan and Keith, 1995), other forms of expert opinion, and scientific and model uncertainty. Subjective methods will be required in many cases because objective methods are not available or may only cover part of the range. Because experts disagree, one strategy is to determine how different world views, policy views and scientific views affect input ranges and responses (Reilly *et al.*, 2001). Transparency should be maintained at each step so that how the choice of method affects the results is communicated along with the results.

When two or more ranges of uncertainty are combined, the resulting probability distribution will favour the central tendencies of the input probability distribution functions (PDFs) subject to the method of combination.[88] Monte Carlo analysis (repeated random sampling) and Bayesian analysis (testing the impact of prior

---

[88] See for example Jones, 2000b; Schneider, 2001; Wigley and Raper, 2001; Jacoby 2003; Wigley, 2003 on combining ranges of uncertainty for two or more variables.

information on the results) are two of the methods commonly used.[89] Bayesian reasoning can be used to test different input assumptions to determine whether the results are sensitive to those assumptions.[90] If they are, then the structure of input uncertainties is important and further research in that area is warranted. If the output is insensitive to input assumptions, then that aspect of the analysis is robust under uncertainty.

Bayesian methods are more flexible and parsimonious than predictive methods for 'forecasting' climate change. Alternatively, as a variation on predictive methods Allen and Stainforth (2002) propose the use of multiple runs of different climate models sampling a range of plausible climate sensitivities and feedbacks, and taking into account different natural climate forcings, and land-use change and emission futures have been proposed. These schemes have the potential to construct PDFs where the choice of priors has less influence on the results than Bayesian or expert methods (though such an exercise can never be truly 'objective'). However, this would require several orders of magnitude more model runs than assessed through other methods (Allen and Stainforth, 2002).[91] Such methods will have more success exploring biophysical uncertainties but will have more difficulty with human agency in selecting different socioeconomic pathways than Bayesian or Monte Carlo methods.

Human agency can alter ultimate outcomes by concentrating on different futures and by anticipating and responding to situations as they unfold (e.g. Risbey, submitted). It unclear how forecasting methods could be used to assess different policy options, unless a different forecast is made for each set of starting assumptions. Forecasts are largely exploratory methods, based on how the future might unfold given a starting set of assumptions. Normative methods are based on a description of how the future might be according to different moral or ethical concerns or due to particular ideologies. For example, sustainability is a normative condition, whereas business-as-usual scenarios projecting current trends are exploratory or forecasts. Stabilisation scenarios that are designed to explore issues of dangerous climate change are normative with regard to the UNFCCC.

Risk assessment, drawing on Bayesian and Monte Carlo methods, are more flexible in their ability to take explicit account of the role of human agency and to evaluate both exploratory and normative futures. In a recent example, Webster *et al.* (2003) contrast the effect of the non greenhouse gas policy SRES scenarios with stabilisation scenarios on the likelihood of mean global warming in 2100. Key questions

---

89 The need to apply different constructions of probability and likelihood within a Bayesian framework is discussed by Jones (2000b), New and Hulme (2000), Moss and Schneider (2000), Schneider and Kuntz-Duriseti (2002) and Webster (2003).

90 For example, New and Hulme, 2000; Jones and Page, 2001; Dessai, 2003; Webster *et al.*, 2003.

91 For example, Allen and Stainforth (2002) describe the results of the Climate Model Intercomparison Program which assessed over 2,000 simulations.

are: What happens if no climate change policies are applied? What is the impact of different policy regimes?

2.2 *Spatial issues and scale*

The dependence of impacts on scale is critical for assessing impacts and adaptation. In most sectors impacts vary widely between locations and groups of people. Impacts will vary widely as a response to climatic interactions with the physical environment, vulnerability will vary widely in response to the socio-economic environment and adaptation will be subject to the relationship between both aspects. A major impediment to large scale or top-down impact assessments is that the results from one location often cannot be usefully applied to another. Local knowledge is critical. The experiential aspect of adaptation where people have to "learn by doing" means that local stakeholder involvement is critical to both assessment and application (e.g. Jones *et al.*, in review; Conde and Lonsdale, 2004). Stakeholders are also important in the development and understanding of critical thresholds (Jones, 2001).

Most of the top-down methods of impact assessment summarised by Hitz and Smith (this volume) contain a degree of spatial representation to account for differences between regions (e.g. Parry *et al.*, 2001). This heterogeneity is due to both the nature of the climate changes and to the distribution of impacts and vulnerability, which will vary from place to place. However, circumstances often dictate that single relationships are utilised widely in global impact models, even though the investigators are aware of widely varying relationships between locations. Assessments in developed countries with relatively good data, e.g. for agriculture in the Europe and the USA, cannot be immediately transferred to Africa or Asia where data is not readily available, because the results will misrepresent those countries' adaptive capacities. The main benefit of global assessments is that they provide a broad picture of outcomes but may be limited in spatial and temporal detail. In some cases this may significantly constrain estimates of damage costs.

Bottom-up assessments that explicitly represent local conditions can manage these details much better but need to be aggregated in some way if a global picture is to be obtained. Presently, this is done through IPCC assessments or meta-analyses that integrate a larger number of studies (e.g. Thomas *et al.*, 2004). Site specific assessments can also provide local detail that can be used to inform global impact models. For example, transfer relationships based on local information can be used to scale up impact assessments in larger models, but are needed for many regions, not just the few where assessments have been carried out.

Large-scale impact models that are physically and/or spatially complex can usually only be run with a limited number of scenarios. A major advantage of local assessments is that scale and complexity can be constrained, thus allowing the development of models that can be run with a large number of scenarios. The results can then be used to explore input uncertainties and construct probability distributions (Jones, 2001). However, the results from local assessments driven with variables critical to a specific activity cannot be readily compared with those from another activity. For example, a crop assessment driven by mean changes in temperature, rainfall and atmospheric CO_2 cannot be easily contrasted with an assessment of storm damage.

Expressing the results in terms of global mean warming will allow damages from different activities to be compared (e.g. the impacts expected with 1, 2 and 3°C warming), as is being demonstrated in a number of recent studies (e.g. Leemans and Eickhout 2003). Despite this, it will only be possible to attach likelihoods to a subset of potential impacts (where uncertainty can be constrained to a sufficient degree). The examples given in section 4 below address situations where such an approach is feasible. It remains to be seen how broadly applicable such an approach may be in terms of different sectors, regions and levels of uncertainty.

2.3 Temporal issues and inertia

The delay between action and response is a key contributor to climate change uncertainty. Although greenhouse gases will mix in the atmosphere within one to two years, the resultant impacts may not manifest for decades to centuries (IPCC, 2001a). Following the stabilisation of greenhouse gases in the atmosphere, warming will continue for some decades at a gradually slowing rate, providing there are no abrupt shifts in the global climate system. Delays in warming the deep ocean mean that sea level rise may continue for centuries. This gradualist model also fails to account for catastrophic changes, or singularities that, while currently being at very low probabilities, become increasingly more likely as warming continues and accelerates. The delay between gradual forcing and sudden response is a plausible scenario because this type of response has been a feature of previous climate changes (Schneider and Lane, this volume). Therefore, a wait and see approach is highly risky because the climate system could be subject to irreversible changes in future without showing any sign of such changes today.

The delay between radiative forcing and the climatic response also affects planned adaptation. If a current activities or investments, such as the building of fixed infrastructure like bridges and dams, then climate change over that period may need to be anticipated and adapted to in the design of that investment. Irreversible impacts over different timeframes that are deemed critical also need to be addressed (see Schneider and Lane, this volume). Therefore, adaptation needs are closely linked to planning horizons.

Planning horizons will be of different lengths, whereas policy horizons mostly focus on the short to medium term (Jones *et al.*, in review). This requires outlooks to be long, whereas subsequent actions based on those outlooks can have a limited lifetime, to be re-adjusted as new information becomes available or as a policy sunset clause is exercised. This model suits a conceptual model of adaptation where adaptation is based on managing the risks posed by existing climate variability and extremes in the near-term. In terms of planning adaptation, decisions must also anticipate how those risks may change in the future, or new risks that may emerge. For example, short-term activities, such as crop choice in agriculture, can be useful adaptation options for climate change but are implemented incrementally, so can respond to climate variability on a seasonal basis. In contrast, mitigating the impact of agricultural activity on the landscape and its effects on processes such as salinisation, will require very long-term planning horizons. The latter will prove a challenge for shorter-term policy horizons. Thus single activities may contain both short- and long-term adaptation planning horizons and both planning horizons may affect nearer term adaptation policy decisions.

In a similar manner, the pathway to the stabilisation of climate (or mitigation pathway) is likely to be determined by a series of shorter-term policy actions moving towards a longer term goal.

3. Elements of climate risk assessment

This chapter frames risk as the likelihood of exceeding the ability to cope with climate change. At the local scale, the ability to cope is measured using impact and location-specific criteria. At the global scale, the ability to cope is guided by Article 2 of the UNFCCC. The case studies described in this chapter explore methods for linking these scales in order to manage risk through both adaptation and mitigation consistent with the UNFCCC. The starting point begins with risk under current climate, then utilises methods to asses how these risks may change with climate change.

Over time, societies have developed an understanding of climate variability in order to manage climate risk. People have learnt to modify their behaviour and their environment to reduce the harmful impacts of climate hazards and to take advantage of their local climatic conditions. The range of coping mechanisms within a given exposure unit or system that can be described in terms of a range of climate or climate-related phenomena is called the coping range (Hewitt and Burton, 1971; Smit and Pilofosova, 2001). Technological and social developments can be used to expand the coping range. Climate-related outcomes are beneficial or neutral when they are well within the coping range, display a tolerable level of harm towards the margins and exceed tolerable levels of harm beyond the margins, where a system becomes vulnerable (Jones and Boer, 2004). This limit of tolerable harm is used to define the critical threshold (Pittock and Jones, 2000). In an agricultural system, outcomes may range from bountiful crops in the coping range, to marginal crops at the edge to crop failure beyond the range. Sometimes, the threshold for criticality is clear, as in the failure of flood protection; but often it is indistinct, where the outcomes are contingent on socio-economic influences with varying groups showing different levels of vulnerability (Smit and Pilofosova, 2001).

The coping range is a largely heuristic device that reflects the human experience to climate variability and extremes. Practitioners have a range of responses to climate which is based on their experience and includes a sense of when events become critical. It serves as a frame of reference based on current and past experience of adaptation that can be used to assess the need to manage future risks. Different individuals and groups undertaking the same activity will have different coping ranges and critical thresholds based on their current capacity to cope. In this sense, stakeholder-derived critical thresholds can be strongly normative because they represent the values held by the practitioners. Other thresholds may be more widely held and imposed externally (e.g. formal definitions of disaster and loss from extreme events). The development of the coping range has not yet accounted for critical thresholds that mark the transition of one activity to another or how such transitions should be managed (e.g. Schneider and Lane this volume). Both coping ranges and critical thresholds should be constructed with this possibility in mind.

In most climate-related systems, the further one moves beyond the coping range at any given time, the greater the vulnerability will be. Adaptation will expand the coping range and mitigation may limit the likelihood of that range being exceeded under climate change. Adaptive capacity refers to the potential to adapt and measures the ability to expand the coping range in response to anticipated or experienced hazards (Brooks *et al.*, 2004).

The coping range is most easily defined for an activity, group and/or sector, although sectoral and national coping ranges have been proposed (Yohe and Tol, 2002). Critical thresholds for these coping ranges will be context specific and can be expressed using a wide range of measures. A global coping range is implied by the structure of Article 2 of the UNFCCC where 'dangerous' climate change forms a global critical threshold. However, this threshold is difficult to define and will be interpreted variously according to different levels of risk perception, risk tolerance and the criteria used to measure damages and benefits (Dessai *et al.*, 2004).

3.1 Global scale

The IPCC Third Assessment Report concluded that as global warming increases, the number of activities damaged by climate change will increase, as will the damage suffered by individual activities (Smith *et al.*, 2001). This allows the magnitude of global warming to be linked with the severity of consequences (IPCC, 2001a; Parry *et al.*, 2001; Swart *et al.*, 2002). If a large number of impacts can be expressed as a function of the increase in global mean temperature, then it becomes possible to integrate their results. At low levels of warming there may be many positive and fewer negative impacts, but those negative impacts will affect the poor disproportionately. At high levels of warming the impacts will be more negative and more widespread (Smith *et al.*, 2001). The likelihood of reaching dangerous climate change is therefore low at low levels of global warming, increasing at ever higher levels, although the decision of what is dangerous is a policy-related rather than scientific question (Azar and Rodhe, 1997). This conclusion does not invalidate the role of research in exploring the concept of dangerous climate change, but makes it clear that researchers will not provide the answer independently of investigations involving wider communities (Dessai *et al.*, 2004).

In terms of exceeding given levels of global warming, the likelihood of exceeding the lower limit of the potential range of warming will always be higher than the likelihood of exceeding the upper limit. We assume that the range of 1.4°C to 5.8°C (IPCC, 2001b) forms the limits of the well-calibrated range of possible global warming by 2100, the lower limit of 1.4°C is highly likely to be exceeded and the upper limit is highly unlikely to be exceeded. This outcome holds even though the objective probability distribution of global warming remains unknown and will hold for all subjective distributions.

A cumulative probability distribution function (CDF) provides a measure of the likelihood of exceeding a given level of warming (probability over threshold) from low to high levels of warming. For example, the likelihood of exceeding a particular threshold moves from 100% at the lowest limit of warming to 0% at the upper limit (assuming the whole range of uncertainty has been captured).[92] Therefore a CDF provides more information about potential risk than a peaked probability distribution assessing the "most likely" value of global warming.

Using this risk analysis structure, the adverse consequences of climate change increase with global warming, while the likelihood of reaching successively higher levels of damages decreases. However, the product of probability × consequence remains unknown. For damage functions that increase slowly, risk may diminish as probabilities reduce; for damage functions that increase steeply, risk may increase non-linearly and peak towards the upper limit of the warming range; for damage functions that are step-like, especially global-scale singularities such as ice-sheet collapse, risk may be extremely high at very low probabilities occurrence.

3.2 Local scale

At the local scale, damages will increase as climate change takes a system beyond its coping range. Most damages will occur in response to altered variability and extremes. For example, increases in the rate and magnitude of coral bleaching, inundation of low coasts during storm surges, fire frequency and drought and floods may exceed rates that allow system recovery. For each system and location, this relationship needs to be understood under past, present and future conditions.

The coping range can be used as a conceptual model to operationalise and communicate risk. The climatic stimuli and their responses for a particular locale, activity or social grouping can be used to construct a coping range if sufficient information is available (e.g., Smit and Pilifosova, 2001). Current climate is an important reference for the coping range, where the current ability to cope can be reviewed under both changing climate and changing socio-economic conditions (Jones and Boer, 2004). Changes in climate hazards may lead to critical thresholds being exceeded more frequently (Pittock and Jones, 2000).

---

92 All the CDFs shown in subsequent figures follow this format.

Ideally a critical threshold marks a known level of vulnerability, broadly defined as the outcome of climate-related hazards in terms of cost or any other value-based measure (Jones and Boer, 2004).[93] Society will also change over time, altering its capacity to cope with climate hazards. It is therefore possible to compare changing hazards with current capacity or to alter the coping capacity of society in line with projected socio-economic change at some time in the future (Jones and Mearns, 2004). This can be in the form of an anticipated change, for example as the result of policy or planned development, or a desired change, for example a future sustainable state. An example of how the ability to cope can totally alter the response to climate-driven hazards is provided by Tol and Dowlatabadi (2001) who describe how economic development in an emissions scenario can negate changed exposure to malaria vectors caused by the accompanying projected climate change. Changes in climate hazards can be modified by mitigation and the width of the coping range modified by adaptation. Both will alter projected risks.

4. Application of risk analysis

The following sections show how to formulate both local and global scale risk assessments in a manner consistent with Article 2 of the UNFCCC. Cumulative probability distribution functions are created to assess the likelihood of exceeding critical thresholds: 1) using generic examples of global warming and sea level rise, 2) an example of local coral bleaching 3) an example using regional water supply and 4) assessing the likelihood of being able to stabilise atmospheric CO_2 below a given global average temperature at equilibrium.

Risk analysis can be applied to climate change impacts in two ways, depending on whether the starting point focuses on the climate hazard or on the socio-economic outcome. The first method assesses the likelihood of a given hazard then of the ensuing consequences. This is the approach usually used in natural hazards research and is consistent with the standard approach to impact assessment as described by Carter *et al.* (1994) with the addition of probabilistic methods as described by Jones (2001). It can be applied to simple and complex climate hazards and to secondary hazards closely aligned with climate, such as fire and flood, where climate can be closely aligned with outcomes. The second method focuses on the outcomes, setting criteria based on possible desirable or undesirable future states, such as critical thresholds, and then analysing the likelihood of exceeding those criteria. These criteria can be based on current or plausible future levels of adaptation. This is largely a normative method, where thresholds can measure a level of criticality that is to be avoided, or a set of desirable future conditions (i.e. that achieve sustainability). Both methods – assessing

---

93 The consequences of climate change for a particular activity can be assessed using a critical threshold marking the tolerable limit of harm (Parry *et al.*, 1996), or as a threshold based on a continuous relationship that can be partitioned according to given levels of success or failure (Jones and Boer, 2004).

risk as a function of climate hazard and as a function of climate change consequences - are applied in the following sections.

4.1 Global temperature and sea level rise

The following example shows how probabilities of exceedance can be applied to thresholds for warming and sea level rise set based on global average values. The use of global values for location-specific thresholds is a low-precision method best used when regional climate scenarios or quantitative links between climate and impacts are not available.

The example compares two hypothetical critical thresholds for global warming of 1.0°C and 2.5°C mean temperature increase. Of those two thresholds, a global average warming of 1.0°C is much more likely to be exceeded. For example, coral reefs that bleach above a warming threshold of 1.0°C face a far greater risk than those that will not bleach below 2.5°C. (In other words, it is more likely that a 1°C threshold will be exceeded than a 2.5°C threshold when looking across all possible climate outcomes.) The most southern permafrost zones in Europe, Asia and North America will experience more seasonal melting at lower temperature rises than those zones located further north. Similarly, alpine ecosystems currently close to their marginal limits will be more severely affected than those ecosystems at higher altitudes. The same principle holds for sea level rise. For example, we intuitively know that the lowest areas of coast are the most likely to be inundated, irrespective of the level ultimately reached. Applying the IPCC (2001b) range of sea level rise of 0.09 to 0.88 m at 2100, a section of coast vulnerable to a rise of 0.25 m is much likelier to be affected than a section vulnerable to a rise of 0.75 m.[94] This principle extends to all activities where critical thresholds or other risk-based criteria can be characterised as a function of mean global warming or of sea level rise.

Figure 2 shows how these thresholds can be related to probability distributions of global warming and sea level rise. On the upper and lower left are ranges for temperature and sea level rise from the IPCC TAR (2001b). The central panels show PDFs for 2100 constructed by multiplying two randomly sampled ranges of component uncertainties for both temperature and sea level rise. These are framed as conditional forecasts, showing the 'most likely' value of global mean warming and sea level rise.

For temperature, it is difficult to associate each pair of thresholds in the central panels with its relevant PDF. In fact, the lower temperature threshold of 1°C falls outside the range of probabilities, but common sense tells us by 2100 this threshold will be exceeded under all outcomes (based on priors drawn from the SRES scenarios and IPCC, 2001b). This is also the case when a uniform PDF is applied. For sea level rise,

---

[94] Both mean global warming and sea level rise are suitable for this type of treatment because they move in a single direction. The critical thresholds chosen are arbitrary and unrelated between the two variables.

the two thresholds have a similar likelihood of being met – the domination of thermal expansion on the outcomes means that the Monte Carlo sampled joint PDF is similar to the uniform PDF.[95]

In the right-hand panels, we have moved from conditional forecasts to risk analysis. The PDFs are recast as CDFs, illustrating the likelihood of threshold exceedance. The lower thresholds for temperature and sea level rise both are much more likely to be exceeded than the upper thresholds. Therefore, the locations where the lower threshold applies face a much greater risk. By comparing the relative likelihoods of exceeding a range of thresholds, locations facing higher risks can be identified and prioritised for the purposes of risk management.

Casting likelihood as the probability of exceeding a given threshold is also more robust to changes in input assumptions, such as the shape of the probability distribution function. Figure 2 shows that a substantial difference when relying on uniform and non-uniform PDFs to predict the most likely global warming. Yet even though the likelihood of predicting a particular outcome in terms of climate is very low, the probability of calculating threshold exceedance is much less sensitive to different input assumptions. For example, the uniform CDF (all points in the range of global warming and sea level rise being equally likely) shows similar probabilities of exceedance to the non-uniform CDF. Thus, different prior distributions of the component uncertainties may alter both PDF and CDFs but will not alter the basic conclusion that framing probabilities in terms of risk offers much more utility than attempting forecasts.[96]

4.2 Local coral bleaching

This example deals with the risk of thermal bleaching of corals at a single location. The thresholds used are bleaching thresholds measured in degree days of bleaching using a temperature–duration relationship determined from observations (Berkelmans, 2001). This is the point at which corals become bleached due to the expulsion of symbiotic algal species known as *xoozanthellae*. Also used are a series of thresholds where local sea surface temperature exceeds a temperature–duration curve set at 0.5°C increments above the bleaching threshold to bleaching +2.5°C. As each successively hotter threshold is breached, both coral mortality and the subsequent recovery time will be increased. Risk assessments at the local scale require estimates of local changes in key climate variables and locally relevant criteria such as critical thresholds for bleaching and mortality. The site used in this paper is Magnetic Island on the Great Barrier Reef, Australia but the thermal bleaching thresholds have been

---

95 Bear in mind, that if regional sea level rise was being sought, a PDF constructed from joint input uncertainties would be very different, due to the large model to model differences in estimating regional sea level rise.

96 This is consistent with the findings of Patt and Schrag (2003) who find that if the public are provided with climate change 'forecasts', they interpret them in terms of risk.

calibrated for over a dozen sites (Berkelmans, 2001) and tested for climate change at three sites (Done *et al.*, 2003). After assessing the risk at the local scale, this risk is then framed as a function of global warming.

Figure 2. Relating threshold exceedance to the likelihood of climate change

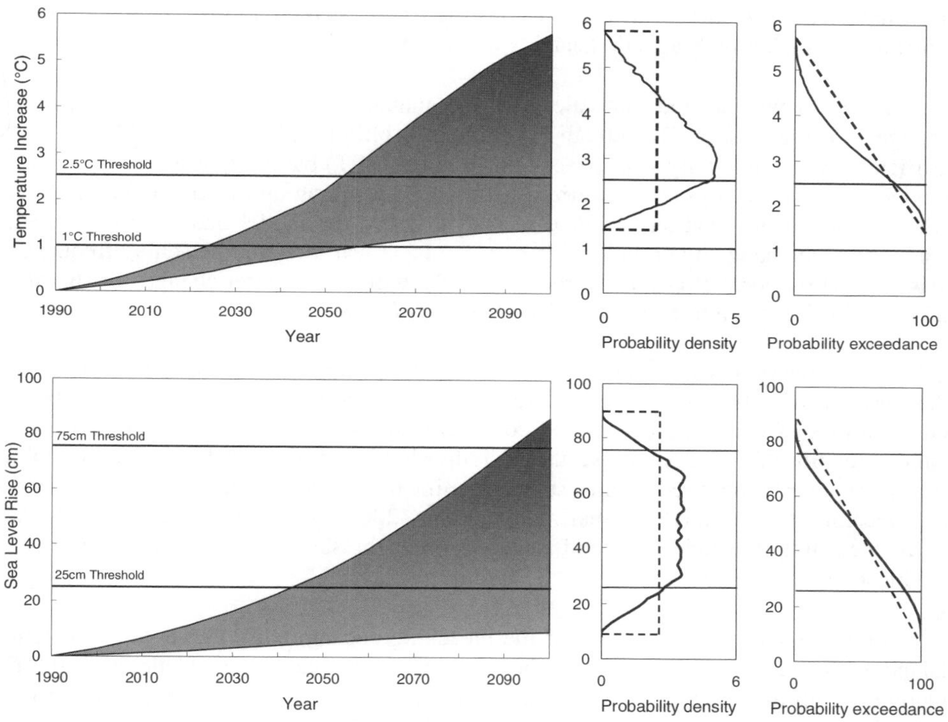

Note:

Temperature was constructed from the range of radiative forcing in Wm^{-2} for the SRES scenarios (4.2 to 9.2 Wm^{-2}) and the range of climate sensitivity (1.7 to 4.2°C at $2{\times}CO_2$) taken from IPCC (2001b), both sampled using a uniform distribution. The ranges for sea level rise are also produced from two sources: the range of thermal expansion as one range of input uncertainty, here taken to be 75% of the input uncertainty (an approximation of IPCC 2001b results), and all other sources of uncertainty as the other range, again sampled using uniform distributions.

The left-hand panels show ranges of temperature increase (upper charts) and sea level rise (lower charts) 1990–2100, showing a 1°C and 2.5°C global warming threshold and a 0.25 m and 0.75 m sea level rise threshold, respectively. The centre panels illustrate the 'most likely' outcomes (shown as percent) for temperature and sea level rise in 2100 based on prior assumptions of input uncertainties. These probability distributions each combine two ranges of uncertainty, randomly sampled and multiplied in a manner consistent with Schneider (2001; see also text section 3). They are also shown with uniform probability density. The right-hand panels show the same probability distributions recast as likelihood of threshold exceedance. The dashed line represents a uniform probability distribution (all points are equally likely).

Projected rises in sea surface temperature are likely to increase the frequency and severity of coral bleaching under greenhouse warming (Hoegh-Guldberg, 1999). Coral thermal bleaching occurs when local sea surface temperature rises above a threshold described by a duration–temperature relationship constructed from observed bleaching events (Berkelmans, 2001).[97] If the number of days over the Austral summer (December–March) exceeds a given temperature, then corals are expected to bleach. Thermal bleaching curves have been constructed for several locations on the Great Barrier Reef including Magnetic Island (Berkelmans, 2001).

A model was created that calculates the annual risk of bleaching under climate change, using the Berkelmans' thermal bleaching curves (2001) and artificially-generated daily sea surface temperatures (SST) based on observed data. This model was used to assess the severity of the 2002 bleaching on the Great Barrier Reef which matched observations at three reef sites where the model had been calibrated. Under climate change, the relationship between local warming and bleaching frequency was created by perturbing a daily record of SST with incremental changes in summer SST obtained from eight climate models (Done *et al.*, 2003).

The vulnerability of corals is related to the frequency and degree of mortality and the duration of subsequent recovery, rather than bleaching itself (Done *et al.*, 2003). Observations of bleaching events and coral relocations suggest that mortality shows a similar relationship to days above the SST threshold as does bleaching. For example, Acropora, an important reef genus, suffers significant mortality at 0.5°C to 1.0°C above the bleaching threshold (Berkelmans, pers. comm. 2002). This example therefore looks at the bleaching threshold itself and a succession of thresholds at 0.5°C intervals above the bleaching threshold.

Figure 3 shows the structure of the bleaching and mortality curves for Magnetic Island with the bleaching curve on the far left and a succession of curves at 0.5ºC increments up to bleaching+2.5°C. The most severe observed bleaching events on the Great Barrier Reef exceeded the bleaching+0.5ºC threshold in 1998 and 2002 (the latter event was used to validate the model). Although significant mortality was observed in some hot-spots, recovery has been widespread. Many bleached corals recovered and fast-growing species are recolonising heavily affected areas – positive signs of a resilient ecosystem (McClanahan *et al.*, 2002). However, as one moves to the right of Figure 3, the numbers and species of coral affected will increase, as will aesthetic and ecological recovery times. Sustained higher temperatures will kill both fast-growing (sensitive) and slow-growing (tolerant) species. The fast-growing species are generally more sensitive to temperature, so if the slow-growing species are being killed in some years, the fast-growing species are likely to be affected in most years. The recovery period from a bleaching+2.5ºC event is uncertain, but is likely to be decades.

---

97 The temperature-bleaching relationship is extremely precise and is sensitive to errors in input temperatures of 0.1–0.2°C (unpublished data).

The next step is to estimate the frequency of bleaching events at current and increased levels of SST. Figure 4 is a bleaching-temperature response surface showing the annual likelihood of threshold exceedance across different levels of increase in (local) sea-surface temperatures for all thresholds from bleaching to bleaching+2.5°C. The current risk of bleaching at Magnetic Island is about 40% and the risk of exceeding bleaching+0.5°C is about 25%. This estimate is consistent with recent events; during the period 1990–2002 these probabilities were 30% and 15% respectively. Figure 4 can be read in two directions. If read from the vertical axis across to the right then down, the chart links temperature increase to a given bleaching frequency. Read from the horizontal axis up then left, the chart shows the annual probabilities of exceedance for a given level of warming.

Figure 3. Coral bleaching and mortality curves for Magnetic Island.

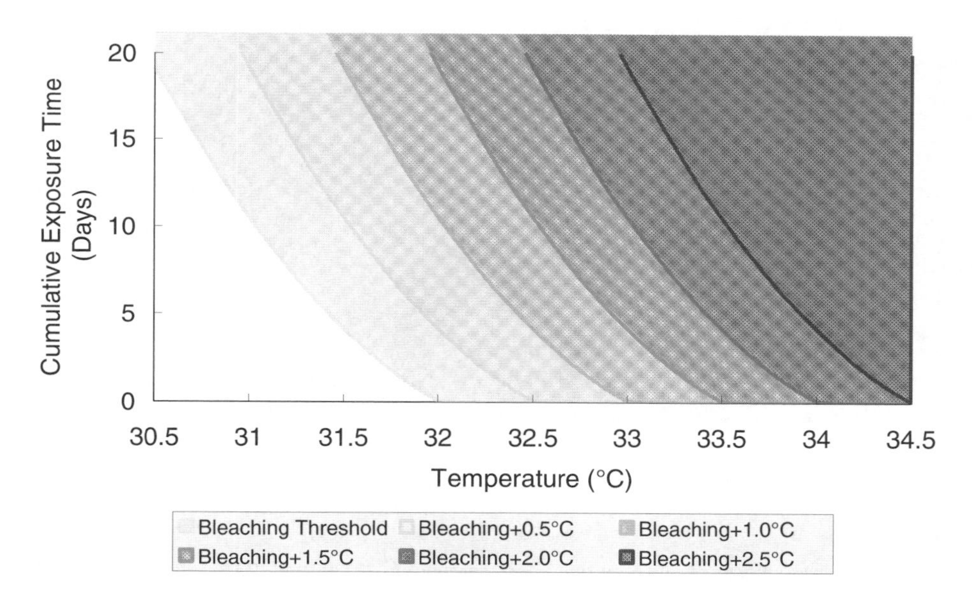

Note:

Temperature is increase in sea surface temperature. Bleaching curves are stepped up in increments of 0.5°C. The most severe observed events have exceeded the bleaching+0.5°C threshold. As successively more severe events occur, the recovery time will be longer. Events reaching temperatures on the right-hand side of the graph are likely to have a recovery time of decades.

CDFs for local increases in SST were then created and superimposed on the response surface in Figure 4. Two component ranges of uncertainty contributed to these CDFs: local increase in SST provided by eight climate models, calculated as change per degree of global warming, multiplied by ranges of global warming for 2030 and 3070 from the IPCC TAR (2001b). Uncertainty analysis showed that global warming contributes to 60% of the range of uncertainty in local SST change and local warming contributes about 40%.[98]

The vulnerability of coral depends on the severity and frequency of bleaching. If the recovery time between bleaching events is too brief, then the reef ecosystem will degrade, initially altering the composition of the coral community. With higher frequencies of severe bleaching, the environment will ultimately become unsuitable for coral growth (Done *et al.*, 2003). The level of damage that constitutes a critical threshold for coral is highly uncertain and needs to be explored.

Figure 4. Response surface for warming and annual coral bleaching risk for Magnetic Island

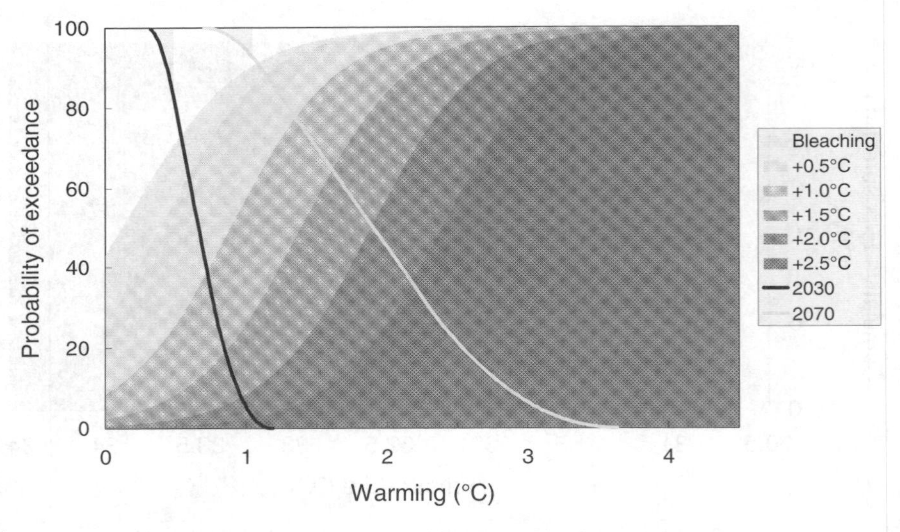

Note:

Probability distributions of average summer warming in 2030 and 2070 superimposed. The response curves for bleaching to bleaching+2.5°C thresholds are expressed as the annual risk of bleaching on the vertical axis. The warming curves are expressed as the likelihood of exceeding a given mean summer warming by 2030 or 2070. Warming is measured from a 1990 baseline, the bleaching baseline is 1990–2002.

The graph also shows CDFs for 2030 and 2070 local sea surface temperature increases as a function of global warming. The CDF was constructed by randomly sampling and multiplying these two uncertainties assuming a uniform probability distribution of the components. The range of local warming for Magnetic Island on the Great Barrier Reef for the summer period is 0.66°C to 1.07°C per degree of global

98 Using a non-linear PDF for global warming taken from Wigley and Raper (2001) did not significantly change the results.

warming; ranges of global warming are 0.54°C to 1.24°C in 2030 and 1.17°C to 3.77°C in 2070.

During the 1990s there was sufficient recovery time between bleaching events (Done *et al.*, 2003), but increasing the annual frequency of exceeding the bleaching+0.5°C threshold to 50% (equivalent to a 1998 or 2002 bleaching every second year) would place coral ecosystems under severe stress. From Figure 4, by 2030 this frequency would be exceeded by over half of the projected range of warming. By 2070 the entire range of projected warming would exceed the bleaching+0.5°C threshold of every second year. Also by 2070, 95% of all outcomes exceed a 1°C warming and half of the possible outcomes exceed the bleaching+2.0°C threshold more frequently than one in every two years. Although spatial differences in bleaching thresholds suggest that thermal adaptation does occur, we do not know what adaptation rates may be, and adaptation has not yet been observed in response to past bleaching events. However, at an arbitrary rate of 0.05°C per decade, a cushion of 0.15°C and 0.3°C would be provided by 2030 and 2070 respectively. This is only a small part of the total range of warming but would reduce both the lower and upper limits slightly.

The next step is to estimate the consequences of coral bleaching as a function of global warming (see Figure 5 note). Outcomes in Figure 5 are based on exceeding a succession of bleaching and bleaching/mortality thresholds one year in every two. By 2025, temperatures are likely to be causing bleaching at unprecedented levels (the bleaching+1.0°C threshold being exceeded in 50% of years), resulting in unprecedented levels of damage. At a global warming of 2.5°C the bleaching+2.5°C threshold will be exceeded in 50% of years, at 4.5 °C bleaching+2.5°C threshold will be exceeded every year. Using a spatial model of bleaching risk based on the 1998 and 2002 bleaching events on the Great Barrier Reef, Berkelmans *et al.* (2004) estimate that 82% of the GBR would bleach with a warming event equivalent to a bleaching+1.5°C event, 97% of the GBR would bleach during a bleaching+2.5°C event and 100% of the GBR during a bleaching+3.5°C event. Although not shown on Figure 4, a bleaching+3.5°C event affecting 100% of the Great Barrier Reef has a 1 in 100 possibility with a local increase in SST of about 2°C, increasing to 50% of frequency years with a local warming of 3.5°C. From Figure 5, these latter two outcomes would become possible from 2040 and 2060 respectively.

Figure 5 shows bleaching risk at single site expressed as a function of global warming. The following information is needed to expand this to a global risk assessment for coral reefs: 1) a better understanding of critical thresholds related to levels of ecosystem vulnerability, including concepts of resilience and recovery, 2) better developed spatial models of bleaching risk, 3) a larger number of site-based models of bleaching risk, 4) an understanding of other stresses contributing to bleaching or ecosystem resilience including freshwater, turbidity, over-exploitation and nutrient pollution. This can then be used to ask questions like "What level of climate change looks dangerous for coral reefs?" and "How do the risks faced by coral reefs compare with risks faced by other areas of concern?"

Figure 5. Bleaching threshold exceedance as a function of local SST change and global warming

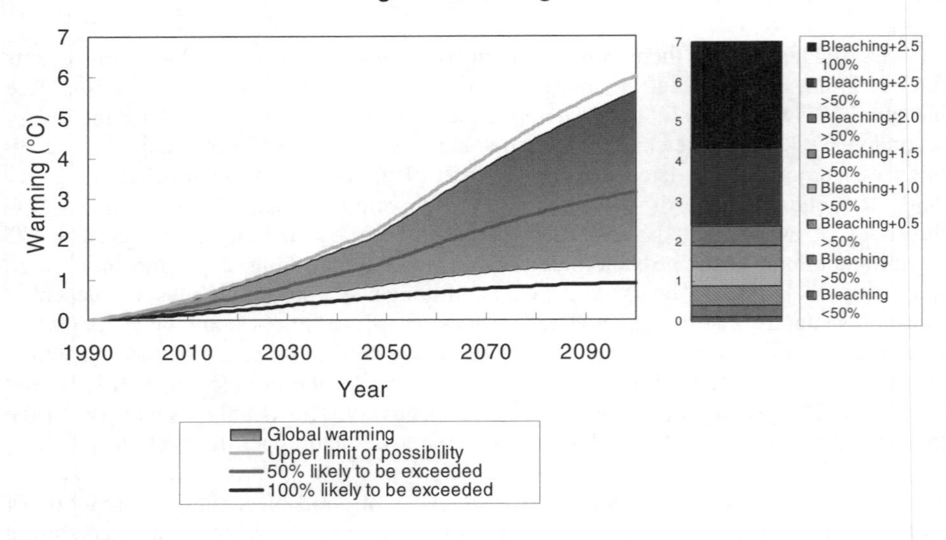

Note:

Global warming based on the range of SRES marker scenarios and climate sensitivity consistent with IPCC (2001) with super-imposed local warming ranges of SST for Magnetic Island, Great Barrier Reef, on the left. On the right is the range of consequences portrayed as thresholds based on exceeding the 50% levels (one year in every two) of a succession of bleaching and bleaching/mortality thresholds (and 100% of the bleaching+2.5°C threshold).

4.3 Catchment Water Supply

Impacts where rainfall is a major driver will have a large uncertainty when presented as a function of global warming because of uncertainties surrounding the direction and magnitude of rainfall change. Despite this, hydrological risks are far too important to be ignored. This next example describes water supply risks in an Australian catchment, the Macquarie River in eastern New South Wales (Jones and Page, 2001). This analysis considers three outcomes to assess risk: storage in the Burrendong Dam (the major water storage), environmental flows to the Macquarie Marshes (nesting events for the breeding of colonial waterbirds), and proportion of bulk irrigation allocations met over time.

Figure 6a shows the results for 2030 that project the most likely outcomes in terms of change to mean annual supply. Although there is an increased flood risk with constant and increased flows, the drier outcomes are considered worse in terms of lost productivity and environmental services. The driest and wettest extremes are less likely than the central outcomes where the line is steepest. The extremes of the range are about +10% to -30% in 2030 and about +25% to -60% in 2070, but the most likely outcomes range from about 0% to -15% in 2030 and -0% to -35% in 2070.

Figure 6a. Probability distribution for changes to mean annual Burrendong Dam storage, Macquarie Marsh inflows and irrigation allocations in 2030

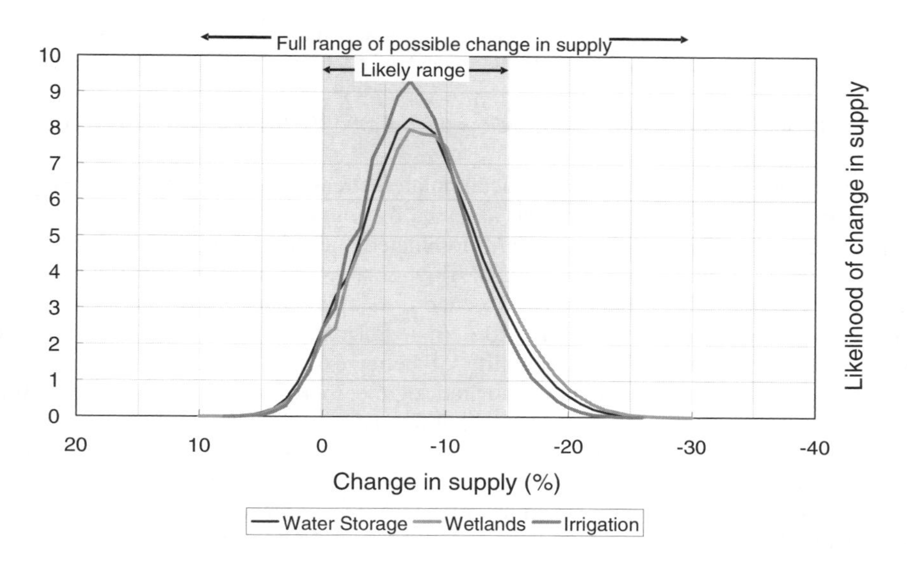

Figure 6b. Cumulative probability distribution showing likelihood of exceeding critical threshold in drought-dominated, normal and flood-dominated rainfall regime in 2030

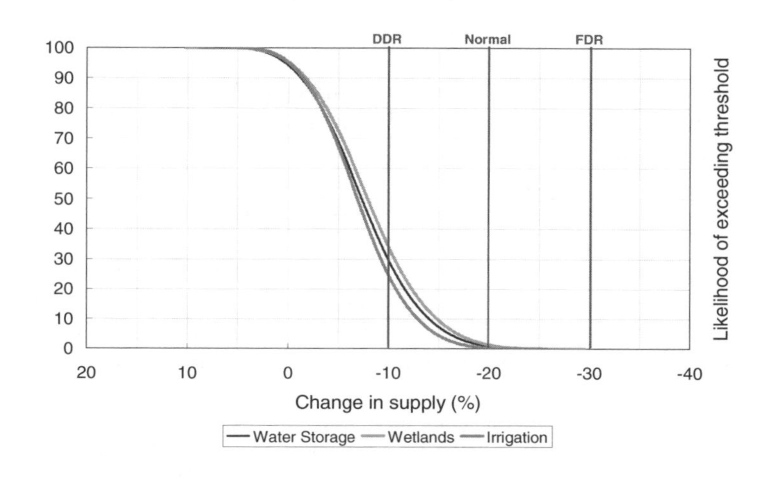

Note:

Based on Monte Carlo sampling of input ranges of global warming, δP and δEp in 2030. See Annex for details of the calculations.

To assess vulnerability, two critical thresholds for the system were considered:

1. Bird breeding events in the Macquarie Marshes, taken as 10 consecutive years of inflows below 350 GL.

2. Irrigation allocations falling below a level of 50% for five consecutive years.

Both thresholds are a measure of accumulated stress rather than a single extreme event. Critical threshold exceedance was assessed by comparing changes from the entire historical period (normal climate), drought-dominated and flood-dominated components as a function of change in mean annual flow. From the sample of fifty-six runs described above, both thresholds were exceeded if mean annual flow declined by >10% in a drought-dominated climate, by >20% in a normal climate and by >30% in a flood-dominated climate. Their probability of being exceeded by 2030 is shown in Figure 6b. If climate is in a drought-dominated regime, there is a 30% probability of the two thresholds being exceeded, but the likelihoods are small if rainfall is close to normal or in a flood-dominated regime. The probabilities of critical threshold exceedance in 2070 are much higher (Jones and Page, 2001). This shows that for rainfall dominated systems risk is a combination of low frequency variations in climate that are naturally occurring (i.e. flood-dominated or drought-dominated), combined with human-induced climate variations.

Uncertainty analysis was carried out to understand how each of the component uncertainties contributed to the range of outcomes. The contribution of individual ranges of input uncertainty (global warming and local changes in precipitation (P) and potential evaporation (Ep)) on the total outcome was assessed as a percentage contribution.[99] The results reveal that in both 2030 and 2070, δP provides almost two-thirds of the total uncertainty, global warming about 25% and δEp just over 10% (where δ is change; Jones and Page, 2001).

The effects of different ranges, sampling strategies and probability distributions within each range was also explored through Bayesian analysis (e.g. New and Hulme, 2000) by altering the input PDFs of δP, δEp and global warming and using different sampling strategies. The likelihood of exceeding the two critical thresholds do not change markedly (usually <10% for the original distribution; Jones and Page, 2001). The 'most likely' parts of the ranges were largely unaffected by increasing the ranges of input uncertainty, even though the breadth of the range increased. Some of the tests included adding different weighting on individual model-based projections of regional rainfall change, sampling beyond the range of regional rainfall change obtained from the models, sampling on quarterly, six-monthly and a seasonal basis, and altering the

99 Each input was held constant constant in turn within a Monte Carlo assessment while allowing the others free play (e.g., Visser *et al.*, 2000; Wigley, 2003). Global warming was held at 0.91°C in 2030 and 2.09°C in 2070. δP was taken as the average of the nine models in percent change per °C global warming for each quarter. δEp was linearly regressed from δP, omitting the sampling of a standard deviation.

input PDF of global warming. This suggests that the uncertainty is more dominated by the central tendencies of combining the three ranges of uncertainty, than of the internal structure of any particular range. However, adding a large new uncertainty, such as variability in mean decadal rainfall, has a significant affect on the results.

Figure 7. Change to annual water storage and supply for the Macquarie River shown as a function of global warming

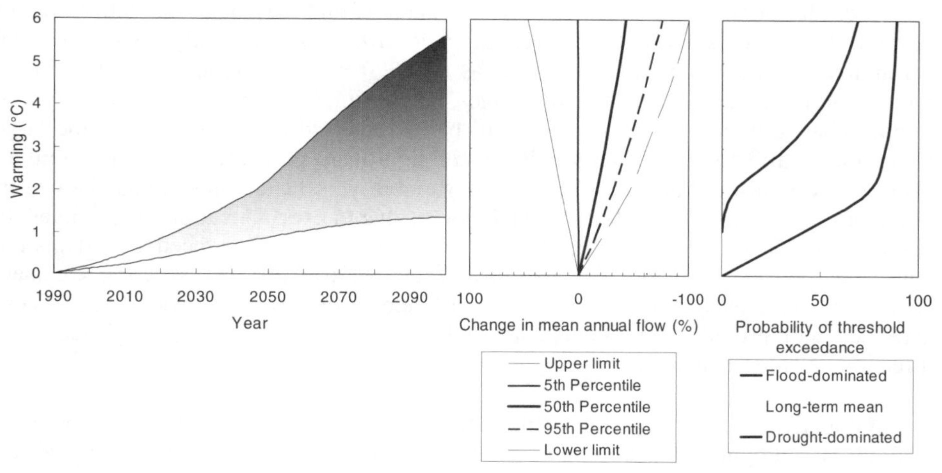

Note:

The results were converted to express changes in mean annual water storage as a function of global warming, while keeping decadal rainfall regimes separate. This was undertaken by making incremental changes in global warming (in the range 1°C to 6°C at 1°C intervals) and holding that level constant while sampling the full ranges of δP and δEp in a manner similar to the uncertainty analysis carried out above (see Annex for full details). The left-hand temperature access applies to all three panels. The central panel shows the total range of uncertainty as a function of global mean warming (left-hand panel). It also shows the range of change in flow using the upper and lower limits, 5th and 95th percentile and the 50th percentile. The likeliest 90% of outcomes lies within the 5th to 95th percentiles extending from neutral to strongly negative. The right-hand panel shows the likelihood of exceeding critical thresholds as a function of both the decadal rainfall regime and global mean warming.

The range of uncertainty for mean annual flow increases markedly with global warming although the most likely 90% of outcomes form a substantially smaller range (the area lying within the 5th and 95th percentiles in the central panel of Figure 7). The likelihood of exceeding the two critical thresholds also increases with global warming. With a warming of 2°C there is approximately a one in three chance of exceeding the critical thresholds if decadal rainfall variability is close to the long-term mean; this is less likely if rainfall variability is in a flood-dominated regime and more likely if rainfall is in a drought-dominated regime.

The largest uncertainty is the direction and magnitude of rainfall change which, if known, would reduce total uncertainty enormously. Climate models indicate that once a direction of rainfall change is established, the magnitude increases with global warming (at least until stabilisation; Cai *et al.*, 2003), therefore the identification of rainfall change is critical for impacts dependent on the volume of precipitation. Once the direction of rainfall change is established, drought and/or floods will intensify over time, requiring adaptation commensurate with the magnitude of global warming. The regions where small changes in climate are likely to breach a critical threshold are those that face the greatest risk.

Decadal rainfall variability complicates matters, and may delay the attribution of rainfall change for some decades (Hulme *et al.*, 1999). In the study region, decadal variability can vary rainfall by as much as ±20% from the long-term mean. Without understanding the dynamics of such changes, diagnosing the magnitude and direction of rainfall change under global warming will be extremely difficult. Therefore, the large uncertainties affecting the results will remain significant until the direction of rainfall change is established and decadal rainfall variability is better understood. However, much of the vulnerability within the catchment is due to over-allocation of resources. In the case of Australia, this vulnerability can be significantly reduced by adaptation consistent with carrying out the National Water Reform process, securing adequate environmental flows and increasing the efficiency of irrigation water use. These activities will provide benefits that are independent of the ultimate magnitude and direction of rainfall change.

Globally, aggregating information from different water supply systems is a difficult task. Populations exposed to water stress is the major criterion used by Arnell *et al.* (2002) in a global assessment using model output from a single GCM forced by two stabilization profiles. However, few large assessments have involved inputs from more than one climate model, and the results are highly dependent on rainfall changes which are very model specific. The technique used here has been to explore uncertainty using a very simple model constructed from a more complex system specific model. To demonstrate a wider utility, similar methods need to be applied in a larger number of river basins around the world.

4.4 Greenhouse gas stabilisation

This section looks at managing climate risks by stabilising greenhouse gases at a level that aims to avoid 'dangerous' climate change consistent with the UNFCCC. Based on a series of input assumptions, the conditional probabilities of meeting joint equilibrium temperature and stabilisation targets are explored. Note that this section is exploratory, serving to provide examples only. In a genuine risk assessment, a wide range of underlying assumptions should be explored.

The previous examples illustrate analytic frameworks that, by expressing local climate risks as a function of global warming, can be placed within a global context. Although these risks increase appreciably with global warming, the critical thresholds that occur at the lowest levels of warming will have the highest probability of exceedance. Damages occurring at low levels of global warming display the greatest need for adaptation. However, where adaptation is not feasible, where serious and irreversible outcomes are possible, or where aggregated damages from a large number of impacts become significant, stabilisation of greenhouse gases at safe levels will be required thus calling for significant mitigation of greenhouse gas emissions.

The SRES scenarios are non greenhouse-gas-policy scenarios that have no specific allowance for mitigation of greenhouse gases (Swart *et al.*, 2002), although some do have normative environmental elements. One robust aspect of risk management through mitigation is that if reductions in greenhouse gases are sustained over time, the potential range of global mean warming will reduce from the top down. The warmest outcomes become progressively less likely while the probabilities of threshold exceedance remain largely unchanged (unless stabilisation leads to the reduction of the lower limit of warming compared to that produced by the SRES B1 scenario). Probabilities can be used to explore how mitigation can be used to manage risk by stabilising greenhouse gas emissions at different levels and thus limit warming (Webster *et al.*, 2003; Wigley, 2003).

Figure 8 shows two constructions of probability for temperature at stabilisation: the probability density function and likelihood of exceedance distributions as a function of increasing temperature. Figure 8a shows that (according to the input assumptions) the most likely outcome is close to 3°C but that warming at stabilisation ranges from <1°C to >11°C. Figure 8b presents the same distribution as a probability of exceedance function, showing that the 50th percentile is about 4°C. Therefore, based on this example, if policy makers decided that it was desirable to remain below a stabilisation level of 1,000 ppm CO_2 in the atmosphere, there would be a 50% probability stabilising below 4°C and 50% probability of being above. Limits of 1.5°C and 6°C are also shown. These limits are used to explore 'what if' questions about establishing the level of dangerous climate change, shown in Figure 9.

Having information on risk as a function of global warming on a global or a local basis allows us to ask the question: "What is the level of greenhouse gas stabilisation needed to stay below critical outcomes?" The probability distribution in Figure 8 can be partitioned to provide guidance on this question. Assuming that it is not possible to identify the level of dangerous climate change *a priori* then the likelihood of reaching different levels can be investigated by relating global warming to an aggregate of local criteria, or to global criteria. At this stage, given the limited information linking global warming to levels of damage or to damage functions, it seems prudent to investigate the likelihood of reaching a wide range of targets. The probability of exceeding different levels of global warming was investigated as a function of stabilised atmospheric CO_2. The likelihood of stabilising between 1.5°C and 6°C was also assessed.

Figure 9 shows the joint probabilities of reaching different levels of warming under a range of atmospheric CO_2 at stabilisation. The curves each represent the range of temperatures between 1.5°C and 6°C in increments of 0.5°C. The vertical axis indicates the likelihood of being below a given temperature. The horizontal dotted, solid and dashed lines denote a 2 in 3, 50/50 and 1 in 3 probability respectively, of being below a given temperature and atmospheric CO_2 concentration, denoted as the point where a temperature curve crosses each line.

Figure 8. Probabilities of equilibrium global mean warming at stabilisation (with stabilisation below 1000 ppm CO_2)

Panel A. Probability distribution function for the global average temperature at stabilisation

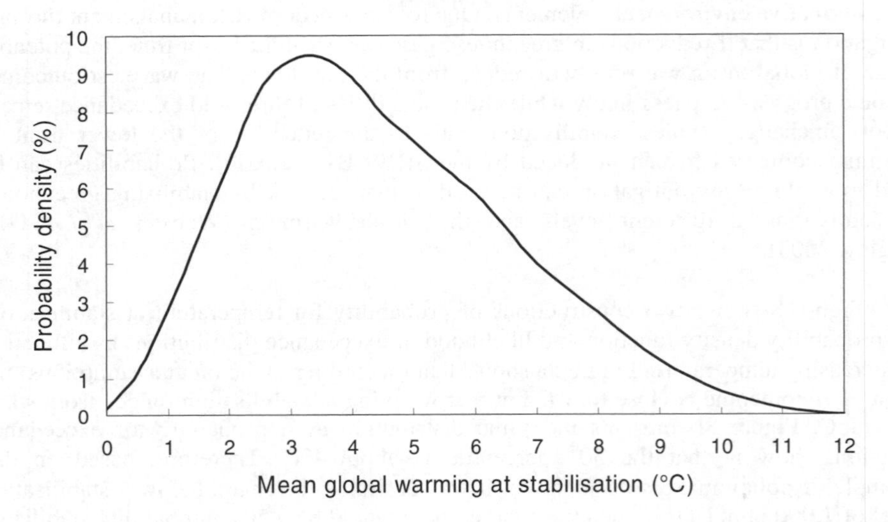

Panel B. Probability of exceeding a given level of warming at stabilisation

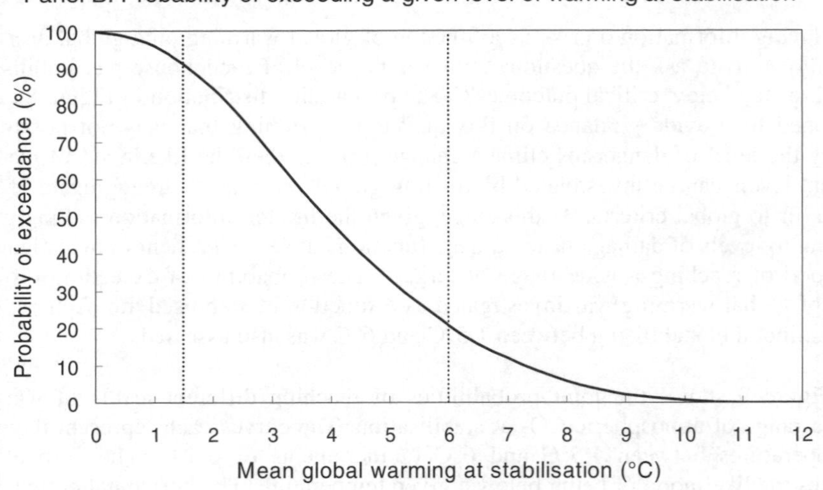

Figure 9. Likelihood of meeting joint global warming and atmospheric CO_2 concentration targets

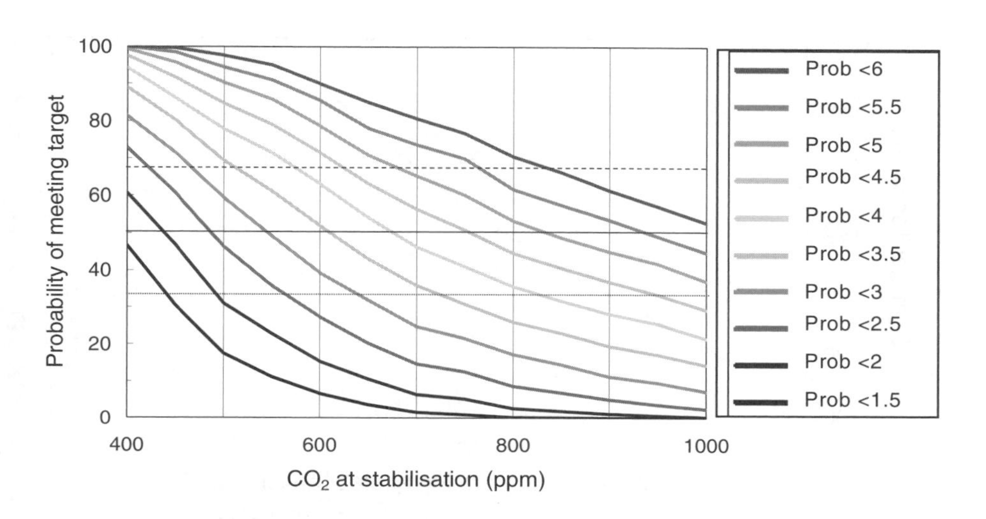

Note:

Three horizontal dotted, solid and dashed lines denote a 2 in 3, 50/50 and 1 in 3 probability of stabilising below a series of equilibrium temperature increases ranging from 1.5°C to 6°C at a given level of CO_2 stabilisation, are shown.

For example, if we wish to investigate whether stabilisation below 2°C is possible, then we have a 50/50 chance of being below that level at about 440 ppm CO_2. If we wish to stabilise below 3.5°C, there is 2 in 3 chance of success by stabilising at 500 ppm CO_2. At 650 ppm this falls to a 40% chance of being below 3.5°C. Conversely, if we wish to stabilise CO_2 at 750 ppm, then we have a 1 in 3 chance of being below 4°C, a 50/50 chance of being below 5°C and a 2 in 3 chance of being below 6°C.

Alternatively, we may be interested in the levels of risk hedging. Figure 10 summarises results from Figure 9 in terms of hedging risk at 1 in 3, 50/50 and 2 in 3 probabilities. What levels of hedging would policymakers prefer to utilise if they wish to stabilising emission in order to avoid damages? This area of policy research is yet to be explored in any detail.

The axes in Figure 10 can also be viewed in terms of knowledge about increasing costs of mitigation down the vertical axis and increasing cost of damages along the horizontal axis. A reasonable cost-benefit outcome would occur in the zone where the costs on both axes are similar, or where net benefits outweigh costs. However, even if the socio-economic outcomes could be controlled through risk management, the scientific uncertainties may remain sufficiently large to limit the likelihood of being able to reach a positive cost-benefit outcome. Utilising uncertainty analysis as detailed here and in Wigley (this volume) would help to constrain such uncertainties.

Figure 10. Risk hedging for joint targets of global warming and atmospheric CO_2

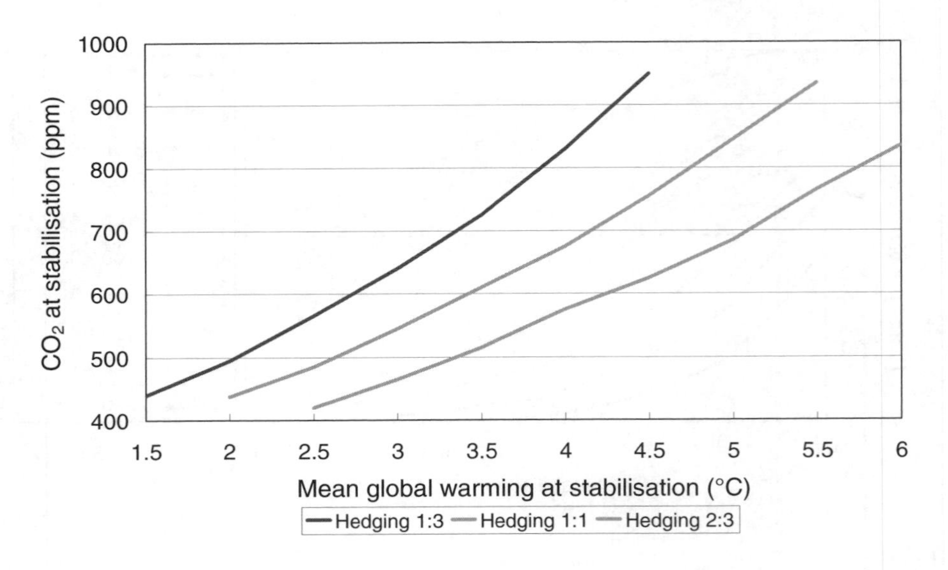

Note:

Lines of likelihood, hedging a 1 in 3, even probability and 2 in 3 chance of stabilising below a given temperature at different levels of CO_2 stabilisation derived from Figure 9. For example, to have a 2 in 3 chance of being below an equilibrium temperature of 3°C, would require stabilisation of CO_2 at about 450 ppm. At 5°C, with the same level of hedging, 700 ppm would suffice.

The outcomes of such risk analysis is very much at odds with models that optimise damages in order to assess costs (e.g. scenario-based cost benefit analysis; see Tol, 2003). By taking risk into account and analysing a multitude of possible outcomes, the likelihood of reaching a balanced outcome is limited, whereas models that optimise costs always reach a balanced outcome. The requirement within the UNFCCC to use the precautionary principle in managing climate uncertainties suggests that explicitly dealing with risk is the more appropriate approach. Therefore, this example shows that the probability of being able to meet a 'balanced' outcome on the basis of cost-benefit needs to be factored into cost-benefit analyses.

The analysis presented here does not necessarily favour any particular decision analytic approach, such as cost-benefit analysis or a safe minimum standard (see Tol, 2003), but assumes that both will be utilised by researchers, and that insights gained from assessing risk across scales will benefit whatever approach is used. Any global assessment will benefit from the aggregation of local benefits developed through bottom-up assessments that build on those described here, and through increasingly sophisticated models that can assess damages at the global scale. This is an area that requires considerable further work.

5. Discussion and conclusions

5.1 Managing risk across spatial scale

This chapter proposes risk assessment methods structured around the objectives of Article 2 in the UNFCCC that aim to manage the uncertainty, scale and delays between implementation and response under climate change. Two of the key linkages concern aspects of spatial scale:

- the scale of risk analysis in the form of top-down and bottom-up methods; and
- the scale of the prescribed risk management options of adaptation and mitigation.

Risk can be analysed through assessing climate–society relationships at these different scales. Relationships between climate hazards and coping strategies are largely specific to activities, communities and locations and range from local to global scale risks such as large ice-sheet collapse and the cessation of deep ocean-water formation. If, by increasing the number and magnitude of hazards, global warming moves a system beyond its coping range, that system will become vulnerable to risk unless that risk is managed. The critical threshold is located where the level of vulnerability exceeds a tolerable limit; criticality is a normative condition that may be derived from a range of locally specific criteria to globally accepted criteria. Many of the potential damages of climate change span the local to global scale, but direct incidence of damages is mainly on smaller scales.

Successful adaptation modifies the coping range of a system by expanding its breadth, whereas the mitigation of greenhouse gases modifies the climate hazards themselves. Therefore, adaptation and mitigation are complementary, because they treat different aspects of climate risk. Although adaptation and mitigation cannot be directly mapped onto top-down and bottom-up methods, adaptation is better suited to bottom-up methods of assessment, and mitigation, because it acts directly on the magnitude of climate change, is better suited to top-down methods. Although the impact of mitigation on climate is best measured at the global scale, mitigation can be implemented across a range of scales (e.g. industry, enterprise and regional scales). The ancillary benefits of mitigation will also accrue at these scales while benefits of reduced damages will occur on impact-relevant scales. Therefore linkages between different spatial scales are needed (bottom-up and top-down) as well as within a single scale of assessment. There is a great deal of work that needs to be done before adaptation and mitigation options can be implemented within a single sector or region. Whether such work will lend itself to the use of probabilities at this regional (or sectoral) scale remains to be seen.

In the case studies presented here, probabilities have been attached mainly to risk analysis rather than risk management. Likelihoods of reaching or exceeding specific outcomes are assessed by sampling and combining the ranges of uncertainty contributing to those outcomes. The two case studies of coral bleaching and water supply show this in terms of key climate variables and thresholds specific to each situation.

Thermal coral bleaching was assessed in terms of local sea surface temperature, with the hazard measured as degree days above a series of bleaching and mortality thresholds. Although the coping range for coral under thermal beaching is uncertain, recent bleaching events indicate that coral is close to its limits in many locations. The analysis in Section 4.2 shows that severe impacts will occur on the Great Barrier Reef under modest increases in local warming (c. ~2 °C). Further work needs to determine critical thresholds for coral based on both biophysical and social criteria, for example the ecology and aesthetics of reef systems as proposed by Done *et al.* (2003) or in terms of its impacts on tourism (Hoegh-Guldberg and Hoegh-Guldberg, 2003).

The results from the catchment water supply assessment were presented in two ways: as 'most likely' outcomes in terms of water supply, and in terms of critical threshold exceedance. These thresholds were subject to changes in both mean rainfall and decadal rainfall variability imposed on interannual rainfall variability. If mean decadal rainfall is in a drought-dominated regime, then a further decrease in mean annual streamflow of 10% would exceed the catchment-wide coping range in terms of irrigation supply and environmental flow. These thresholds would be -20% in a normal climate and -30% in a flood-dominated climate. Thus planning in a drought-dominated regime will need to be more sensitive to the critical threshold which has a higher likelihood of being exceeded under all future scenarios of climate change. This shows that assessments for planning adaptation must account for all future states in climate whether they be natural or greenhouse induced.

Both case studies presented their results in terms of a locally specific coping range. This structure minimises the input uncertainties providing information for risk assessment at the scale most suitable for assessing vulnerability and adaptation needs. Because both impact assessments were couched in impact-specific terms, they had no common metric that could be used for aggregation. Converting their results to a factor of global warming increases their uncertainty, because global warming comprises only 60% and 25% of the joint ranges of input uncertainty for the coral bleaching and water supply cases respectively. However, this procedure allows different local assessments to be compared and, eventually, their results to be aggregated to larger scales.

The task of aggregation will be made easier if common measures specific to each type of impact are developed to allow results to be aggregated at the global scale. For example, a common level of frequency and/or severity above a given bleaching threshold could be used to assess potential damages to coral reefs worldwide, allowing the global proportion of coral reefs under threat to be assessed. For water resources, water quality and supply under stress at the catchment scale and aggregated to number of people (as in Arnell, 1999), or areas of wetlands (e.g. Nicholls and Lowe 2004), may provide suitable global measures.

Recasting impacts as function of global warming will work best for activities where temperature is a very strong driver, and less well for activities affected by moisture availability or multiple independent drivers. For example, assessing the likelihoods of hydrological impacts such as flood and drought, and biodiversity impacts will be more complex to manage because of the number of climatic and non-climatic drivers. However, a recent bottom-up assessment on extinction risk did manage to analyse risks to a wide range of species within a common framework (Thomas *et al.*, 2004).

The relative risks between different activities, sectors and regions can also be assessed to determine which faced the greatest risks or benefits under climate change. The examples used here merely talk about criticality – obviously the relative magnitudes of critical outcomes will vary depending on the activity, scale and across different measures of criticality. At some stage, a series of measures providing links across sectors will also need to be developed.

The structure of the coping range can also be applied to global scale risks on a conceptual level, using the UNFCCC as a guide. Climate change threatens to global mean temperature beyond levels previously experienced by humans, by human systems and by many other species. Dangerous climate change forms a threshold which is strongly normative. Consensus on such a threshold is unlikely to be achieved in the short term and cannot be predicted *a priori.* Thus it would be necessary to define any such threshold in policy terms rather than objectively through scientific means. However, scientific research can contribute to its definition through an understanding of risk under successively higher levels of global warming. This understanding will evolve over time as new information emerges.

5.2 Managing risk over time

Sections 4.1, 4.2 and 4.3 provide information on the changing probabilities of exceeding thresholds attached to individual activities over the course of this century. Section 4.4 deals with stabilising climate at an indeterminate time in the future, rather than assessing evolving risks over time. To be effective, risk management needs to act on robust information as it becomes available. The role of policy is crucial in deciding when and how to act. Near-term action in mitigation and adaptation would reduce risks (IPCC, 2001a) but the framework for how specific actions should be decided has not yet been established.

The critical thresholds presented in this paper are static, but thresholds based on rates of change, such as those associated with forest migration (Leemans and Eickhout, 2003) or anticipated through socio-economic development can also be used. The case studies showed that the critical thresholds exceeded under low levels of global warming were likely to be exceeded earlier rather than later, and under a greater range of emission scenarios. Activities with critical thresholds that are sensitive to low levels of climate change are often also subject to significant risks under current climate – in such cases adaptation will yield both short and longer-term benefits.

Using these general principles it is possible to use likelihoods of critical threshold exceedance under climate change to prioritise the need for adaptation. Activities likely to be affected within their planning horizons would receive a high priority for adaptation. Adaptation measures can then be prioritised according to their feasibility and net and ancillary benefits. If autonomous and planned adaptation cannot sufficiently reduce the risk of critical outcomes, mitigation will become the primary mechanism to manage these risks.

Adaptation does not alter the frequency and magnitude of primary climate hazards, but increases the ability of an activity to cope with climate change. This will either delay critical threshold exceedance or if adequately integrated with mitigation, may allow the rate of climate change to slow and stabilise before a given critical threshold is exceeded. Secondary climate hazards such as flooding or fire are influenced by human activities, so can be partially modified by adaptation in addition to mitigation. Because stakeholders are not necessarily engaged in the UNFCCC, they may not be interested in the distinction between adaptation and mitigation at the local scale and may just want to reduce climate risks while being productive over the longer term.

Elsewhere, we have assessed adaptation in two probabilistic impact assessments in the agricultural sector: milk loss under heat stress (Jones and Hennessy, 2000) and wheat production (Howden and Jones, 2001). In both cases, the net benefits of currently practised adaptations increased with global warming even though those benefits did not keep up with increasing loss rates in all cases, showing that eventually, losses can outstrip benefits under larger climate changes. For activities where damages increase with global warming, adaptation will become more difficult and expensive with increasing climate changes and at higher levels of global warming more activities will also require adaptation. This suggests that the ability to adapt is limited and is best suited to modest changes in climate (e.g. IPCC, 2001a).

Policy-directed mitigation will reduce the risks of climate damages below those that would otherwise have occurred.[100] For example, the range of warming from the non-policy SRES scenarios at 2100 is 1.4–5.8°C, whereas the range of warming from stabilisation scenarios with targets ranging from 450 ppm to 1,000 ppm CO_2 has been estimated to be about 1.2–3.6°C at 2100 by the IPCC (2001a; see also Wigley, this volume). This implies that successive efforts to mitigate greenhouse gases will reduce the likelihood of reaching levels of global warming from the top down, with the highest potential temperatures being avoided first. Successive mitigation efforts will produce a progressively cooler range of global warming over time. This would suggest that the avoided economic costs associated with the adverse impacts of climate change (IPCC, 2001a, para 9.27) are highest with the first cuts, becoming successively smaller as mitigation efforts continue. The sum of these avoided costs will depend on the

100 If we accept the range of scientifically-based uncertainties as quantified by the IPCC (IPCC, 2001b) as being realistic, greenhouse gas mitigation will reduce the level of radiative forcing in the atmosphere independently of those scientific uncertainties.

sensitivity of the climatic response and the cumulative damage function of climate impacts.

Figure 11. Synthesis of risk assessment approach to global warming

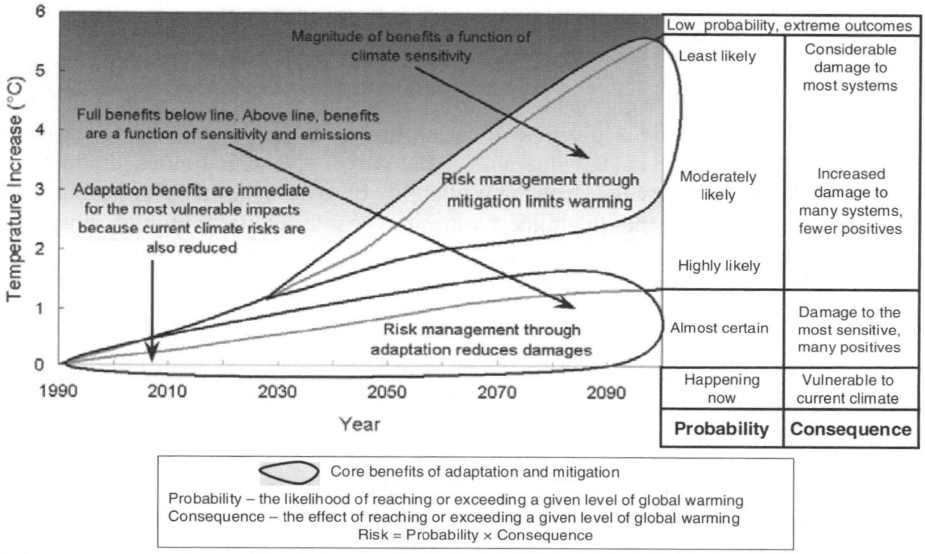

Magnitude of benefits a function of climate sensitivity
Full benefits below line. Above line, benefits are a function of sensitivity and emissions
Adaptation benefits are immediate for the most vulnerable impacts because current climate risks are also reduced
Risk management through mitigation limits warming
Risk management through adaptation reduces damages
Temperature Increase (°C)
Year

Low probability, extreme outcomes	
Least likely	Considerable damage to most systems
Moderately likely	Increased damage to many systems, fewer positives
Highly likely	
Almost certain	Damage to the most sensitive, many positives
Happening now	Vulnerable to current climate
Probability	**Consequence**

Core benefits of adaptation and mitigation

Probability – the likelihood of reaching or exceeding a given level of global warming
Consequence – the effect of reaching or exceeding a given level of global warming
Risk = Probability × Consequence

Note:

The left part of the figure shows global warming based on the six SRES greenhouse gas emission marker scenarios with the zones of maximum benefit for adaptation and mitigation. The right side shows likelihood based on threshold exceedance as a function of global warming and the consequences of global warming reaching that particular level based on the conclusions of IPCC WG II (Smith *et al.*, 2001). Risk is a function of probability and consequence.

Figure 11 shows how risk management options that are linked to levels of global warming can be assessed under the current state of knowledge. These are robust outcomes that are insensitive to the exact nature of known contributing uncertainties. The range of mean global warming under the non-greenhouse gas policy SRES scenarios is shown in the left-hand graph, while likelihood and consequences are shown on the right-hand side. Adaptation will be most beneficial to activities that are vulnerable to current climate and likely to be worsened under climate change and those that are likely to be affected under small to modest increases in global warming. Adaptations to larger warmings will be difficult and costly, needing to cover a large number of activities and a large range of change in any single activity. Adaptation for critical outcomes exceeded under larger warmings could only be contemplated if the benefits without climate change were large and/or the consequences of not adapting are severe. The most optimal range of warming for adaptation is the lower shaded zone.

The mitigation of greenhouse gases will act from the top down, reducing the highest possible temperatures within the possible range of uncertainty. This will also reduce the likelihood of the most extreme consequences, therefore the benefits of avoided damages would be correspondingly high, reducing over time with increasing mitigation measures. Mitigation is unlikely to limit global warming to very low increases; this has led IPCC to conclude that some climate change is inevitable (IPCC, 2001a). Inevitable climate change risks thus occurring are best managed by adaptation, if at all possible. The most optimal range of warming for mitigation is in the upper shaded zone.

The Table on the right-hand side of Figure 11 shows the probability and consequences of exceeding a given level of warming. Where it is feasible, targeted adaptation will allow the most vulnerable systems to cope with limited amounts of warming while mitigation will reduce the probability of extreme levels of warming from occurring (bearing in mid that the highest warmings would require both high emissions and a high climate sensitivity). Thus, adaptation and mitigation are complementary in terms of scale and in managing risk over the potential range of global warming (and sea level rise).

There are many situations dominated by human agency where probabilities cannot be attached to outcomes as described in the examples provided in this paper (e.g. Barnett, 2001), such as those activities where climate is a contributing factor but where vulnerability is dominated by a combination of socio-economic factors such as drought-related famine. In these cases, development-based methods focussing on the socio-economic aspects of vulnerability which assume that the natural hazard will recur may be the best course (Adger, 1999; Barnett, 2001; Callaway, this volume). Such cases will fit into the framework outlined in Figure 11 as long as qualified likelihoods of change linked to global warming can be estimated. For example, areas vulnerable to current drought and where drought is expected to persist or increase with small increases in global warming but where famine is related to political instability and market failure.

5.3 Conclusion

In this chapter I argue that probabilities can be attached to climate change assessments and that this is best carried out using a risk assessment framework. Expectations amongst some scientists and policymakers that climate change can be forecast and responded to by decree are both unrealistic and unworkable. The previous section showed that some measures can be prioritised and applied in the current environment of uncertainty with a high probability of success. Scientists and policymakers do not share a clear picture as to how climate change risks should be weighed up against policy risks, including risks of high policy costs. This is a difficult area for the IPCC who endeavour to be policy-relevant without being policy-prescriptive. Risk assessment will allow different policy options to be tested. This has not been carried out in the analyses presented here, but Bayesian assessments showing the benefits of different policy options are beginning to be explored (Webster *et al.*, 2003; Wigley, submitted).

Policy-related risks include the risk of doing too little, the risk of doing too much, the risk of moving ahead without sufficient information and the risk of delaying and suffering irreversible change. The certainty surrounding climate change is too large and the system too complex to manage a command and control situation, where climate change is forecast, a target is chosen and then policymakers act to meet that target. Some of the existing uncertainties may take years to narrow down to a sufficient level and some will be irreducible (Schneider, 2002; Dessai *et al.*, 2004). Although this chapter has concentrated on analysing the risks of climate change, these need to be weighed up with policy risks. Risk assessment offers a flexibility and robustness that forecasting does not. Critical levels of impact can be assessed independently of climate change scenarios to serve as criteria for assessing outcomes. The likelihood of exceeding these critical thresholds can then be assessed under different policy regimes. It is in this context that the benefits of climate policy can be assessed.

A number of insights can be gained by moving from assessments based a limited number of scenarios to those dealing with ranges of uncertainty:

1. Ranges of quantified uncertainties assessing probability of exceeding given criteria of harm (benefit) can be utilised to identify the activities facing the greatest risk under climate change rather than relying on single, unrelated scenarios.

2. The assessment of critical thresholds allows vulnerability assessment to be conducted independently of its likelihood of occurrence.[101]

3. Dangerous levels of anthropogenic greenhouse gases do not need to be predicted before the prioritisation of adaptation and mitigation options can begin. Adaptation and mitigation to manage the activities identified as most at risk can proceed while ongoing risk assessments at the local and global scale are pursued.

4. Bayesian methods of constructing priors for different ranges of uncertainty can be used to determine which are robust and which are sensitive to input uncertainties.

5. The sustained mitigation of greenhouse gases will reduce the likelihood of the highest potential warming occurring, irrespective of the ultimate value of climate sensitivity. While the magnitude of net benefits will be a function of climate sensitivity and the damage curve, the earliest mitigation efforts will always yield the largest economic benefits in terms of damage reduction (unless delayed mitigation is cheaper than accrued damages from that delay). Short-term ancillary benefits of mitigation, such as reduced pollution or reduced energy costs, will ensure both short- and long-term returns.

---

101 Thus fulfilling the conditions of Grübler and Nakicenovic (2001).

6. Optimised cost-benefit outcomes may only have a limited probability of being achieved.

The framework presented in this chapter is consistent with Article 2 of the United Nations Framework Convention on Climate Change and can be used to investigate the risks of climate change using both top-down and bottom-up methods. By expressing the outcomes of individual assessments as a function of global warming, it is possible to aggregate (or at least look) across scale for a single activity using common thresholds and eventually between activities expressing the outcomes as a function of global warming. A risk assessment framework provides better management of uncertainty than does linear assessments of climate change. It allows the prioritisation of adaptation and mitigation options according to the greatest need and can be refined as new information becomes available. It also provides a synthesis consistent with the aims of the UNFCCC that can unite the interests of the IPCC Working Groups I, II and III in preparing for the Fourth Assessment Report.

Annex

This annex contains a brief explanation of the calculations and models used to develop the results presented above. First is the water supply model which was constructed to relate regional climatic change parameters to global warming (measured through global mean temperature increase). This model is the basis for results presented in section 4.3 above. Second is the model used to the risk of exceeding a wide range of global mean temperature thresholds, the results of which are outlined in section 4.4.

Catchment water supply risk assessment (section 4.3).

The risk assessment of water supply in the Macquarie River basin in Australia is based on a river management model whose output based on historical observations is then used as input to a Monte Carlo analysis that explores the linkage between between regional hydrological change and global warming.

First, regional changes to potential evaporation (Ep) and precipitation (P) were used to perturb historical daily records of P and Ep in the Macquarie basin from the period 1890–1996 which served as input into a river management model. The historical time series was separated into a drought-dominated (dry) period (1890–1947) and a flood-dominated (wet) period (1948–1996) allowing different modes of decadal rainfall variability to be assessed along with greenhouse-induced rainfall change. Three outputs were considered for risk assessment: storage in the Burrendong Dam (the major water storage), environmental flows to the Macquarie Marshes (nesting events for the breeding of colonial waterbirds), and the proportion of bulk irrigation allocations met over time.

Second, a transfer function summarising the results of individual models runs was created to investigate probability distributions using Monte Carlo sampling. Fifty-six simulations using a range of scenarios exploring the IPCC (2001b) range of global warming, and regional changes in P and Ep from nine climate models were analysed. The results were then used to create the following transfer function:

$$\delta flow = a \times (atan (\delta Ep / \delta P) - b)$$

where atan is the inverse tan function, δEp and δP are in mm yr^{-1}, δflow is mean annual flow in gigalitres (GL) yr^{-1} for water storage and environmental flow and percent of a capped allocation for irrigation, and a and b are constants. The results have an r^2 value of 0.98 and standard error in mean annual flow ranging from 1 to 2%, allowing this simple function to substitute for the more complex river management model for the purposes of risk and uncertainty analysis.

Three ranges of input uncertainty contributed to the analysis: global warming, and regional δP and δEp expressed as percentage change per degree of global warming. Monte Carlo methods were used to sample the IPCC (2001b) ranges of global warming for 2030 and 2070. These were then used to scale a range of change per °C of global warming on a quarterly basis for P. Ep was then sampled using a relationship between P and Ep established from climate model output. Finally, quarterly changes for P and Ep were totalled to determine annual δP and δEp which was then applied to the above transfer function. The following assumptions were applied to the analysis:

- The range of global warming in 2030 was 0.55–1.27°C with a uniform distribution. The range of change in 2070 was 1.16–3.02°C (Note these ranges are slightly different to those used in sections 4.1 and 4.2 – they were undertaken with earlier, provisional data preceding IPCC (2001b)).
- Changes in P were taken from the full range of change for each quarter from the sample of nine climate models. The annual range of change in P was about ±4% per degree of global warming.
- Changes in P for each quarter were assumed to be independent of each other (dependence between seasonal changes could not be found).
- The difference between samples in any consecutive quarter could not exceed the largest difference observed in the sample of nine climate models.
- δEp was co-dependent with δP and sampled accordingly (δEp = 5.75 – 0.53δP, standard error = 2.00, randomly sampled using a Gaussian distribution, units in percent change).

Risk Analysis of Global Mean Temperature Change (based on SRES)

Section 4.4 uses Monte Carlo sampling to explore the input uncertainties of the temperature at stabilisation, measured as the change in temperature since 1990, using the relationship:

$$T_{stab} = -0.7 + \partial T_{2\times} \times Ln(CO_{2\,stab} / 278) / Ln(2) + \partial Q_{non\text{-}CO2} / (3.71 \times \partial T_{2\times})$$

Where -0.7 allows for warming already experienced, T_{stab} is mean global warming in °C at stabilisation, $\partial T_{2\times}$ is temperature sensitivity, CO_{2stab} is the atmospheric concentration of CO_2 in ppm and $\partial Q_{non\text{-}CO2}$ is the radiative forcing of non-CO_2 elements (greenhouse gases and aerosols) in Wm^{-2}. The ranges of change were 354 ppm to 1000 ppm for CO_{2stab}, -0.5 Wm^{-2} to 3.5 Wm^{-2} $\partial Q_{non\text{-}CO2}$ and 1.5°C to 4.5°C $\partial T_{2\times}$. The range of CO_2 stabilisation extends from the concentration in 1990 to 1,000 ppm; the range of non-CO_2 forcing is 1.5 Wm^{-2}, close to the average of all SRES scenarios to 2100, with uncertainty bounds of ±2 Wm^{-2}; the climate sensitivity has been unchanged since 1PCC (1990) and the forcing relationships can be found in Appendix 2 of IPCC (1997). Each of these ranges was sampled independently assuming a uniform probability across the range, for a total of about 65,000 samples. The inverse of this relationship has been used by Wigley (submitted) in applying a PDF for 'dangerous' warming at stabilisation to determine targets for stabilising atmospheric CO_2.

Acknowledgements

This paper was greatly influenced by presentations at the OECD Benefits of Climate Policy Workshop in Paris, December 2003 and the OECD's support in commissioning a more complete version is gratefully acknowledged. The projects that fuelled the several years' research into risk assessment summarised in this paper, were funded by various Australian state governments, the United Nations Development Programme, the Australian Greenhouse Office and CSIRO. Ray Berkelmans, Terry Done and colleagues at the Australian Institute of Marine Science are thanked for bleaching thresholds and the data that was used to build the coral bleaching model. The analysis in Section 4.4 was based on provisional analyses by Tom Wigley that I turned around to provide the inverse version. Tom Wigley, Steve Schneider, Jan Corfee-Morlot, Jake Jacoby, Barrie Pittock, Granger Morgan, Suraje Dessai and Bill Hare provided valuable comments.

References

Adger, W.N. (1999), "Social vulnerability to climate change and extremes in coastal Vietnam", *World Development* **27**, 249-269.

Allen, M.R. and D.A. Stainforth (2002), "Towards objective probabilistic climate forecasting", *Nature*, 419, 228.

Arnell, N.W. (1999), "Climate change and global water resources", *Global Environmental Change – Human And Policy Dimensions*, **9**, S31–S49.

Arnell, N.W., M.G.R. Cannell, M. Hulme, R.S. Kovats, J.M.B. Mitchell, R.J. Nicholls, L.M. Parry, M.T.J. Livermore and A. White (2002), "The consequences of CO_2 stabilisation for the impacts of climate change", *Climatic Change*, **53**, 413-446.

Azar, C. and H. Rohde (1997), "Targets for stabilisation of greenhouse atmospheric CO_2", *Science*, **276**, 1818-1819.

Barnett, J. (2001), "Adapting to climate change in Pacific Island Countries: The problem of uncertainty", *World Development*, **29**, 977-993.

Beer, T. and F. Ziolkowski (1995), "Environmental risk assessment – an Australian perspective", in T.W. Norton, T. Beer and S.R. Dovers, *Risk and Uncertainty in Environmental Management*, Centre for Resource and Environmental Studies, Canberra, 3-13.

Beer, T. (1997), "Strategic risk management: a case study of climate change", *World. Res. Rev.* **9**, 113-126.

Berkelmans, R. (2001), Time-integrated thermal bleaching thresholds of reefs and their variation on the Great Barrier Reef, *Marine Ecology – Progress Series*, **229**, 73-82.

Berkelmans, R. (2002), personal communication, Australian Institute of Marine Science, Townsville, Australia.

Berkelmans, R., G. De'ath, S. Kininmonth and W.J. Skirving (2004), "A comparison of the 1998 and 2002 coral bleaching events on the Great Barrier Reef: spatial correlation, patterns, and predictions", *Coral Reefs*, **23**, 74-83.

Brooks, N., W.N. Adger and S.R. Khan (2004), *Measuring and Enhancing Adaptive Capacity*, in B. Lim, I. Burton and S. al Huq (eds.), Adaptation Policy Framework, Technical Paper 2, United Nations Development Programme, New York (in press).

Cai, W.J., P.H. Whetton and D.J. Karoly (2003), "The response of the Antarctic Oscillation to increasing and stabilized atmospheric CO_2", *Journal of Climate*, **16**, 1525-1538.

Callaway, J.M. (2004), "Assessing and linking the benefits and costs of adapting to climate variability and climate change", in *The Benefits of Climate Change Policies*, OECD, Paris.

Carter, T.R. and E.L. La Rovere (2001), "Developing and applying scenarios", in J.J. McCarthy, O.F. Canziani, N.A. Leary, D.J. Dokken and K.S. White (eds.) *Climate Change 2001: Impacts, Adaptation, and Vulnerability*, Contribution of

Working Group II to the Third Assessment Report of the Intergovernmental Panel on Climate Change, Cambridge University Press, Cambridge, 145–190.

Carter, T.R., M.L. Parry, H. Harasawa and S. Nishioka (1994), *IPCC Technical Guidelines for Assessing Climate Change Impacts and Adaptations*, University College, London and Centre for Global Environmental Research, Japan, 59 pp.

Conde, C. and Lonsdale, K. (2004), *Stakeholder engagement in the adaptation process*, in B. Lim, I. Burton and S. al Huq (eds), Adaptation Policy Framework, Technical Paper 2, United Nations Development Programme, New York (in press),

Dessai, S. (2003), "Heat stress and mortality in Lisbon Part II: An assessment of the potential impacts of climate change", *International Journal of Biometeorology*, **48**, 37–44.

Dessai, S. and M. Hulme (2003), "Does climate policy need probabilities?" *Tyndall Centre Working Paper No. 34*, Tyndall Centre for Climate Change Research, Norwich UK, 42 pp.

Dessai, S., W.N. Adger, M. Hulme, J. Turnpenny, J. Köhler1 and R. Warren (2004), "Defining and experiencing dangerous climate change", *Climatic Change*. **64**, 11-25.

Done, T., P.H. Whetton, R.N. Jones, R. Berkelmans, J. Lough, W. Skirving and S. Wooldridge (2003), *Global Climate Change and Coral Bleaching on the Great Barrier Reef*, Australian Institute of Marine Science, Townsville, 54 pp.

Giles, J. (2002), "Scientific uncertainty: When doubt is a sure thing", *Nature*, **418**, 476-478.

Grubler, A. and N. Nakicenovic (eds.) (2001) "Identifying dangers in an uncertain climate", *Nature*, **415**, 15

Hewitt, K. and I. Burton (1971), *The Hazardousness of a Place: A Regional Ecology of Damaging Events*, University of Toronto, Toronto.

Hitz, S. and J. Smith (2004), "Estimating global impacts from climate change", in *The Benefits of Climate Change Policies*, OECD, Paris.

Hoegh-Guldberg, O. (1999), "Climate change, coral bleaching and the future of the world's coral reefs", *Marine and Freshwater Research*, **50**, 839–866.

Hoegh-Guldberg, H. and O. Hoegh-Guldberg (2003), *Climate Change, Coral Bleaching and the Future of the Great Barrier Reef*, World Wildlife Fund Australia, http://www.greenpeaceusa.org/media/publications/coral_bleaching.pdf

Howden, S.M. and R.N. Jones (2001), "Costs and benefits of CO_2 increase and climate change on the Australian wheat industry", Australian Greenhouse Office, Canberra, Australia, http://www.greenhouse.gov.au/science/wheat/

Hulme, M., E.M. Barrow, N.W. Arnett, P.A. Harrison, T.C. Johns. and T.E. Downing, (1999), "Relative impacts of human-induced climate change and natural climate variability", *Nature*, **397**, 688-691.

Hulme, M. and T.R. Carter (1999), "Representing uncertainty in climate change scenarios and impact studies", in T.R. Carter, M. Hulme and D. Viner. (eds.), *Representing Uncertainty in Climate Change Scenarios and Impact Studies*, ECLAT-2 Workshop Report Number 1, Climatic Research Unit, Norwich, UK, 11–37.

IPCC (1990), *Climate Change: The IPCC Scientific Assessment*, J.T. Houghton, G.J. Jenkins, J.J. Ephraums (eds.), Cambridge University Press, Cambridge, UK, 365 pp.

IPCC (1997), *An Introduction to Simple Climate Models used in the IPCC Second Assessment Report*, J.T. Houghton, L.G. Meira Filho, D.J. Griggs and K. Maskell (eds.), Intergovernmental Panel on Climate Change, Geneva, 47 pp.

IPCC (2001a), *Climate Change 2001: Synthesis Report*, R.T. Watson, D.L. Albritton, T. Barker, I.A. Bashmakov, O. Canziani, R. Christ, U. Cubasch, O. Davidson, H. Gitay, D. Griggs, J. Houghton, J. House, Z. Kundzewicz, M. Lal, N. Leary, C. Magadza, J.J. McCarthy, J.F.B. Mitchell, J.R. Moreira, M. Munasinghe, I. Noble, R. Pachauri, A.B. Pittock, M. Prather, R.G. Richels, J.B. Robinson, J. Sathaye, S.H. Schneider, R. Scholes, T. Stocker, N. Sundararaman, R. Swart, T. Taniguchi and D. Zhou (eds), Cambridge University Press, Cambridge, 397 pp.

IPCC (2001b), "Summary for Policymakers", in J.T. Houghton, Y. Ding, D.J. Griggs, M. Noguer, P.J. Van Der Linden and D. Xiaosu (eds.) *Climate Change 2001: The Scientific Basis*, Contribution of Working Group I to the Third Assessment Report of the Intergovernmental Panel on Climate Change, Cambridge University Press, Cambridge, 1-29.

IPCC (2001c), Summary for Policymakers, in J.J. McCarthy, O.F. Canziani, N.A. Leary, D.J. Dokken and K.S. White (eds.) (2001) *Climate Change 2001: Impacts, Adaptation, and Vulnerability*, Contribution of Working Group II to the Third Assessment Report of the Intergovernmental Panel on Climate Change, Cambridge University Press, Cambridge, 1-17.

Jacoby, H.D. (2003), "Informing climate policy given incommensurable benefits estimates", Working paper, OECD Workshop on Benefits of Climate Policy:

Improving Information for Policymakers, Organisation for Economic Co-operation and Development, Paris, 22 pp.

Jacoby, H.D. (2004), "Toward a framework for climate benefits estimation", in *The Benefits of Climate Change Policies*, OECD, Paris.

Jones, R.N. (2000a), "Analysing the risk of climate change using an irrigation demand model", *Climate Research*, **14**, 89-100.

Jones, R.N. (2000b), "Managing uncertainty in climate change projections – issues for impact assessment", *Climatic Change* **45**, 403-419.

Jones, R.N. and K.J. Hennessy (2000), *Climate change impacts in the Hunter Valley: a risk assessment of heat stress affecting dairy cattle*, CSIRO Atmospheric Research, Melbourne, 23 pp. http://www.dar.csiro.au/publications/Jones_2000a.pdf.

Jones, R.N. (2001), "An environmental risk assessment/management framework for climate change impact assessments", *Natural Hazards*, **23**, 197-230.

Jones, R.N. and C.M. Page (2001), "Assessing the risk of climate change on the water resources of the Macquarie River Catchment", in F. Ghassemi, P. Whetton, R. Little and M. Littleboy (eds.), *Integrating Models for Natural Resources Management across Disciplines, issues and scales* (Part 2), Modsim 2001 International Congress on Modelling and Simulation, Modelling and Simulation Society of Australia and New Zealand, Canberra, 673-678.

Jones, R.N. and Boer, R. (2004), *Assessing Current Climate Risks*, in Lim, B., Burton, I. and al Huq, S. (eds), Adaptation Policy Framework, Technical Paper 4, United Nations Development Programme, New York (in press),

Jones, R.N. and L.O. Mearns (2004), *Assessing Future Climate Risks*, in B. Lim, I. Burton and S. al Huq (eds), "Adaptation Policy Framework", Technical Paper 5, United Nations Development Programme, New York (in press).

Jones, R.N., B. Lim and I. Burton (in review), "Using coping ranges and risk assessment to inform adaptation", *Climatic Change*.

Leemans, R. and B. Eickhout (2003), "Analysing changes in ecosystems for different levels of climate change", OECD Workshop on the Benefits of Climate Policy: Improving Information for Policy Makers, Organisation for Economic Co-operation and Development, Paris, 27 pp.

Lempert, R. and M.E. Schlesinger (2001), "Climate-change strategy needs to be robust," *Nature*, **412**, 375 pp.

McClanahan, T., N. Polunin and T. Done (2002), "Ecological states and the resilience of coral reefs", *Conservation Ecology*, **6**, Art. No. 18.

Morgan, M.G. and M. Henrion (1990), *Uncertainty: A Guide to Dealing with Uncertainty in Quantitative Risk and Policy Analysis.* Cambridge University Press, Cambridge, 332 pp.

Morgan, M.G. and D. Keith (1995), "Subjective Judgments by Climate Experts", *Environmental Science & Technology*, **29**(10), 468-476, October 1995.

Moss, R.H. and S.H. Schneider (2000), *Uncertainties in the IPCC TAR: Recommendations to lead authors for more consistent assessment and reporting*, in *Third Assessment Report, Cross-Cutting Issues Guidance Papers*, R. Pachauri, T. Taniguchi and K. Tanaka (eds.), World Meteorological Organisation, Geneva, Switzerland.

Nakicenovic, N. and R. Swart (eds.) (2000), *Emissions Scenarios*, Special Report on Emissions Scenarios, the Intergovernmental Panel on Climate Change, Cambridge University Press, Cambridge UK, 570 pp.

New, M. and M. Hulme (2000), "Representing uncertainty in climate change scenarios: a Bayesian Monte-Carlo approach", *Integrated Assessment* **1**, 203–213.

Nicholls, R. and J. Lowe (2004), "Benefits of mitigation of climate change for coastal areas", *Global Environmental Change: Special Edition on the Benefits of Climate Policy* **14**: 229-244.

Nordhaus, W.D. (1994), "Expert opinion on climate change", *American Scientist*, **82**, 45-51.

Parry, M.L., T.R. Carter and M. Hulme (1996), "What is a dangerous climate change?", *Global Environ. Change*, **6**, 1–6.

Parry, M., N. Arnell, T. McMichael, R. Nicholls, P. Martens, S. Kovats, M. Livermore, C. Rosenzweig, A. Iglesias and G. Fisher (2001), "Millions at risk: defining critical climate change threats and targets", *Global Environmental Change*, **11**, 1-3.

Patt, A. and D.P. Schrag (2003), "Using Specific Language to Describe Risk and Probability", *Climatic Change*, **60**, 17–30.

Pittock, A.B. and R.N. Jones (2000), "Adaptation to what and why?", *Environmental Monitoring and Research*, **61**, 9-35.

Reilly, J., P.H. Stone, C.E. Forest, M.D. Webster, H.D. Jacoby and R.G. Prinn. (2001), "Uncertainty and Climate Change Assessments", *Science*, **293**, 430-433.

Risbey, J.S. (submitted), "Agency and the Assignment of Probabilities to Greenhouse Emissions Scenarios", *Climatic Change*.

Schneider, S.H. (2001), "What is 'dangerous' climate change?" *Nature*, **411**, 17–19.

Schneider, S.H. (2002), "Can we estimate the likelihood of climatic changes at 2100?" *Climatic Change*, **52**, 441–451.

Schneider, S.H. (2003), "Abrupt non-linear climate change, irreversibility and surprise", Working paper, OECD Workshop on Benefits of Climate Policy: Improving Information for Policymakers, Organisation for Economic Co-operation and Development, Paris.

Schneider, S.H. and K. Kuntz-Duriseti (2002), "Uncertainty and Climate Change Policy", in S.H. Schneider, A. Rosencranz, and J.O. Niles, (eds.), *Climate Change Policy: A Survey*, Island Press, Washington D.C., 53-88.

Schneider, S.H. and J. Lane (2004), "Abrupt non-linear climate change and climate policy", in *The Benefits of Climate Change Policies*, OECD, Paris.

Smit, B. and O. Pilifosova (2001), "Adaptation to climate change in the context of sustainable development and equity", in J.J. McCarthy, O.F. Canziani, N.A. Leary, D.J. Dokken and K.S. White (eds.) *Climate Change 2001: Impacts, Adaptation, and Vulnerability*, Contribution of Working Group II to the Third Assessment Report of the Intergovernmental Panel on Climate Change, Cambridge University Press, Cambridge, 877-912.

Smith, J.B., J-J Schellnhuber and M.M.Q. Mirza (2001), "Vulnerability to climate change and reason for concern: a synthesis", in J.J. McCarthy, O.F. Canziani, N.A. Leary, D.J. Dokken and K.S. White (eds.) (2001) *Climate Change 2001: Impacts, Adaptation, and Vulnerability*, Contribution of Working Group II to the Third Assessment Report of the Intergovernmental Panel on Climate Change, Cambridge University Press, Cambridge, 913-967.

Swart, R., J. Mitchell, T. Morita and S. Raper (2002), "Stabilisation scenarios for climate impact assessment", *Global Environmental Change*, **12**, 155-166.

Thomas, C.D., A. Cameron, R.E. Green, M. Bakkenes, L.J. Beaumont, Y.C. Collingham, B.F.N. Erasmus, B.F.N., M.F. de Siqueira, A. Grainger, L Hannah, L. Hughes, B. Huntley, A.S. van Jaarsveld, G.F. Midgley, L. Miles, M.A. Ortega-Huerta, A.T. Peterson, O.L. Phillips and S.E. Williams (2004), "Extinction risk from climate change", *Nature*, **427**, 145–148.

Tol, R.S.J. (2003), "Is the uncertainty about climate change too large for expected cost-benefit analysis?" *Climatic Change*, **56**, 265–289.

Tol, R.S.J. and H. Dowlatabadi (2001), Vector-borne disease, development and climate change, *Integrated Assessment*, **2**, 173–181.

USPCC RARM (1997), "Framework for Environmental Health Risk Management", Final Report Volumes 1 & 2, US Presidential/Congressional Commission on Risk Assessment and Risk Management, Washington DC.

Visser H., R.J.M. Folkert, J. Hoekstra and J.J. de Wolff (2000), Identifying key sources of uncertainty in climate change projections, *Climatic Change*, **45**, 421–457.

Webster, M. (2003), "Communicating climate change uncertainty to policy-makers and the public", *Climatic Change*, **61**, 1–8.

Webster, M., C. Forest, J. Reilly, M. Babiker, D. Kicklighter, M. Mayer, R. Prinn, M. Sarofim, A. Sokolov, P. Stone and C. Wang (2003), "Uncertainty analysis of climate change and policy response", *Climatic Change*, **61**, 295–320.

Wigley, T.M.L. (2003), "Modeling climate change under no-policy and policy emissions pathways", Working paper, OECD Workshop on Benefits of Climate Policy: Improving Information for Policymakers, Organisation for Economic Co-operation and Development, Paris.

Wigley, T.M.L. (2004), "Modelling climate change under no-policy and policy emission pathways", in *The Benefits of Climate Change Policies*, OECD, Paris.

Wigley, T.M.L. (submitted), "Choosing a stabilisation target for CO_2", editorial essay submitted to *Climatic Change*.

Wigley, T.M.L. and S.C.B. Raper (2001), "Interpretation of high projections for global-mean warming", *Science*, **293**, 451–454.

Yohe, G. and R.S.J. Tol, (2002), "Indicators for social and economic coping capacity - moving toward a working definition of adaptive capacity", *Global Environmental Change*, **12**, 25–40.

Chapter 9

TOWARD A FRAMEWORK FOR CLIMATE BENEFITS ESTIMATION

by Henry D. Jacoby,

Massachusetts Institute of Technology, United States

Public discussion of the threat of global climate change, and policymaking regarding a societal response, require a widely shared conception of what is at stake. Unfortunately, there is no single measure of the benefits of avoiding anthropogenic change that can provide a commonly accepted basis for judgment. Part of the difficulty stems from weaknesses in the underlying science, but also available global estimates are rendered incommensurable because of different attitudes to risk, problems in valuing non-market effects, and disagreements about aggregation across rich and poor nations. The needed information can be provided through the development of a portfolio of measures and its maintenance over time. It is recommended that such a portfolio include global variables that can be analyzed in probabilistic terms, regional impacts stated in natural units, and integrated monetary valuation. Creation of such a portfolio is a research task, and elements of the required program of work are summarized.

ISBN 92-64-10831-9
THE BENEFITS OF CLIMATE CHANGE POLICIES

CHAPTER 9. TOWARD A FRAMEWORK FOR CLIMATE BENEFITS ESTIMATION

by Henry D. Jacoby[102]

1. The challenge

By its initiative on the Benefits of Climate Policies, recorded in this volume, the OECD has taken up one of the most difficult and contentious issues in the environmental policy domain. All nations face a century-long search for appropriate responses the climate threat, but the issue is now most salient for OECD member governments. Because of the wealth of these nations, and their dominant contribution to current atmospheric greenhouse gas concentrations, they are the natural focus of efforts at emissions mitigation. Underlying the resulting policy debates are assessments of the benefits to be gained by such efforts. For example, what level of restriction of human emissions is called for given our understanding of the value of climate impacts avoided? What actions are justified to ease adaptation to climate change that we may experience in any event? Responses to such questions may reflect the viewpoint of a single country, they may encompass a group like the Annex B nations, or they may be framed in terms of a global total. Assessments may incorporate uncertainty in various ways, and include different assumptions about future behavior as it influences the benefits of action today. But, however they are formulated, answers to these types of questions imply a weighing-up of the benefits expected, for comparison with the costs to be borne. Sometimes explicit but more often implicit, such estimates are an inescapable component of any conclusion about what nations should do about this issue.

An ability to communicate about perceived benefits thus is an essential need of authorities seeking a common response to the threat of human-caused change. They need some shared conception of what is at stake in the choice of one level of effort or another, and a common terminology for incorporating these considerations into international negotiations and domestic decision-making. Essays produced originally for the December 2002 OECD Workshop explore the various aspects of this task—showing possible ways forward but sometimes simultaneously revealing just how daunting the

[102] Joint Program on the Science and Policy of Global Change, Massachusetts Institute of Technology.

challenge is. Drawing on these papers, and in particular on my earlier contribution (Jacoby, 2004), the text below explores a possible framework for organizing efforts at benefits estimation. It is natural to seek a single estimation procedure, with all benefits converted to a common monetary unit, to allow direct comparison with estimates of the costs of emissions control. Unfortunately, the complexities of the climate issue conspire against any single, widely accepted measure of this type. Inevitably, governments will be confronted with sets of benefits estimates that are incommensurable—i.e., they will share no common basis for comparison. Therefore a portfolio of benefits measures is recommended, structured to provide transparency when viewing alternative estimates.

The development of such a portfolio is a research task, and an effort is made here to outline the work needed. To limit the scope of the discussion several issues are given less attention than they surely deserve. In keeping with the OECD Workshop objectives, the focus is on direct benefits to be gained by limiting climate damage (net of any positive effects) by means of emissions mitigation. Such estimates must consider opportunities for adaptation and the proper accounting for associated costs, but the difficult issues attending the proper analysis of adaptation are touched on only briefly. Because of the focus on direct benefits, questions of secondary or ancillary benefits of mitigation actions also are pushed aside. This omission is unfortunate, for many of the issues raised about (net) climate damage apply as well to ancillary benefits and costs. Finally, although distributional issues will emerge, the discussion does not pretend to cover the range of concerns of developing countries or of sustainable development more broadly. Again, these issues are important, but they only add more dimensions to the problem of incommensurability explored here.

Exploration of this complex topic begins in Section 2 with a quick survey of the role that benefits estimates play in long-term strategy development and in the formulation of near-term policy. Different issues arise depending on whether the task is to justify a long-term stabilization target, to inform the setting of a current level of mitigation effort, or to provide information about possible regional effects and guidance for adaptation. A summary follows in Section 3 of those characteristics of the climate issue that combine to limit our ability to develop commonly accepted, comprehensive measures of climate benefits. Three are given special attention: the handling of uncertainty and risk preferences, problems of valuation of non-market impacts, and the lack of accepted means of aggregating welfare across human populations. This view of the challenges of benefit estimation leads to a conclusion that no single benefits measure is going to be universally acceptable, and to a Section 4 discussion of the design of a portfolio of measures to meet this need. Development of this approach is naturally a substantial research task, and Section 5 begins the work of laying out the tasks that would have to be pursued to realize the advantages of the approach suggested.

2. How benefit estimates are used in policy formation

Projections of the economic and environmental effects of climate change enter the policy process in a variety of ways—from pictures of specific consequences of change (endangered polar bears, shriveled crops) intended to stir public interest in the issue, to projections of climate effects at regional scale used to inform public and private managers about opportunities for adaptation. Here, however, the focus is on quantitative measures of benefits attributable to policies for restricting human emissions, limiting their contribution to global radiative forcing. Further, the view taken here is that a useful "framework" for such discussions must yield information that is widely understood and accepted among the diverse sets of parties that must ultimately take a role in mitigation policymakng, now and in future decades. These participants include not just members of the OECD (diverse enough in itself) but also developing countries and economies in transition.

Benefit (damage) information has an influence on the long-term strategy taken in response to the risk of climate change, and the choice of institutions to be employed. Such an effect can be seen in the structure of the Climate Convention, discussed below. But the main purpose of any analysis of benefits and costs of policy is to inform decisions about actions to be taken *now*. Nations have a very limited capacity to commit to actions in the distant future. Moreover, the climate issue is characterized by the stock pollutant character of the greenhouse gases, by long lags in the climate system, and by the prospect that some uncertainties will be resolved in the next decade or two. Under these conditions, nations will decide and re-decide their global response over time. The key decision to be informed, then, is what to do in the near term. Actions intended over longer time horizons are important, but mainly for what they may imply about desired activity today.

In this circumstance, benefit information comes into play in policy evaluation at two levels: directly as a guide to near-term mitigation effort, and indirectly as it informs the setting of long-term goals, which in turn have implications for the adequacy of efforts to date and committed in the short term.

2.1 Informing the level of near-term effort

Usually the benefit side of climate change policy assessments is implicit. Various categories of climate impacts may be presented, but their integration to some overall benefit impression takes place in the mind of the observer. Less often, aggregate benefit assessments are made explicit, sometimes leading to a calculation of the marginal value of future impacts avoided by additional mitigation undertaken today. These latter studies have a feature in common: in order to identify the desired level of current effort, cost and benefit data are converted to some common measure. Almost universally costs are estimated in monetary units, so estimates of avoided climate damage are similarly expressed. These are then summed across diverse climate effects and the various components of the decision unit of interest (e.g., the sector, the nation, the globe).

In these attempts at explicit analysis, impact estimates usually are summarized by a (net) damage function, stated in terms of a projected change in global average temperature (e.g., Nordhaus and Boyer, 2000; Tol *et al.*, 2004). Sometimes, an optimal path of greenhouse gas (GHG) emissions is computed, along with the associated marginal cost or emissions penalty stated in dollars per ton of CO_2 equivalent.

Schneider and Lane (this volume) apply just such a global damage function in their exploration of emissions paths leading to abrupt change. Such analyses typically assume that all future mitigation efforts are carried out in an optimal manner, considering the estimated costs and benefits of actions along the way. Almost always they are applied at a global level of aggregation, calculating a path of emissions penalties that is assumed to apply to all nations. Occasionally such assessments include features affecting policy choice such as the presence of uncertainty and the possibility if its resolution (e.g., Webster, 2002) and differentiation of mitigation effort among countries. Whatever the method, the main focus of these studies is the optimal level of current effort, and the associated pattern of stringency in the future.

Later this discussion will return to insights to be drawn from such studies. The relevant observation at this point, however, is that this explicit benefit-cost framing of policy choice requires agreement on a single measuring rod, and a means of converting all climate damage effects into its units. Any individual analyst may do this, of course, and draw insight from the results. The difficulty arises when agreement is needed on such a procedure (and its underlying values) among analysts, representatives of national governments, and members of diverse interest groups. The problem is aptly illustrated by the review of available studies carried out by the UK Department for Environment, Food and Rural Affairs (DEFRA), seeking guidance in setting a social cost of carbon emitted today for use in policy formation. As discussed in the summary and analysis of this effort by Pittini and Rahman (this volume), the available estimates differ by a factor of twenty or more!

2.2 Guiding the Choice of Long-Term Goals

Anticipating the difficulty of direct estimation of the benefits of GHG mitigation, as might be required to support the goal of a socially beneficial mix of mitigation and adaptation over time, the diplomats drafting the Framework Convention on Climate Change (FCCC) stated its objective (in Article 2) in terms of a constraint: the avoidance of "dangerous" levels of GHG concentrations in the atmosphere. This formulation directs the policy debate to a comparison of the advantages of alternative concentration levels compared to the estimated costs of various paths to their achievement. For any atmospheric target, the resulting cost-effectiveness analysis leads to a recommendation of the efficient path of mitigation effort over time, and thus to the level of effort (frequently stated in terms of dollars per ton CO_2 equivalent) that should be undertaken now (e.g., Wigley, Richels and Edmonds, 1996; Manne and Richels, 1997). Also, Article 2 connects to a provision of Article 4 of the FCCC that requires nations to report periodically on the adequacy of efforts "until the objective of the Convention is met". Thus the very notion of an Article 2 target provides a basis for debate about whether current efforts are consistent with some particular atmospheric goal. This provision has led to various methods, such as tolerable windows analysis (Toth *et al.*, 1998; Schellnhuber *et al.*, this volume) for analyzing what must be done in the short term if an assumed Article 2 goal is to remain within the realm of economic and political feasibility.

Underlying the language of Article 2, and the analyses (formal and otherwise) that follow from it, is the facilitating myth that some level of GHG concentrations can be identified above which there is "danger", and below which there is not.[103] The absence of widely-agreed, summary *scientific* evidence of such a threshold leads to a form of meta benefit-cost assessment about what constitutes the "danger" level, with the debate ranging across levels from 450 ppmv to 750 ppmv.[104] Observers who believe that emissions reductions will be cheap, but climate damage severe, argue that 450 ppmv (or lower) is the correct target. Others who think mitigation will be costly, but climate benefits questionable, argue against any target lower than 650 ppmv. One key objective of benefits work, then, is guidance regarding the marginal gains of moving from a loose atmospheric target to a more stringent one.

A brief look at a couple of recent efforts to represent the climate damage function, one by the IPCC and the other prepared for this OECD project, provide a useful prelude to discussion of the difficulties of clarifying this debate. Both use temperature change as the variable representing climate, and each tries to organize available information in a simple expression or functional form. First, Working Group II of the IPCC's Third Assessment Report (TAR) summarizes a huge body of data, research and analysis on the impacts of potential anthropogenic climate change (McCarthy *et al.*, 2001). One of their tasks of was to help answer the question above: what can scientific, technical, and socio-economic analysis contribute to the determination of what constitutes dangerous anthropogenic interference with the climate system? The TAR's Synthesis Report states that the IPCC could not produce a coherent answer to the question: "Comprehensive, quantitative estimates of the benefits of stabilization at various levels of atmospheric concentrations of greenhouse gases do not *yet* exist" [emphasis added] (Watson, R., *et al.*, 2001, p. 22).

In part the TARs inability to meet this objective results from uncertainties in the links between atmospheric concentrations and radiative forcing, climate response, etc. But more relevant for this discussion, the Synthesis Report states,

> .... impacts such as the changes in the composition and function of ecological systems, species extinction, and changes in human health, and disparity in the distribution of impacts across different populations, are not readily expressed in monetary or other common units.

---

103 In analysis under certainty, GHG levels associated with abrupt change might be identified, such as substantial slowdown of the thermohaline circulation discussed by Schneider and Lane (this volume). Many analysts would argue for a limit far stricter than the levels so calculated, however, and even abrupt-change thresholds lose clear definition in the face of uncertainty.

104 For simplicity, the problem of defining an atmospheric target if multiple greenhouse gases are considered is ignored in this discussion. See Sarofim *et al.* (2004) and Reilly, Jacoby and Prinn (2003).

Nonetheless the authors made an attempt at a summary representation of what is at stake. Apparently hoping that colors have the same meaning across cultures, they employed a palate—shown in Figure 1 of the chapter by Corfee Morlot and Agrawala (this volume)—to express in a qualitative way where "danger" lies. It is a creative attempt to deal with a daunting problem, but literally the interpretation lies in the eye of the beholder. Efforts are made to use this IPCC result as a basis for numerical definition of dangerous atmospheric influence, such as the one by Schneider and Lane (this volume) who define the danger level in terms of a particular shade of red in the diagram. The approach may have some value in demonstrating methodology, but there is in this representation scant guidance for setting a "danger" threshold for policy purposes, or for estimating a marginal benefit function to be applied in analytical work.

The second example, the OECD-sponsored study by Hitz and Smith (this volume) draws on analyses reviewed in the TAR and others published since the TAR was completed. The studies that were available to these authors were few and several of them used global average temperature as the only climate change indicator, even for types of effects where precipitation and dryness, and change in variability (droughts and floods), are more important. Also, most of the available studies were based on changes in equilibrium climate, offering little insight into the effects of transient change.[105] Like the TAR, Hitz and Smith make no attempt to convert the diverse results to a common metric. Instead, they seek to reveal the shape of the various damage relations (e.g., damage increasing monotonically with temperature change, showing first some net benefits then increasing damages, or unknown) and to identify the level of temperature change where damages seem to increase significantly. Available studies allow them to so classify some sectors, but not others.

It is a carefully prepared survey, with studious qualification of its results. But even so there remains a danger of over-interpretation of its conclusions. One of the key conclusions is that:

> .... by an approximate 3 to 4°C increase in global mean temperature, all of the studies we examined, with the possible exception of those on forestry, suggest adverse impacts. It appears likely that temperatures exceed this range, impacts in a vast majority of sectors will become increasingly adverse. Although below this temperature level . . . [some] are negative and some are positive.

When setting an environmental constraint in the face of poorly quantified benefits, it is natural to look for an "elbow" in the benefits relation. Even if the level of benefits is not known, it is a good place to consider setting a target, because there seems a better chance there than elsewhere in the possible span of control that marginal benefit (lower below the elbow, higher above) will intersect the marginal cost of mitigation. Hitz and

105 As emphasized below, any effort to meet the OECD's objective of improved benefits estimates should begin with an increased allocation of resources to the correction of these inadequacies in the fundamental science.

Smith warn against this interpretation, citing the limits of the studies available to them, but it is there to be made by the unsuspecting reader. Since most sectors covered in the study are not quantified using a common metric, there is no way to weigh up the relative damage among sectors. Even if such an "elbow" could be identified within a range of hypothesized temperature changes, it would vanish with the introduction of uncertainty in the relations of emissions and atmospheric concentrations to climate outcomes. Thus while the survey provides insight into the effect of temperature on individual sectors, it cannot provide a basis for saying what the aggregate damage function looks like.

Similar concerns can be raised about any effort to identify a elbow in the aggregate benefit function by building up from a set of individual damage areas represented by sigmoid functions, each with a zone of critical change (Schellnhuber *et al.*, this volume), or about attempts to extract such a relation from direct assessments of critical thresholds defined at the level of local effects (Jones, this volume; Yohe, 2004). Such analyses may provide useful information about local vulnerability and targets for adaptation, but (assuming a common metric can be constructed) their aggregation is not likely to reveal a substantial discontinuity in the aggregate benefit relation.

Thus, even after the earnest effort that went into these studies, and others not cited here, summaries of the climate effects studies now available offer little quantitative guidance to the determination of national or global policy, either for the choice of an atmospheric target or for the level of near-term effort. Moreover, fundamental barriers would impede the provision of better aggregate benefit estimates even if the needed work on the underlying science were available. It is not just that such analysis is not available "yet", as in the quote above from the TAR Synthesis Report. It is unlikely that a single, widely understood and commonly accepted benefit estimate will become available *ever*. This expectation leads to the portfolio approach to the benefits issue, outlined in Section 4.

3. The challenge of aggregate benefit estimation

At the outset it was argued that benefit-cost considerations are inescapable in any policy choice. Decisions about a climate response—stringent or relaxed policies now, tight or loose atmospheric constraints for the future—imply some weighing-up of the likely climate benefits. That proposition reflects a viewpoint underlying this essay: that the concepts and analysis methods of welfare economics, while not the only input to decision, do provide the best approach to clear thinking about choices in this domain. Within this approach "value" is seen as human mitigated, notions of willingness to pay or willingness to accept are adopted as an appropriate foundation for thinking about benefits, and value is usefully parsed into different categories (use, option, existence).[106] These concepts provide a common language, which is one of the essential elements

[106] For an introduction to the concepts at issue, see Kolstad (2000).

required for coherent domestic and international debate about a response to the climate threat.

The challenge presented by the climate issue is that these ideas must be adapted to choices of a scope, magnitude, and complexity never foreseen by earlier generations of thinkers about social choice. The underlying economic theory is sound, but severe problems of empirical estimation arise in application to the climate issue, as will be evident in the discussion below. Of course, any one analyst or analytical group, or the authorities of any one national party to the FCCC, can always agree on benefit estimates, and on a functional form for their inclusion in assessments of mitigation policy. Unfortunately, the challenge presented by the climate issue is not just to guide a single decision-maker or interest group. It is rather to inform multiple-nation negotiations where participants have different understandings of the world, and to support national decision-making where different segments of the population hold widely varying views of the seriousness of the issue and what should be done about it. Some of these difficulties are implicit in the discussion to this point. It is, however, useful to review three of these challenges in particular, to bolster an argument made later that much of the scientific guidance about benefits that governments justifiably want is not to be had, at least not in a single comprehensive, quantitative measure.

3.1 Uncertainty and risk preferences

Considered choices about climate policy require the formation of a linkage between actions that could be taken and the climate change effects they might prevent. Unfortunately, a lengthy chain connects the two. Fundamental to the understanding of climate is the interaction between the physics, chemistry and biology of the atmosphere, oceans and terrestrial biosphere and human and natural systems with which they interact. Uncertainty in these phenomena creates serious challenges to the construction of a measure of the benefits of efforts to reduce human interference. The difficulty is not just in quantifying the physical[107] aspects of climate change, but also in differences among relevant groups in their perceptions of the risks that such analysis may reveal.

The choice of an atmospheric concentration target under FCCC Article 2 illustrates the challenge. At our current level of understanding, we cannot specify the precise degree of emissions control that would achieve a particular "danger" level of interference, even if the climate effect were to be defined simply in terms of global average temperature. Figure 1, adapted from work by Webster *et al.* (2003), illustrates the problem. It shows two PDFs of global temperature change. One is an estimate of the distribution of change assuming no emissions control, taking account of uncertainty in both human emissions and the response of the climate system. The other is the same distribution under a profile of global emissions control that would (under central

---

107 The term "physical" is used to describe impacts of climate change including physical, chemical and biological measures. That is, it refers to impacts stated in their own natural units, and not converted into a monetary measure.

tendency estimates) lead to a stabilization of greenhouse gases at roughly 550 ppmv. The insight to be drawn from the figure is that there is no one-to-one link between policy action over time and the climate change avoided. At best the policy outcome can be stated in terms of a confidence interval, or the odds that a particular climate result will be achieved. The same result is illustrated by Wigley (this volume).

Figure 1. Temperature change under no-policy and stabilization cases

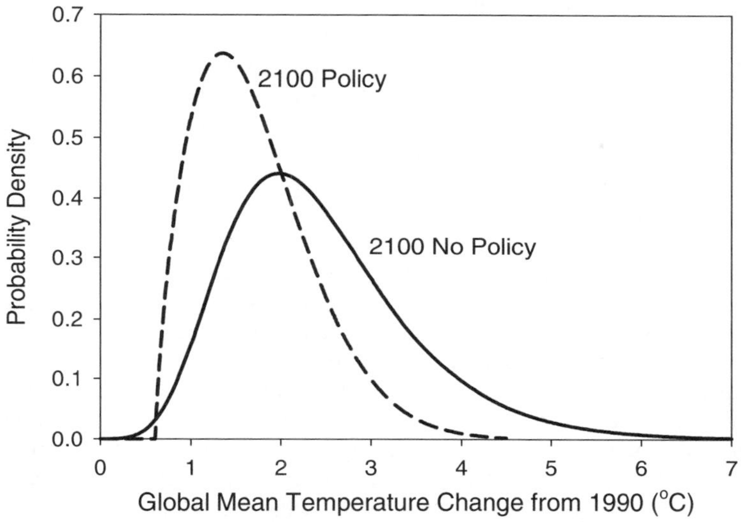

Source: Adapted from Webster *et al.*, 2003.

More troublesome for benefit estimation is the need to represent climate effects at the regional level, where most of the consequences of climate change must be studied. Uncertainty grows substantially at regional scale, even for climate variables such as temperature but particularly for precipitation, dryness, etc. One can still think in terms of the odds of outcomes, but existing climate models cannot even crudely quantify the uncertainty at regional scale. The difficulty can be seen in another result shown by Wigley (this volume) who demonstrates a method for gaining insight about the confidence that should be attributed to regional projections by a comparison of results from a group of seventeen ocean-atmosphere global climate models. Such inter-model comparisons can yield only the crudest impression of uncertainty in projections. Accepting this limitation, however, he argues that, for regional temperature, the mean of these estimates is statistically significant at regional scale (i.e., its magnitude is substantially greater than the noise of the differences among the models). For regional precipitation, however, this is not true: the noise drowns out the signal. Only with a substantial relaxation of the standards of policy useful evidence is it argued that the models can yet tell much about precipitation, which is critical to many climate change effects.

Finally, even with certain knowledge of future regional climate conditions, our knowledge of adaptive response and ultimate impact would remain severely limited. The lack of understanding is particularly troubling for unmanaged ecosystems, but it also complicates analysis of managed, human systems. The damage from a particular pattern of climate change should be assessed net of anticipated adaptation measures, taking account of their costs. However, the difficulty in properly estimating this net damage can be seen in the analysis by Callaway (this volume). As he points out, most systems have some built-in flexibility (his example is a water supply reservoir) to deal with existing natural variability on a short (daily to decadal) time scale—"weather" in the climate scientist's terminology. The climate damage ultimately realized depends on the ability of managers to distinguish natural variability from climate change, and on the interplay of the flexibility built into human systems at any time (and the costs of their use) and shifts in capital stock appropriate to a changed long-term circumstance (and their costs). Callaway shows the complexity of such analysis even for relatively simple systems like the one used as an illustration, although other effect areas, like sea level rise (Nichols and Lowe, 2004), should be easier to assess.

Thus although the benefits of emissions mitigation policy may be most appropriately thought of in probabilistic terms, the weakness of available uncertainty analysis at regional scale, and the difficulty of adequately representing what is known and not known at this level of detail, leads to the use of distant proxies like global average temperature and to a collapsing of the analysis to "reference" values of uncertain emissions and climate system parameters. Given that there is no "solution" to the threat of anthropogenic climate change, mitigation policy is most appropriately formulated as seeking the appropriate level of risk reduction. But substantial barriers remain to be overcome to allow a clear statement of what the risk is.

Furthermore, even where such estimates of the risk-reducing effect of a policy path can be calculated, the construction of a commonly accepted single benefit function faces another challenge. Attitudes to risk differ among cultures and across individuals within a culture or nation (Renn and Rohrmann, 2000; Slovic, 2002). People have different views of what it would be worth to reduce a particular risk, even if they agree on the magnitude of the effect under various outcomes. Thus it is difficult to imagine a single benefit function—say, expressing the benefit of reducing the future atmospheric concentration from 650 to 550 ppmv—that will communicate across the diverse parties to climate policy negotiations. There may be many ways to summarize such information, each with meaning to a particular party, but there may be no way to reach agreement on a common estimate, and perhaps even a difficulty in achieving a common measuring rod.

3.2 Valuation of Non-Market Impacts

The next link in the chain from mitigation action to benefit is the conversion of the many physical, chemical and biological effects into a common measure that can be compared with cost. The issue is "valuation", and the task falls into two familiar categories: effects that can be reasonably represented by calculations using market prices (or near-market analogies) and those that cannot. Some climate impacts can be credibly formulated in monetary terms, because market prices are available to value the physical changes that may be estimated to occur. Examples include the effects of environmental change on agriculture and commercial forestry. When prices are not available, valuation can sometimes be achieved by appeal to the revealed preference of consumers—monetary value being imputed from observed behavior in markets that do exist. Data for such analysis may be found in closely related markets, as when the value of clean air is estimated from variations in property values. The value of an environmental bad also may be imputed from expenditure on defensive measures (sound proofing to avoid noise pollution), or of an environmental good by analysis of spending that allows full use of the good (travel expenditure to enjoy a park).[108]

Unfortunately, only a small number of the myriad climate change impacts are candidates for revealed preference treatment, and the ones where data are available for many countries are still more limited. Where direct market-origin data are absent, another approach is to apply contingent valuation—seeking an estimate of what consumers would pay for an environmental good if a market for it did exist. A couple of techniques are in use. In one, people are surveyed to try to determine their willingness to pay for an environmental improvement, or what they would have to be paid to be willing to accept the loss of some aspect of environmental quality they already enjoy (Hanemann, 1994). In a similar approach, laboratory experiments are conducted wherein subjects are put into an artificial market where a non-market good is actually traded, and behavior is observed in the search for underlying preferences.

Continuing and sharp controversy surrounds the use of these methods, and serious questions are raised whether responses to imaginary situations can ever yield information consistent with that revealed by real choices (e.g., see Diamond and Hausman, 1994). Beyond these difficulties, Pittini and Rahman (this volume) call attention to yet another category of concern, what they call "socially contingent effects", that falls outside the domain of conventional economic valuation. These effects include knock-on consequences of climate change that may appear in the social domain—such as migration and associated stresses, economic and political instability, and promotion of conflict. Thus whatever one's taste in these analytical methods, the scope of the impacts of climate change, and poor respondent understanding of the implications of changes at regional and global scale, means that any such applications are going to be partial in coverage. It is hard to imagine the application of any available method to a relevant question such as, "What would you be willing to pay to prevent the loss of all arctic tundra?" More important for this discussion, it is questionable whether such applications can produce estimates that are commensurable across regions or realms of climate change impact.

---

[108] For a summary of method used for this type of analysis, see Kolstad (2000) or Smith (1993).

3.3 Aggregation across effects, peoples and circumstances

The aggregate damage from climate change is the sum of myriad local effects, as human and natural systems are pushed outside the "coping range" to which they have adapted over millennia within a relatively stable climate regime. The overwhelming difficulty of aggregation across local effects can be seen in the two examples provided by Jones (this volume): coral bleaching of the Great Barrier Reef and changes in flows within an Australian river basin. Such studies of specific local effects are valuable in that they can identify key vulnerabilities and help make the climate change threat concrete, and they can inform decisions about anticipatory adaptation. However, it likely is not possible to develop the bottom-up analyses that would be needed to span from this local scale and narrow scope to national or global totals, nor is it possible to imagine adaptation analysis being carried out at the level of detail suggested by the example presented by Callaway (this volume).[109] Even if such detailed local assessments were available, the challenge of proceeding from results in natural units to some common measure, discussed above, would remain.

Further, however local or national climate damages are estimated and valued, different participants in the search for a climate solution will add them up in different ways. The aggregation task is the same as that faced in valuing any public good, but it is particularly problematic for climate change. In the classic solution to public goods valuation, benefits are determined by eliciting estimates of willingness to pay (or accept) from each party affected, then summing them up. The approach is consistent in theory; the difficulties arise in practical application. The procedure assumes that the income distribution underlying the estimates is optimal, or at least acceptable. But in application to climate change, with its conflict between North and South, this is not the case, as amply illustrated by the exploration of distributional issues by Tol *et al.* (2004). The issue was painfully explored in preparation of the IPCC's Second Assessment Report, in the debate over the use of willingness-to-pay measures of the value of human life (Pearce *et al.*, 1996). Without correction, the same human loss in a rich country receives much greater weight than in poor a one. Pittini and Rahman (this volume) discuss the literature on the correction of estimates by some form of global welfare function--a procedure that usually leads to the computation of weights by region. In the climate context, however, this approach only pushes the issue back one step to the selection of the weights, for which there is no widely accepted process. Different parties will have different weighting schemes, and their summary estimates will be incommensurable.

A further difficulty arises because benefits estimates are not constructed by the textbook method of summing individual valuations. Rather, analysis groups in one country or another prepare estimates for the whole, applying willingness-to-pay (or accept) estimates and relative weights as they see fit. Even with the best intentions,

---

[109] Attempts at such estimates are made a regional scale (e.g., Fankhauser, 1995; Mendelsohn *et al.*, 1998; Nordhaus and Boyer, 2000; Tol *et al.*, 2004) but they do not pretend to deal with information at the detail explored by Jones (this volume).

aggregation methods will differ, and alternative estimates will be incommensurable for this reason alone. This aggregation issue arises at all scales from the individual upward to various social aggregates, and in all benefit estimation problems. But it is particularly troublesome for climate change, which is so fraught with inter-country equity issues, not to mention the need to aggregate the (imagined) preferences of future generations.

4. A framework for future benefit studies

A conclusion to be drawn from the discussion thus far is that no single benefits measure is going to be universally applicable.[110] By extension, no single method for calculating the magnitude of the marginal benefits of greenhouse gas mitigation, stated in the same units as marginal costs, is going to be widely accepted. As a result, there is no single analytical basis for selecting a target level of atmospheric concentrations or other climate variables like global mean temperature rise. As in other areas of public decision, governments face the limits of neutral assessments as a guide to political decision, and they have little choice but to prepare to deal with this situation by structuring benefits research so that the results are transparent, and as universal as possible in their acceptance.

In this circumstance the desired "framework" for benefits estimation will involve a portfolio of estimates, related to one another but at different scales, in different units, and with alternative degrees of aggregation. Discussion of this approach begins with information needed to inform the debate over long-term atmospheric goals. The proposed set of benefits measures includes global physical variables, regional indicators in natural units, and monetary aggregates. Their development involves a set of parallel tasks, because the different indicators need to be coordinated with one another. Thus while the discussion to follow makes some recommendations regarding the content of such a portfolio of measures, what is proposed here is really the outline of a research agenda.

4.1 Informing the selection of long-term atmospheric goals

A benefits portfolio would include estimates at a global aggregate level, with global variables that can be analyzed in probabilistic terms, impacts that can be measured at regional scale (most likely in natural units), and exercises in monetary valuation. With this structure, variables that might gain broad acceptance would be given prominence, and a foundation of common information would be laid for the more problematic matters of non-market valuation and aggregation. Each is explored in turn.

110 For a similar argument, see Yohe (2004) or Schneider, Kuntz-Duriseti and Azar (2000).

4.1.1 Global physical variables, uncertainty and risk reduction

Because of the uncertainty that pervades the climate issue, the ideal would be to formulate climate policy as a way of reducing the risk of damage, in analysis at all levels of aggregation. Unfortunately, as noted earlier, climate models cannot support this type of analysis at a regional level. The risk reduction perspective is nonetheless very important to convey, and thus the first level of benefit representation should be at a global level, expressed in those variables that can be analyzed in probabilistic terms. Climate variables at this level are usually thought of as drivers of benefits estimates, but here it is proposed that they themselves can serve as a benefits measure.

Several efforts have been made to estimate probability density functions (PDFs) for temperature change under non-policy conditions (e.g., Webster *et al.*, 2003; Wigley and Raper, 2001; Wigley, this volume). These same analyses can be extended to estimate PDFs of climate change under specified policy paths as illustrated by the work of Webster *et al.* (2003), shown in Figure 1. Alternatively, the same probabilistic results can be displayed in the form of cumulative distribution functions. Examples of this approach can be seen in the work of Jones (this volume) who further argues the value of expressing results in terms of the probability of exceeding particular thresholds of change. Also, such approaches could be applied to important variables besides global mean temperature, such as temperature by latitude (e.g., distinguishing risks among polar, temperate and tropical regions) and sea level rise.

A question remains as to where in such measures the prospect of abrupt climate change should be addressed (Alley *et al.*, 2002). Possible events such as a significant slowing of the thermohaline circulation, or the loss of permafrost regions and the associated the release of greenhouse emissions, are too important to be left out of any system. Schneider and Lane (this volume) survey a number of areas of possible non-linear response to human forcing, but focus mainly on the threat of slowdown in the deep ocean circulation as an example. Applying simple models of anthropogenic emissions and earth system response, their study explores ways to represent the effect of different emissions scenarios on the performance of the oceans in taking up heat and CO_2. The greenhouse gas emissions paths that they consider are based on a simulation model that assumes a global climate damage function and other economic factors, but the approach could be applied to any set of assumed trajectories. Analyses attempting to support estimates of the likelihood of such changes in ocean behavior are extremely weak, however, and the potential effects at regional scale are poorly understood. A tentative recommendation is that these possible abrupt changes nonetheless be included in benefit representation at this high level of aggregation, though probably best stated only in qualitative terms.

Research is needed on which ways of representing such results are likely to be most widely understood by target audiences. Alternatives include distribution functions like the ones in Figure 1 or the ones illustrated by Jones (this volume), statements of the odds of avoiding certain specific conditions (e.g., sea level rise exceeding 30 cm), or big-loss vs. small-loss segments of roulette wheels. However these measures are constructed, the objective at this stage of benefits analysis would be to express results in ways that do not run afoul of the controversies over valuation and aggregation, and that cast benefits in a proper risk-reduction framework. In addition, a framework might be provided so that differences in risk perception and tolerance could at least be discussed.

4.1.2 Effects at regional scale, stated in natural units

A second component of a benefits portfolio would include measures of climate change impact at regional scale. For most effects these measures would be stated in natural units, short of monetary valuation, although in some cases monetary valuation may in fact provide the most natural metric. For example, useful market measures have been achieved for sectors like agriculture, commercial forestry and energy use, and these monetary estimates would be preferred to physical measures like tons of grain, board-feet of lumber or barrels of oil. Uncertainty analysis is desirable at this regional level as well but, absent dramatic improvement in regional climate forecasting, regional estimates likely would have to be made on an expected magnitude basis, a common approach among analysts of climate impacts.[111] As an example, Jones (this volume) develops two cases wherein human and natural systems are modeled (at regional scale) as adapted to a "coping range" of existing variability (loosely, weather), and regional climate change (itself treated as uncertain) is modeled as shifting the zone of short-term variability away from this range of familiar conditions. The approach shows promise for well-documented local systems, but the challenge is to develop a small number of indicators, in natural units, that together convey as comprehensive a picture as possible of what different levels of climate change might involve.

There are several aspects to the task of developing a set of indicators at regional level. First is research on the best regional aggregation for this purpose. Most studies use an aggregation chosen to be consistent with integrated assessment models, and so their structure is strongly influenced by the data available for cost analysis and the assessments at hand for market-price based impacts studies. For example, for physical indicators (to be discussed below) the identification of polar regions would seem important, although they are usually missing from such studies. Next is the identification of physical measures or indices that are mutually consistent (in terms defined below) and that have high information content. Several examples can be seen in the papers prepared for this OECD benefits project. For example, a number of studies use numbers of people subjected to one or another hardship (risk of hunger, flooding, water stress, disease) as seen the summary by Hitz and Smith (this volume). And studies have used terrestrial vegetation models to calculate measures of the disruption of particular ecosystems, illustrated by the work of Leemans and Eickhout (2004) and applications by the Potsdam Institute (Toth, Cramer and Hizsnyik, 2000), the latter described briefly by Schellnhuber *et al.* (this volume).

---

111 For example, see Fankhauser (1995), Nordhaus and Boyer (2000), Fussel and van Minnen (2001), and a summary of work as of 1995 by the IPCC (Pearce *et al.*, 1996).

A number of criteria are important for the usefulness of a set of such effects measures at regional scale. First, there should not be too many measures in the portfolio else they become indigestible, and each element selected should capture a circumstance where climate change is a major stress compared to other influences. Each measure should have a clear definition, to allow consistent measurement across regions, and each should be applicable to all or most areas of the world. Furthermore, they will be most useful if they can be applied at different scales. To the extent possible the measures should be independent of one another, to avoid double counting and to preserve additivity in the case of subsequent incorporation into monetary estimates and wide-scale aggregation. Finally, for each there needs to be some baseline basis for comparison, so observers have an idea whether a change is large or small. For example, the measures of change in terrestrial ecosystems might be calibrated on a regional basis with an estimate of the last 100 years of anthropogenic change, caused by the conversion of natural ecosystems to agriculture and urban use.

Clearly there is no fully satisfactory approach to this task, only more or less disappointing compromises. But recall the need: it is for indicators at regional level that lie somewhere in between the color diagram of Figure 1 reproduced by Corfee Morlot and Agrawala (this volume) and a thousand measurements of effects like hectares of coral reef, numbers of alpine meadows, and days of trout fishing.

4.1.3 Monetary valuation of market and non-market effects and aggregation

A third component of the portfolio would be the construction of benefit functions, at regional and global level, aggregated in monetary units. Estimates of this type bring together two rough categories of analysis and judgment: market and non-market effects. Useful market estimation has been achieved for several sectors, as noted above. Others, like recreation, have simulated-market analogies that may prove widely acceptable.[112] More problematic are health impacts, where economic estimates of pain and suffering and value of human life may be common practice in some countries and anathema in others. Estimation of effects even further removed from market measures—like ecosystem change, species loss, and amenity values—present still greater difficulties, as noted above.

However, despite the difficulties, estimates of monetary aggregates belong in the portfolio of measures because they serve the important function of imposing discipline on benefit estimates and indicating efficient time patterns of response. Without some means of adding up effects—which requires a common measuring rod—it is difficult to limit the claims for one or another physical effect of change, or the mitigation cost justified in preventing them. Whatever one's view of the valuation issue, there are tradeoffs to be faced. The construction of aggregate benefit functions provides a framework for testing the reasonableness of the total of benefit claims, including a

---

[112] For discussion of the boundaries, see Nordhaus and Kokkelenberg (1999).

transparent accounting of possible substitutes for good and services or environmental conditions that may be lost or displaced due to climate change.

4.2 Informing short-term effort

Even if analysis can be based on a single monetary benefit function, or the benefit analysis yields an atmospheric constraint, the distance between estimates of climate change effect and the choices made today is great. The condition is, unfortunately, characteristic of a stock pollutant emitted by many sources, where the direct guidance to political decisions about current mitigation effort is at best indirect. Many of the considerations and uncertainties that intrude when concern shifts to the benefits attributable to action taken today (the most important issue at stake) are discussed by Pittini and Rahman (this volume) in their summary of the UK Government's attempt to determine the social cost of carbon. And still other difficulties have to be faced in these analyses. For example, the effects on climate risk of actions taken today are not independent of what future generations of policymakers may decide to do along the way. Many calculations of desired near-term effort are based on the assumption that all future generations will make choices that are socially optimal given the assumed cost and benefit functions—examples including Nordhaus and Boyer (2000) or the calculation of emissions paths by Schneider and Lane (this volume). Even those calculating the minimal effort required now to keep open the option of meeting of some future goal (e.g., Toth *et al.*, 1998) must impose assumptions about future behavior. Further complicating the analysis of these choices at each step is the fact that the options that future decision makers will face are influenced by actions of their predecessors. Today's actions influence not only the available set of future technological options, but also the inheritance of institutional capability and public understanding.

Still there are important insights that economic benefit analysis can contribute to climate policy, even if no more than the general shape of a global benefit function is accepted and only rough assumptions are made about future behavior. For example, almost all studies (e.g., Nordhaus and Boyer, 2000) conclude that an efficient mitigation response to the climate threat will start small and increase in stringency over time. The precise level of current effort may still be left unresolved, but the nature of the desired path is not, and it is an insight missing from much current climate discussion and policymaking.

5. Research directions

The components of a framework for benefit estimation proposed here are not new. Work is under way on almost every topic, building on years of research and analysis stimulated by the climate change issue. A couple of points from the discussion above should be emphasized however. First, a main suggestion is that a more formal structuring of climate effects estimates should be prepared, according to region and the level of detail, applying the criteria laid out at the end of Section 4.1.2. The objective is a portfolio of information that can be further analyzed and aggregated according to the abilities and values of the various participants in climate policy discussions. Second, the development of such a structure is a research task—requiring a focusing of effort and redirection of available resources toward work on climate change impacts and the benefits of emissions mitigation. The overriding priority in meeting these needs is research on the fundamental science of climate change impacts, with a special focus on natural ecosystems, their ability to adapt, and potential damage when they cannot. The discussion above and the survey by Hitz and Smith (this volume) highlight the inadequacy of current knowledge. In part the lack is attributable to the great complexity of these systems, but unfortunately our ignorance also results from the inadequate allocation of funds to research in these areas by the major national and regional organizations. Furthermore, climate change is a century-scale problem, and the development of a limited portfolio of measures to help inform the policy process should be seen as requiring continuous, long-term effort.

If such a portfolio concept is to be pursued, several other areas of research beyond the basic science are of high priority.

Develop Regional Impact Indicators. A research program should be formulated with the purpose of developing the best set of impacts indicators that can be supported by current research and anticipated future results. Market measures may be included, but of particular importance is the development of indicators of non-market impact—such as changes in natural ecosystems and effects on human systems that defy easy market valuation. The research would begin with an effort to identify those regions and effects where available analysis shows that climate change effects would be substantial, and where the resulting social and/or environmental consequences are most important (the latter criterion being necessarily subjective in the absence of market measures of impact). The assessment of climate change effects would need to take account of the other stresses to which systems are subject, including natural variability and other human insults. Where possible such a sorting process would call for a comprehensive risk analysis (e.g., see Jones, this volume). Once a selection of effects was so identified, then the task would be to develop indicators of impact that meet as many as possible of the criteria suggested earlier: clear definition, wide applicability, ability to scale up and down, independence, and existence of a baseline for comparison.

Advance the Methods of Analysis and Communication of Uncertainty. Analysis of uncertainty in climate projections, and in the effects of policy measures, is being pursued, as noted above. These efforts should be encouraged, and supported. Also, methods for incorporating this work into benefits measures at the level of global climate need to be further developed, to aid in informing choices about long-term climate goals and analysis of the implied level of current mitigation effort. This research should involve not only challenges of modeling and estimation but also issues of communication with a lay public and its political representatives. Too little is known about how to serve constituencies who are not trained in the frequently-used concepts and terminology of uncertainty analysis.

Improve the Estimation of Market Impacts in Developing Countries. Most analyses of the effects of climate change, particularly the ones stated in market terms, have been carried out in the richer countries, and very crude methods have been applied to extend the results to the developing world. Research directed specifically at impacts in developing countries should be increased, in order to fill out those regional indicators where monetary quantities are a natural unit of measure. Particularly important in such an effort will be consideration of potential improvements over time in adaptive capacity.

Improve the Formulation Global Monetary Damage Functions. Global damage functions are an essential input to many forms of analysis applied to the climate issue. Yet the number of efforts to prepare such estimates is very small. Moreover, they focus very heavily on available market-based estimates, with insufficient research on ways of incorporating effects on unmanaged ecosystems. Research to improve these aggregate functions should be better supported, particularly the representation of developing country impacts and the inclusion of non-market, ecosystem effects.

Design and Maintain a Portfolio of Benefits Measures. Based on analysis of available information and a forecast of the possible results of the research suggested above, a template should be designed, covering the three elements of a climate benefits portfolio suggested here. It should consist of a limited set of indicators—no more than ten. This might include two or three measures of global climate (e.g., probability of temperature rise greater than 2°C by 2100), four or five summary indicators of regional impact from the work recommended above, and one global monetary measure.

The resulting portfolio of benefit measures would not be the only information generated and made available to aid public discussion and policymaking, but it would be a set of variables continuously maintained and used to describe the results of various policy choices.

6. A final thought

Finally, one message that emerges from this exploration of climate benefits estimation is that policymakers cannot expect the problem of incommensurability to be overcome. No one clear, widely accepted estimate of the benefits of avoiding change is likely to become available to substantially narrow the range of choices regarding what to do today. They can hope for little better than the assistance provided by a well-constructed portfolio of measures for summarizing the diverse forms of information relevant to this issue. The underlying science and economic analysis to support such an effort necessarily will come from a diverse set of sources, but some one institution or group would need to be empowered (and funded) by a sufficiently diverse set of national interests to construct and maintain the summary portfolio. No such group is in place today. If, however, the research and analysis suggested above and elsewhere in this volume were carried out, and such a portfolio could be created and sustained, policymakers could reasonably expect much improved insight into the seriousness of the climate change threat, and guidance for the inescapable judgments about the benefits of emissions mitigation and the climate change it may prevent.

Acknowledgments

For useful comments on an earlier draft, thanks are due to John Reilly, David Reiner, Stephen Schneider and Mort Webster, and to other participants in the OECD project. Any remaining errors of logic and judgment are attributable to the author. Results cited from the MIT Joint Program on the Science and Policy of Global Change were developed with the support of the US Department of Energy, Office of Biological and Environmental Research [BER] (DE-FG02-94ER61937) the US Environmental Protection Agency (X-827703-01-0), the Electric Power Research Institute, and by a consortium of industry and foundation sponsors.

References

Alley, R. et al. (2002), Abrupt Climate Change: Inevitable Surprises, Report of the Committee on Abrupt Climate Change, US National Research Council, Washington D.C., National Academy Press.

Callaway, J.M. (2004), "Assessing and linking the benefits and costs of adapting to climate variability and climate change", in *The Benefits of Climate Change Policies*, OECD, Paris.

Corfee-Morlot, J. and S. Agrawala (2004), "Overview", in *The Benefits of Climate Change Policies*, OECD, Paris.

Diamond, P. and J. Hausman (1994), "Contingent Valuation: Is Some Number Better than No Number?", Journal of Economic Perspectives **8(4)**: 45-64.

Fankhauser, S. (1995), Valuing Climate Change, London, Earthscan.

Fussel, H-M, and J. van Minnen (2001), "Climate impact response functions for terrestrial ecosystems", Integrated Assessment **2**:183-197.

Hanemann, W. (1994), "Valuing the Environment Though Contingent Valuation", Journal of Economic Perspectives **8/4**: 19-43.

Hitz, S. and J. Smith (2004), "Estimating global impacts from climate change", in *The Benefits of Climate Change Policies*, OECD, Paris.

Jacoby, H.D. (2004), "Informing climate policy given incommensurable benefits estimates", Global Environmental Change: Special Edition on the Benefits of Climate Policy **14**: 287-297.

Jones, R. (2004), "Managing climate change risks", in *The Benefits of Climate Change Policies*, OECD, Paris.

Kolstad, C. (2000), Environmental Economics, New York, Oxford University Press.

Leemans, R. and B. Eickhout (2004), "Another reason for concern: regional and global impacts on ecosystems for different levels of climate change", Global Environmental Change: Special Edition on the Benefits of Climate Policy **14**: 219-228.

Manne, A. and R. Richels (1997), "On stabilizing CO2 concentrations—cost-effective emissions reduction strategies", Environmental Modeling and Assessment **2**: 251-265.

McCarthy, J., O. Canziani, N. Leary, D. Dokken and K. White, eds. (2001), Climate Change 2001: Impacts, Adaptation, and Vulnerability, Contribution of Working Group II to the Third Assessment Report of the Intergovernmental Panel on Climate Change, Cambridge, UK, Cambridge University Press.

Mendelsohn, R., W. Morrison, M. Schlessinger and N. Andronova (1998), "Country-Specific Market Impacts of Climate Change, Climatic Change **45** (3-4):553-569.

Nicholls, R. and J. Lowe (2004), "Benefits of mitigation of climate change for coastal areas", Global Environmental Change: Special Edition on the Benefits of Climate Policy **14**: 229-244.

Nordhaus, W. and J. Boyer (2000), Warming the World: Economic Models of Global Warming, Cambridge, MA, MIT Press.

Nordhaus, W., and E. Kokkelenberg, eds. (1999), Nature's Numbers: Expanding the National Economic Accounts to Include the Environment, Report of the Panel on Integrated Environmental and Economic Accounting, Committee on National Statistics, US National Research Council, Washington D.C., National Academy Press.

Pearce, D., W. Cline, A. Achanta, S. Fankhauser, R. Pachauri, R. Toll and P. Vellinga (1996), The Social Costs of Climate Change: Greenhouse Damage and the Benefits of Control, in Climate Change 1995: Economic and Social Dimensions of Climate Change, Contribution of Working Group III to the Second Assessment Report of the Intergovernmental Panel on Climate Change, UK, Cambridge University Press, 179-224.

Pittini, M. and M. Rahman (2004), "The social cost of carbon: key issues arising from a UK review", in *The Benefits of Climate Change Policies*, OECD, Paris.

Reilly, J., H. Jacoby and R. Prinn (2003), "Multi-Gas Contributors to Global Climate Change: Climate Impacts and Mitigation Costs of Non-CO_2 Gases", Washington D.C., Pew Center on Global Climate Change.

Renn, O. and B. Rohrmann (2000), Cross-Cultural Risk Perception: A Survey of Empirical Studies, Kluwer, Dordrecht, The Netherlands.

Sarofim, M., C. Forest, D. Reiner and J. Reilly (2004), Stabilization and Global Climate Policy, MIT Joint Program on the Science and Policy of Global Change, Report 110 (http://mit.edu/globalchange/www/MITJPSPGC_Rpt110.pdf).

Schellnhuber, J., R. Warren, A. Haxeltine, and L. Naylor (2004), "Integrated assessment of benefits of climate policy", in *The Benefits of Climate Change Policies*, OECD, Paris.

Schneider, S. (2002), "Abrupt Non-Linear Climate Change, Irreversibility and Surprise", Paper presented for consideration and review to the participants of the OECD Workshop on the Benefits of Climate Policy, 12-13 December 2002, Paris.

Schneider, S., K. Kuntz-Duriseti and C. Azar (2000), Costing Non-Linearities, Surprises and Irreversible Events, Pacific and Asian Journal of Energy **10(1)**: 81-106.

Schneider, S.H. and J. Lane (2004), "Abrupt non-linear climate change and climate policy", in *The Benefits of Climate Change Policies*, OECD, Paris.

Slovic, P. (2002), "Trust, Emotion, Sex, Politics and Science: Surveying the Risk-assessment Battlefield", P. Slovic (ed), The Perception of Risk, Earthscan, London.

Smith, J. and Hitz, S. (2002), "Background Paper: Estimating Global Impacts from Climate Change", OECD, Paris (ENV/EPOC/GSP(2002)12/REV1).

Smith, V. (1993), "Nonmarket Valuation of Environmental Resources: An Interpretive Appraisal", Land Economics **69(1)**: 1-26.

Tol, R., T. Downing, O. Kuik, and J. Smith (2004), "Distributional aspects of climate change impacts", Global Environmental Change: Special Edition on the Benefits of Climate Policy **14**: 259-272.

Toth, F., L. Bruckner, T. Fussel, H. Leimbach, M. Petschel-Held and H. Schellenhuber (1998), "The Tolerable Windows Approach to Integrated Assessments", in Cameron, Fukuwatari and Morita (eds.), Climate Change and Integrated

Assessment Models [IAMS]—Bridging the Gap, Center for Global Environmental Research, Environmental Agency of Japan, Tsukubaj Japan.

Toth, F., W. Cramer and E. Hizsnyiik (2000), "Climate Impact Response Functions: An Introduction", Climatic Change **46**: 225-246.

Watson, R. *et al.* (2001), Climate Change 2001: Synthesis Report, Summary for Policymakers, An Assessment of the Intergovernmental Panel on Climate Change.

Webster, M. (2002), "The Curious Role of 'Learning' in Climate Policy: Should We Wait For More Data?", *The Energy Journal* **23(2)**: 97-119.

Webster, M., C. Forest, J. Reilly, M. Babiker, D. Kicklighter, M. Mayer, R. Prinn, M. Sarofim, A. Sokolov, P. Stone, C. Wang (2003), "Uncertainty Analysis of Climate Change and Policy Response", Climatic Change, in press.

Wigley, T.M.L. (2004), "Modelling climate change under no-policy and policy emission pathways", in *The Benefits of Climate Change Policies*, OECD, Paris.

Wigley, T, R. Richels and J. Edmonds (1996), "Economic and Environmental Choices in the Stabilization of Atmospheric CO2 Concentrations", *Nature* **379**: 240-243.

Wigley, T. and S. Raper (2001), "Interpretations of High Projections for Global-Mean Warming", *Science* **293**, 451-454.

Yohe, G. (2004), "Some thoughts on perspective", *Global Environmental Change: Special Edition on the Benefits of Climate Policy*, **14**: 283-286.

OECD PUBLICATIONS, 2, rue André-Pascal, 75775 PARIS CEDEX 16
PRINTED IN FRANCE
(97 2004 08 1 P) ISBN 92-64-10831-9 – No. 53623 2004